CUTANEOUS BIOMETRICS

CUTANEOUS BIOMETRICS

EDITED BY

DORIS A. SCHWINDT
Rheinisch–Westfalisch Tecnische Hochschule
Aachen, Germany

HOWARD I. MAIBACH
Department of Dermatology
University of California, San Francisco
San Francisco, California

IN COOPERATION WITH THOMAS DIEPGEN, M.D.

Kluwer Academic / Plenum Publishers
New York, Boston, Dordrecht, London, Moscow

ISBN: 0-306-46159-5

233 Spring Street, New York, New York 10013

http://www.wkap.nl/

10 9 8 7 6 5 4 3 2 1

A C.I.P. record for this book is available from the Library of Congress

Printed in the United States of America

Contributors

Wim M. C. van Aalderen • Department of Pediatric Pulmonology, Emma Children's Hospital Academic Medical Center/University of Amsterdam, 1100 DD Amsterdam, The Netherlands.

Beatrice B. Abrams • Clinical Research, Dermatology, Novartis Pharmaceuticals, Inc., East Hanover, New Jersey 07936-1080.

Dennis E. Babel • Mycology Diagnostics Laboratory, Clinton Township, Michigan 48038.

John Berth-Jones • Department of Dermatology, Walsgrave Hospital, Coventry CV2 2DX, England.

Eric Beyssac • Biopharmaceutic Department, Faculty of Pharmacy, Auvergne University 63000 Clermont-Ferrand, France.

Laura Bolton • ConvaTec, A Bristol-Meyers Squibb Company, Skillman, New Jersey 08558.

C. Bonnabry • Biopharmaceutic Department, Faculty of Pharmacy, Auvergne University, 63000 Clermont-Ferrand, France.

Peter Busch • Henkel KGaA, D-40191, Düsseldorf, Germany.

Caroline A. Cardin • The Procter & Gamble Co., Sharon Woods Technical Center, Cincinnati, Ohio 45241.

J. M. Cardot • Biopharmaceutic Department, Faculty of Pharmacy, Auvergne University, 63000 Clermont-Ferrand, France.

Tsu-Yi Chuang • Department of Dermatology, Indiana University Medical Center, Indianapolis, Indiana 46202-5267.

William J. Cunningham • CU-TECH, Inc., International Cutaneous Technologies & Development, Mountain Lakes, New Jersey 07046.

Björn Edman • Department of Dermatology, University Hospital, Malmö, S-20502 Sweden.

Peter Elsner • Department of Dermatology, University of Jena, Jena, Germany.

Keith D. Ertel • The Procter & Gamble Company, Sharon Woods Technical Center, Cincinnati, Ohio 45241.

Thomas Gassenmeier • Henkel KGaA, D-40191, Düsseldorf, Germany.

Robin A. C. Graham-Brown • Department of Dermatology, Leicester Royal Infirmary, Leicester LE1 5WW, England.

John M. Haigh • Faculty of Pharmacy, Rhodes University, Grahamstown 6140, South Africa.

Helen J. Heacock • Department of Health Care and Epidemiology, University of British Columbia, Vancouver, Canada.

Dimiter Iliev • Department of Dermatology, University Hospital Zurich, CH-8091 Zurich, Switzerland.

Cornelius Kennedy • Department of Health Care and Epidemiology, University of British Columbia, Vancouver, Canada.

Antti I. Lauerma • Department of Dermatology, University of Helsinki, 00250 Helsinki, Finland.

Cheol Heon Lee • Department of Dermatology, Kangnam Sacred Heart Hospital, College of Medicine, Hallym University, Seoul 150-071, Korea.

Howard I. Maibach • Department of Dermatology, School of Medicine, University of California, San Francisco, California 94143-0989.

Otto H. Mills, Jr. • University of Medicine and Dentistry of New Jersey, Robert Wood Johnson Medical School, New Brunswick, New Jersey 08901.

Karyn Monte • ConvaTec, A Bristol-Meyers Squibb Company, Skillman, New Jersey 08558.

Ranjini Natarajan • Center for Statistical Sciences, Brown University, Providence, Rhode Island 02912. Present address: Department of Statistics, Division of Biostatistics, University of Florida, Gainesville, Florida 32611.

Janet E. Pope • Department of Rheumatology, University of Western Ontario, London ON N6A 465, Canada.

Volker W. Rahlfs • idv-Data Analysis and Study Planning, D-82131 Gauting, Germany.

Rukmini Rajagopalan • Glaxo Wellcome, Inc., Research Triangle Park, North Carolina 27709, and University of North Carolina, Chapel Hill, North Carolina 27599

Robert L. Reitschel • Department of Dermatology, Ochsner Clinic and Alton Ochsner Medical Foundation, New Orleans, Louisiana 70121.

Jason K. Rivers • Division of Dermatology, Vancouver Hospital and Health Services Centre, University of British Columbia, and British Columbia Cancer Agency, Vancouver, British Columbia V5Z 4E8, Canada.

Gerhard J. Schmitt • Beauty Care Europe, Procter & Gamble European Service GmbH, D-65823 Schwalbach/Taunus, Germany.

Axel Schnuch • IVDK-Zentrale, Universitätshautklinik, D-37075 Göttingen, Germany.

Doris A. Schwindt • Department of Dermatology, Rheinisch-Westfälisch Technische Hochschule Aachen, D-52074 Aachen, Germany. Present address: Hofman La Roche AG, D-79630 Grenzach, Wyhlen, Germany.

Eric W. Smith • Faculty of Pharmacy, Rhodes University, Grahamstown 6140, South Africa. Present address: College of Pharmacy, Ohio Northern University, Ada, Ohio 45810.

Aline B. Sprikkelman • Department of Pediatric Pulmonology, Emma Children's Hospital Academic Medical Center/University of Amsterdam, 1100 DD Amsterdam, The Netherlands.

Jean François Stalder • Department of Dermatology, CHU Hôtel Dieu, 44035 Nantes, France.

Päivikki Susitaival • Finnish Institute of Occupational Health, FIN-70701 Kuopio, Finland.

Alain Taïeb • Pediatric Dermatology Unit, Hôpital Pellgrin-Enfants, 33076 Bordeaux, France.

Ann-Jeanette Vasko • ConvaTec, A Bristol-Meyers Squibb Company, Skillman, New Jersey 08558.

Roderick B. Walker • Faculty of Pharmacy, Rhodes University, Grahamstown 6140, South Africa.

Martin A. Weinstock • Dermatoepidemiology Unit, VA Medical Center and Department of Dermatology, Rhode Island Hospital and Brown University, Providence, Rhode Island 02912.

Jonathan K. Wilkin • Division of Dermatologic and Dental Drug Products, Food and Drug Administration, Rockville, Maryland 20851.

Mitchell Wortzman • Medicis Pharmaceutical Corporation, Scottsdale, Arizona 85258.

Preface

In the 19th century, dermatology developed into a rich and highly efficient clinical specialty—on the basis of history and morphology.

In the 20th century, the clinical laboratory provided enormous strength in refining the science.

The 21st century will be remembered as the century of cutaneous biometrics. Our clinical judgments are now going to be based upon evidence-based medicine—highly buttressed by metrics.

This slim textbook is our first attempt to pull together some of the rapidly building database on these metrics.

We welcome your comments and suggestions.

Howard Maibach, M.D.
Doris Schwindt, M.D.

Contents

I. Basic Concepts

1. Statistics in Clinical Research

Volker W. Rahlfs

1. Introduction 3
2. The Meaning of the *P*-Value 4
3. The Appropriate Statistical Test 5
   3.1. Parallel Groups 6
   3.2. Correlated Samples 8
4. Effect Size Measures and Confidence Intervals 9
   4.1. Effect Size Measures 9
   4.2. The Confidence Interval 10
5. Multiple Testing and the Control of α 11
6. Tests for Difference and Equivalence 14
7. Two Examples and Their Interpretation 15
8. Conclusions 19
References 20

2. The Case-Control Study

Cornelius Kennedy, Helen J. Heacock, and Jason K. Rivers

1. Introduction 23
2. Reasons for Selecting a Case-Control Design 24
3. Planning and Carrying Out a Case-Control Study 25
   3.1. Developing a Study Protocol 26
   3.2. Data Collection 30
   3.3. Data Analysis and Interpretation of Results 31
4. Criteria Use to Judge Causality 34
5. Conclusion 34
References 35

3. Principles for the Conduct of Nondrug Clinical Studies

Gerhard J. Schmitt

1. Introduction 37
2. World Medical Association Declaration of Helsinki 38
3. Nondrug vs. Drug Studies 38
4. Skin Care Products 39
5. Protection of Trial Subjects: Written Informed Consent 39
6. Study Protocol 39
7. Safety Documentation for Test Products 40
8. Data Handling and Documentation 40
9. Statistics 41
10. Final Report 41
11. Summary 41
References 42

4. Questionnaire Design in Skin Disease Epidemiology

Päivikki Susitaival

1. Questionnaire Methods Used in Skin Disease Epidemiology 43
1.1. Some Epidemiological Concepts 43
1.2. Hand Dermatitis and Other Dermatoses 44
1.3. Atopy 46
1.4. Skin Cancer 47
1.5. Concluding Remarks 47
2. Designing a Questionnaire 48
2.1. Introduction 48
2.2. The Art of Asking Questions 49
2.3. Answer Categories 50
2.4. Final Tasks 50
3. The Tuohilampi Questionnaire 51
References 52

II. Noninvasive Evaluation of Skin Properties

5. Scale Development for Clinical Assessment

Gerhard J. Schmitt

1. Assessment Criteria and Scales 57
2. Case Report Forms 59
3. Validation and Calibration of Investigators 62
4. Relevance of Biophysical and Biochemical Parameters 63

6. *Evaluation of Cosmetics by Sensory Assessment*

Peter Busch and Thomas Gassenmeier

1. Introduction 65
2. The Relationship between Effect and Perception 65
3. Reasons for Sensory Vagueness in Assessing Cosmetic Effects 66
4. The Significance of Language 67
   4.1. Correlation of Experience and Language 67
   4.2. Generation of Verbal Terms 68
5. Quantitative Testing 71
   5.1. Principles of Quantification of Subjective Tests 71
   5.2. Comparison Test vs. Defined References 72
   5.3. Mathematical Aspects 77
   5.4. Structuring of Sensory Profiles 80
6. Areas of Application for Sensory Assessment in the Cosmetic Field . . . 80
References 80

7. *Cutaneous Bioengineering*

Dimiter Iliev, Peter Elsner, and Howard I. Maibach

1. Introduction 81
2. Barrier Function of Stratum Corneum as Measured by Transepidermal Water Loss 81
3. Skin Moisture 82
4. Acid Mantle of the Skin 83
5. Skin Color 83
6. Skin Blood Flow 86
7. Skin Roughness as Measured by Visiometry 86
8. Skin Thickness 87
9. Other Methods and Relations among Different Measurements 87
References 88

III. Skin Diseases and Therapy

8. *Atopic Dermatitis: Severity Scoring*

Alain Taïeb and Jean François Stalder

1. Introduction 93
2. Historical Perspective 94
3. Defining Extent 94
4. Choosing/Defining Severity/Intensity Terms 95
5. Interobserver Variability 96
   5.1. Physician Profile 96
   5.2. Intensity Items 96
   5.3. Extent 98

6. Subjective Items and Quality of Life 98
7. Assessment of Outcome with a Cumulative Compound Index: The SCORAD Index 99
8. Severity and Time Course of the Disease 101
9. Quality Control and Routine Scoring of Atopic Dermatitis 101
10. Scoring with SCORAD for Special Purposes 102
10.1. Localized Assessment 102
10.2. Inclusion Criteria for Clinical Trials 102
10.3. Epidemiological Studies 103
11. Biophysical and Biochemical Evaluation of Severity 103
12. Training 103
Appendix 1. EFFAD Recommendations and Definitions for Grading Intensity Items 103
Appendix 2. A Proposal for Severity Grading of Atopic Dermatitis 105
References 105

9. *A Review of the Use of Essential Fatty Acid Supplementation in Atopic Dermatitis with Emphasis on the Methodology of Trial Design*

John Berth-Jones and Robin A. C. Graham-Brown

1. Introduction 109
2. Rationale for Essential Fatty Acid Supplementation in Atopic Dermatitis . 110
3. Evidence for Efficacy 110
3.1. The *n*6 Series EFA Supplementation 111
3.2. The *n*3 Series EFA Supplementation 112
3.3. The *n*6 and Combined *n*6 and *n*3 Series EFA Supplementation . . . 112
4. Observations on the Methodology of Clinical Trials in Atopic Dermatitis . 114
5. Conclusions 115
References 116

10. *Severity Scoring of Atopic Dermatitis*

Aline B. Sprikkelman and Wim M. C. van Aalderen

1. Introduction 119
2. Atopic Dermatitis Severity Scoring Systems 120
2.1. Severity Scoring Systems for Clinical Practice 120
2.2. Severity Scoring Systems for Clinical Research 121
3. Conclusions and Recommendations 126
References 127

11. *A Metric for Acneiform Eruptions of the Face*

Jonathan K. Wilkin

1. Introduction 129
2. Materials and Methods 131
3. Results 137
4. Comment 139
References 147

12. *Toward a New Psoriasis Severity Scoring Using Bioengineering Techniques*

Doris A. Schwindt and Howard I. Maibach

1. Introduction 149
2. The Psoriasis Area and Severity Index 149
3. Assessing Severity with Bioengineering Techniques 151
   3.1. Infiltration 152
   3.2. Erythema 153
   3.3. Desquamation 154
4. Summary 155
References 156

13. *Epidemic of Melanoma: How Real Is It?*

Tsu-Yi Chuang

1. The Epidemic of Melanoma 159
2. An Analysis of the Epidemic 160
   2.1. Clinical Diagnosis 160
   2.2. Tumor Registry 160
   2.3. Shifting Pathological Diagnostic Criteria 160
   2.4. The Methods of Case Finding and Documentation 161
   2.5. Early Detection and Thin Melanoma 162
   2.6. Increase of Incidence Only in Certain Areas 162
   2.7. Increase of Incidence Only in Certain Types of Melanoma 162
   2.8. Birth Cohort Effect 162
   2.9. The Paradox of Increasing Incidence and Decreasing Mortality . . . 162
3. The Experience of an Epidemic in Kauai, Hawaii, 1981–1990 163
   3.1. Kauai and Its Residents 163
   3.2. Study Design 163
   3.3. Results 164
4. Check the Reality 165
References 166

14. *Prognostic Models in Melanoma*

Ranjini Natarajan and Martin A. Weinstock

1. Introduction 167
2. A Brief Review of Current Methods 168
3. A Mixed-Effects Survival Model 168
3.1. The Model 169
3.2. Assessing Significance of the Model 170
3.3. Reporting Results from a Mixed Model 171
4. Future Directions 171
References 172

15. *Photoaging*

William J. Cunningham

1. Introduction 173
2. Background 173
3. Rationale for Biometrics 174
4. Application of Biometrics 174
4.1. Clinical Aspects 174
4.2. Instrumentation 178
4.3. Histologic Assessments 180
5. Summary 181
References 181

16. *Measurement in Scleroderma Clinical Trials*

Janet E. Pope

1. Introduction 183
2. Statistics and Clinical Trial Guidelines 184
3. Diagnosis of Scleroderma 184
4. Skin Outcome Measurements 187
4.1. Functional Assessment 191
4.2. Global Assessments 192
4.3. Physical Performance Assessments 192
4.4. Visceral Assessments 193
5. Raynaud's Phenomenon in Scleroderma 194
6. Discussion 195
References 198

17. *Quantification of Wound Healing*

Laura Bolton, Ann-Jeanette Vasko, and Karyn Monte

1. Why Measure Healing? 205
2. Overall Healing Measures 206

3. Quantifying the Stages of Healing 206
3.1. Hemostasis 206
3.2. Inflammation 209
3.3. Granulation 211
3.4. Epithelization 211
3.5. Maturation and Complications 211
4. Conclusion 215
References 215

18. Design of Clinical Trials to Evaluate Antimycotic Agents in the Treatment of Dermatophyte Infections
Beatrice B. Abrams and Dennis E. Babel

1. Introduction 221
2. General Trial Design 222
3. Elucidation of the Pathogenic Organism 223
3.1. Specimen Procurement 223
3.2. KOH Examination 224
3.3. Culture 225
4. Clinical Assessments 227
5. Alternative Designs 230
6. Conclusions 230
References 231

19. Sites of Contact Dermatitis in Relationship to Particular Allergens: A Follow-Up Study
Björn Edman

1. Introduction 233
2. Material and Methods 233
3. Statistics 234
4. Results 237
5. Discussion 237
References 241

20. Evaluating Surveillance Systems in Contact Dermatitis
Axel Schnuch

1. Introduction 243
2. Epidemiologic Methods 244
3. Measures of Frequency: The Time Factor 244
4. Data Generation: Active or Passive 245
5. Study Population (Problem: Selection Bias) 245
6. Diagnostic Procedure (Problem: Information Bias) 246
7. General Aspects of Evaluating a Surveillance System 247

8. Discussion of Several Methods . . . 248
8.1. Active Data Generation . . . 248
8.2. Passive Data Generation . . . 249
9. Conclusions . . . 251
Appendix: Proposal to Establish Guidelines for Evaluating Surveillance Systems in Contact Dermatitis . . . 252
References . . . 253

21. *Sodium Lauryl Sulfate: Water Soluble Irritant Contact Dermatitis Model*

Cheol Heon Lee and Howard I. Maibach

1. Sodium Lauryl Sulfate . . . 257
2. Application Methods . . . 257
2.1. Purity and Carbon Length of SLS . . . 258
2.2. Quantity and Concentration of the Test Solution . . . 258
2.3. Evaporation and Temperature of the Test Solution . . . 259
2.4. Time of Evaluation . . . 259
3. Biologic Endpoints . . . 259
4. Host-Related Factors . . . 261
4.1. Age . . . 261
4.2. Sex . . . 261
4.3. Anatomic Region . . . 262
4.4. Skin Color . . . 262
4.5. Skin Hydration . . . 262
4.6. Sensitive Skin . . . 263
4.7. Hyperirritable Skin (Excited Skin Syndrome) . . . 263
4.8. Atopic Dermatitis . . . 263
4.9. Hand Eczema . . . 264
4.10. Seborrheic Dermatitis . . . 264
5. Conclusion . . . 264
References . . . 264

22. *Quaternium-18 Bentonite in the Prevention of Allergic Contact Dermatitis*

Robert L. Rietschel

1. Introduction . . . 269
2. Time of Occlusion . . . 269
3. When to Read Patch Tests . . . 270
4. What to Grade When Reading Patch Tests . . . 270
References . . . 272

23. Personal Cleansers: Irritation and Moisturization

Keith D. Ertel

1. Introduction . . . 273
2. Irritancy of Personal Cleansing Products . . . 274
2.1. Surfactant Interactions with Skin . . . 274
2.2. Estimating Personal Cleanser Irritation Potential . . . 277
3. Moisturization from Personal Cleansing Products . . . 281
3.1. Aspects of Moisturization from Personal Cleansers . . . 282
3.2. Protocol Considerations for Demonstrating Moisturization from Personal Cleansers . . . 282
3.3. Examples of Moisturizing from Personal Cleansing Products . . . 284
References . . . 284

IV. Pharmacologic Assays

24. Small Plaque Assay

Antti Lauerma and Howard I. Maibach

1. Introduction . . . 289
2. Small Plaque Assay . . . 289
2.1. Why Perform Small Plaque Assay . . . 289
2.2. Selection of Patients, Washout Period, and Pretreatment . . . 290
2.3. Treatment Period . . . 290
2.4. Follow-Up . . . 290
3. Small Plaque Assay Efficacy Parameters . . . 290
3.1. Clinical Scoring . . . 291
3.2. Chromametry . . . 291
3.3. Surface Roughness Assessment . . . 291
3.4. Transepidermal Water Loss . . . 291
3.5. Skin Ultrasound . . . 291
3.6. Histopathology . . . 292
4. Alternative Small Plaque Assay Methods . . . 292
5. Conclusions . . . 292
References . . . 292

25. Application of the Minolta Chromameter to the Assessment of Corticosteroid-Induced Skin Blanching

Roderick B. Walker, John M. Haigh, and Eric W. Smith

1. Introduction . . . 295
2. Methodology . . . 296
2.1. Materials . . . 296
2.2. Methods . . . 296

3. Results 299
3.1. Assessment of Visual Blanching Profiles 299
3.2. Assessment of Chromameter Blanching Profiles 299
4. Conclusions 303
References 303

26. *Assaying for Bioavailability in Acne*

Otto H. Mills, Jr., Caroline W. Cardin, and Mitchell Wortzman

1. Introduction 307
2. Pilot Studies of Benzoyl Peroxide and Topically Applied Erythromycin . 308
3. Comparative Study of Salicylic Acid Levels 310
3.1. Clinical Applications and Sampling Schedule 310
3.2. Analytical Method for Salicylic Acid Analysis (Isocratic HPLC Using UV Detection) 311
4. Conclusion 313
References 314

27. *In Vitro/In Vivo Correlations*

Eric Beyssac, J.-M. Cardot, and C. Bonnabry

1. Introduction 315
2. Topical Availability or Bioavailability 316
3. *In Vitro* and *In Vivo* Methods for Studying Drug Release: Diffusion and Permeation of Drugs from Dermatological Preparations 317
3.1. *In Vitro* Methods 317
3.2. *In Vivo* Methods 322
4. *In Vivo/In Vitro* Correlations or Relationships 324
4.1. Relationships between *In Vitro* Release and Diffusion and *In Vivo* Penetration 325
4.2. Relationship between *In Vitro* Release and Systemic Occurrence of the Product 326
5. Conclusion 327
References 328

V. Pharmacoeconomics

28. *Pharmacoeconomics and Outcomes Research*

Rukmini Rajagopalan

1. Introduction 333
2. Audience for Economic Evaluation and Economic Messages 333
3. Evaluation Methodology 334
3.1. Cost-Benefit Analysis 334

3.2. Cost-Utility Analysis 334
3.3. Cost-Effectiveness Analysis 336
3.4. Cost-Minimization Analysis 336
4. Benefits of Outcomes Research Studies 337
5. Examples of Cost-Effectiveness Studies 337
5.1. An Economic Analysis of Oral Therapies for Onychomycosis 337
5.2. An Economic Evaluation of Patch-Testing in Diagnosis and Management of Allergic Contact Dermatitis 337
5.3. Systematic Analysis of Components of Compounds and Outcomes Data 340
5.4. A Comparison of Outcomes between the Standard Treatment and the Treatment Based on Iterative Feedback and Simulation Modeling . . 342
6. Conclusion 342
References 342

29. The Cost of Atopic Eczema to Society: An International Perspective
Rukmini Rajagopalan

1. Introduction 343
2. Clinical Definition 343
3. Prevalence 344
4. International Scenes 346
4.1. United States 346
4.2. United Kingdom 348
4.3. Australia 349
5. Discussion 351
6. Conclusion 351
References 352

Index 353

Part I

Basic Concepts

Chapter 1

Statistics in Clinical Research

VOLKER W. RAHLFS

1. Introduction

Statistics in medical research is no longer what it was some 15 years ago: some descriptive measures for a collection of numbers and *P*-values based on tests appropriately chosen according to the type of data used.

Current research is based on an enlarged arsenal of statistical tools. Some new methods are: effect size measures and their confidence intervals in addition to *P*-values, tests for equivalence as contrasted with the classical tests for difference, appropriate sample size calculation, and—especially for confirmatory proof of efficacy—control of the "experimentwise" (per study) multiple-level α. The change in methods was accelerated by the appearance of new guidelines for clinical research. In Germany it was the quasi-guideline, *Principles for the Proper Performance of Clinical Studies*, which appeared in 1987 and for the first time stressed the importance of the control of α ("just one efficacy criterion"). In the United States the most important guidelines for clinical research covered many of the new topics (FDA Guideline, 1988). A comprehensive Biostatistics Guideline with requirements for state-of-the-art statistics was compiled by the CPMP working group of the European Union (1994) and was made the core content of the internationally based ICH-Biostatistics Guideline (1998). Some renowned, peer-reviewed, English and American medical journals also issued a concerted guideline, called CONSORT, describing statistical principles that have to be followed in manuscripts in order for them to be acceptable for publication (see, e.g., Altman, 1996).

In the following, basic principles and tools of the new biometry will be explained so that a clinical investigator, a medical writer, or, indeed, anyone who might perform clinical studies can gain some insight into and knowledge of what it is all about.

Volker W. Rahlfs • idv-Data Analysis and Study Planning, D-82131 Gauting, Germany.

Cutaneous Biometrics, edited by Doris A. Schwindt and Howard I. Maibach, Kluwer Academic/Plenum Publishers, New York, 2000.

2. The Meaning of the P-Value

Since the 1960s medical research papers have been required to include at least one P-value. There were times when publications were really crowded with P-values or their symbolic equivalent: an asterisk for $P < 0.5$, two asterisks for $P < 0.01$, and so forth. Yet the concept of the P-value and its interpretation is one of the least well understood by subject matter researchers and cursory users of statistics. Thus the P-value statement derived from a statistical test and the word "significant" are explained in the following.

The P-value is a quantitative measure of probability derived from a statistical test. It is the outcome of a game involving chance. Statisticians explore the distribution of chance results in repeated studies. They assume a distribution of observations in a defined situation—in the classical test for difference this is the "null-hypothesis" of zero-difference—and then calculate the probability of observing the empirical result or something more extreme. If this probability is smaller than a stipulated small value, often 0.05 ($\alpha = 0.05$), they call the empirical result "significant." It should be noted that significant does *not* mean relevant in a medical sense but rather "rare against the background of chance assuming the null hypothesis is true"!

One of the biggest misconceptions about the P-value is that it is the probability of the null hypothesis itself or the probability of the "alternative hypothesis," the latter meaning a difference. This is not the case! Indeed this probability could be calculated using Bayes' theorem and knowing the so-called *a priori* probability of the null hypothesis (see Diamond and Forrester, 1983; Browner and Newman, 1987). This *a priori* probability is, however, generally unknown and thus the reasoning is academic, although it may help to clarify the situation.

We have only the probability of the observed event (or a more extreme one) given the null hypothesis (in general that of a zero-difference). The procedure then is to assert a real (proven) difference if we have a small P-value or no difference if we have a large P-value. Although this reasoning was accepted as good science for several decades (no publication without P-value reasoning) there are some caveats concerning the interpretation:

1. When a chance distribution is calculated for the null hypothesis of zero difference, α is defined so that for a true null hypothesis exactly α (usually 5%) of the observed differences will be declared extreme or significant! This means that in working with real null hypotheses one obtains 1 in 20 significant results. This is not completely academic: In clinical studies when comparing randomized treatment groups on their baseline values the null hypothesis is true and one obtains "significant" results with a 1 in 20 chance. Thus we always risk wrongly inferring a difference: statisticians call this the type I error of statistical testing. In general this type I error must somehow be "controlled." For instance, before registration of a drug the FDA requires that at least two studies, with "significant" results, be submitted. In a single study if several tests are performed the overall α of all tests must be controlled in order that the single test result can be regarded as "statistical proof" or confirmatory.

2. The P-value result is strongly dependent on the sample size. In fact the test statistic of a test, which in turn is "transformed" into the P-value, is the result of a multiplication of an effect size with the size of the study (for this presentation see Rosenthal, 1991). For the t-test, e.g., the equation can be written as:

$$t = [(\bar{x}_1 - \bar{x}_2)/s] * [(1/n_1 + 1/n_2)^{-1/2}]$$

Thus it can easily be seen that a given difference—in this case the standardized difference—can be made "significant" or "not significant" by enhancing or reducing the sample size in the planning stage of a study.

The remedy for this arbitrariness of significant results is the rational calculation of the appropriate sample size for a study, which was already one of the most important requirements of the two cited Guidelines, the German Guideline of 1987 and the FDA Guideline of 1988. The sample size, generally speaking, should not be too small in order not to overlook relevant differences and not be too large in order not to declare trivial differences "significant." It is interesting to note that 10 years ago most clinical studies were clearly undersized or, quantified in numbers, approximately, "quarter-sized" (see Johnson, 1985).

3. Another remedy for misinterpreting the P-value results of a study is the calculation of a measure of the magnitude of a group difference and its confidence interval. A well-sized study has measures with a small confidence interval, while an unduly small study has measures with very large confidence intervals. The requirement of calculating confidence intervals was first imposed by a well-known British medical journal in 1986 (Gardner and Altman, 1986; Gardner *et al.*, 1986) and then again in the 1988 FDA Guideline.

3. The Appropriate Statistical Test

There are two important ways of doing a statistical test to obtain the P-value: the "permutational" and the "asymptotic."

The permutational principle can be understood as follows: Under the null hypothesis of no difference, all possible subdivisions of observed values according to group (treatment groups) are equally likely (of course one subdivision was that given as the study result). Thus the chance distribution can be obtained by calculating the distribution of all possible "reshufflings" or permutations of group configurations of n_1 and n_2 patients. One then obtains the "exact" P-value by counting the number of all results that are as extreme as or more extreme than those observed in the study and then dividing by the number of possible permutations. This is the so called "exact permutational" or briefly "exact" P-value.

There is also the asymptotic principle: For many data types and hypothesis situations statisticians have derived a theoretical distribution from which the P-value may be inferred more directly. In many practical situations both approaches are available in parallel: Thus the well-known t-test for two parallel groups is asymptotic, but there is also a permutational counterpart called the Fisher–Pitman (permutational) test.

Although the permutational tests are mainly counterparts of asymptotic tests they have the definite advantage of making fewer "assumptions" than their asymptotic relatives. Therefore they are generally regarded as the "gold standard" in statistics and should be used whenever they are available in the toolbox of the data analyst. Until recently, however, the necessary number of permutations (up to a billion) was prohibitive even for modern computers. Software engineering in the last decade has led to the invention of completely new permutational algorithms so that these tests are beginning to appear in computer packages. Two packages specialized for in exact tests are TESTIMATE* and StatXact.†

The practical worker has to make a choice among several available tests, each appropriate for a special data type or hypothesis. For this purpose we can define certain classes of data types: (1) normally distributed measurement data or data with another distribution, (2) measurement data with more or less unknown distribution, (3) time-to-event data, (4) ordinal data with a few categories, (5) nominal or categorical data with either two (binary data) or more categories. In addition to this classification a subdivision according to experimental design is necessary. The main classes are: parallel groups (independent groups) and pre-post or half-side experiment (correlated groups). It should be noted that calculating *P*-values for pre-post changes (within-group differences) in general is outdated with the exception perhaps of safety data. The half-side experiment is, however, still popular in dermatological research if the patient's body is symmetrically afflicted with the disease and a topical preparation is to be tested.

In the following some useful and popular tests are described briefly with reference to the subdivision scheme just introduced. The collection of selected tests is far from exhaustive and we recommend that the practical worker consult specialized statistical textbooks on this topic.

3.1. Parallel Groups

(a) Normally Distributed Measurement Data. In general the *t*-test is the method of choice. For a valid result the following conditions must be fulfilled:

1. The difference between groups is one of pure shift, i.e., all patient values are shifted by the same amount.
2. All data of a group are sampled from a Gaussian normal distribution.
3. All data are independent (which in general is guaranteed by design: each patient provides one and only one observation for the test).
4. The two samples are chosen from two homoscedastic populations (i.e., populations with equal variances). There is also a special *t*-test for heterogeneous variances but this is rarely used.

*TESTIMATE, 2000, Software for exact inference, confidence intervals, tests for equivalence/noninferiority, idv-Data Analysis and Study Planning, 82131 Gauting/Munich, Germany, Version 6.0.

†StatXact, Software for exact nonparametric inference, Cytel Software Corporation, Cambridge, MA 02139, USA.

(b) Measurement Data with an Unknown Distribution. For skewed or unknown distributions there are several tests available, which—depending on the situation—may be much more efficient than the t-test. The best-known alternative is the Wilcoxon–Mann–Whitney test, which can also be defined as a t-test applied to rank values (measurements are "transformed" according to their position in an ordered series: 1, 2, 3, 4, etc.), but it is not just an alternative for difficult data:

1. The difference can be one of (a) shift (as with the t-test), but also (b) stochastic superiority (test group for some values equal to control group or better for a certain range of values), or (c) general superiority (better for most values but perhaps worse for some)
2. The groups can have arbitrary distributions.
3. All data are independent [this assumption may only be discarded for the permutational version of the test, see Streitberg (1991)].
4. Appropriately calculated the test does not require the assumption of homoscedasticity (i.e., homogeneity of variances).

(c) Time-to-Event Data. In principle time-to-event data could also be analyzed using the Wilcoxon–Mann–Whitney test (which sometimes is quite correct). But when observing survival data in long-term studies statisticians have noted that the group difference is most frequently one of a constant multiplicator of hazards (risk of dying in a unit of time), the so-called proportional hazard, for which the logrank test (there are several similar tests called logrank) is the most efficient. In long-term studies there are usually patients who drop out before the event can be observed, so the logrank test (and the Wilcoxon test) were constructed anew to include information from "censored" values (observed up to a known time unit, but without the event).

The logrank is now also used outside the field of oncology. For instance, in dermatology one may analyze with the help of the logrank test "time-to-remission" or "time-to-relapse" after remission, and so on.

(d) Ordinal Data with a Few Categories (e.g., Rating Scale). Many data in clinical studies are based on rating scales. An example is the global assessment of efficacy, with categories 0 = none, 1 = slight, 2 = moderate, 3 = good, 4 = very good. Very often these data are analyzed after a dichotomization procedure, i.e., after reduction to a two-category system "no success"/"success" (success: good + very good), but this means discarding the information of about one-third of the patients. Therefore all five ordered categories should be evaluated.

The appropriate test utilizes the order and also takes account of the small number of categories. A good choice is again the Wilcoxon–Mann–Whitney test, either the exact permutational version or the asymptotic version but using a calculation equation that regards the discreteness or discontinuity of data to obtain a correct P-value. (Rahlfs and Zimmermann, 1993). It should be noted that there is a legitimate alternative, namely the "extended Mantel–Haenszel test" (Mantel, 1963), which basically is a modified t-test, with a variance estimator adapted to the restricted number of categories. The choice between the two tests is not critical as long as the data are approximately normally distributed (i.e., well-behaved); for other distributions it does make a difference (see Graubard and Korn, 1987).

(e) Binary Data: Two-by-Two Table. For the binary-data type, the data are usually counted in two classes and doing this for two treatment groups results in the so-called two-by-two table of aggregated data often seen in publications. The most common tests are the "Chi-square" (or the identical "z-test") and "Fisher's exact probability," the latter being an exact permutational test now available in most major program packages. Although the Fisher test is sometimes said to be the "gold standard" against which the other tests have to be evaluated, the two tests are based on rather different philosophies, and statisticians still engage in complete symposia about which of the two (or some other dozen possible tests) should be the test of choice.

During the last 10 years there has been a general trend in favor of the Fisher test, partly because it is now available in computer programs, partly because the odds ratio, a measure of relevance for two-by-two tables, has become more popular, and partly because it has also become more and more esteemed as the "gold standard" of testing.

3.2. Correlated Samples

(a) Normally Distributed Measurement Data. For normally distributed measurement data the paired t-test is still the method of choice, in principle with all assumptions and characteristics of the t-test for parallel groups.

(b) Measurement Data with an Unknown Distribution. For data with an unknown distribution, the Wilcoxon test, also called the "matched-pairs signed ranks test," is the one of choice. Based on rank-transformed differences within pairs, in principle it follows the Wilcoxon rank sum test for parallel groups. It is not generally known that a valid interpretation of results requires the assumption of a symmetric distribution of differences, but, unfortunately, there is nothing better in the major statistics computer program packages.

Another well-known test for this data situation is the sign test, which exploits information from the direction of changes but not the ranks of the quantities. Indeed it makes fewer assumptions but also utilizes less of the information in the data if these are available as measurements. Though being less efficient for "nice" data, the test is a robust tool for more difficult data. Thus it is also used by statisticians when testing laboratory data with unknown distributions and outliers for change.

(c) Ordinal Data with a Few Categories. Although working with differences from ordinal data does not seem to be appropriate, the Wilcoxon test for paired measurements is used for this data type. This is valid for most practical purposes as long as a small number of categories is considered, which the exact permutational P-value does. The sign test is also a reasonable one.

When displaying the data in aggregated form as counted frequencies in a publication one should not only give the univariate results (e.g., pre and post) but also the full cross-classification in the form of a $k \times k$ table with the counted frequencies of all changes (k = number of classes).

(d) Binary Data. The so-called McNemar test is by far the most popular for this situation; basically it is a sign test of frequencies in the two off-diagonal cells in a 2×2 table cross-classification of frequencies.

4. Effect Size Measures and Confidence Intervals

In 1986 a paper appeared in the *British Medical Journal* in which effect size measures and confidence intervals—otherwise not new in statistics—were not only explained to the practical worker but also made a requirement of future submissions to the journal (Gardner and Altman, 1986). Confidence intervals were briefly mentioned in the 1988 FDA Guidelines but only in the EU-Biostatistics Guideline (1994) and the ICH-Biostatistics Guideline (1998) were they made mandatory for pharmaceutical research.

What is it all about? The *P*-value statement is, as has been described, a combined value of effect size and sample size and as such not an indicator for the effect size. Why not take the effect size itself for quantifying the degree of a treatment effect? It is mandatory today to describe observed treatment effects by a measure of relevance or by an "effect size measure."

Some of the most important measures associated with statistical tests are described below.

4.1. Effect Size Measures

(a) One Group. The measures for describing effects in a one-group study are well known: mean value, median value, standard deviation, and—for counting the values of a binary variable—the "proportion" (of units with a certain property). There are many more in statistical textbooks and computer program packages and the practical worker will have to make her/his choice.

(b) Two Parallel Groups. The following are the tests for two parallel groups:

1. t-test: mean difference, but also other measures as, e.g., "standardized difference" or "quotient of mean values."

2. Logrank test: the hazard ratio is a useful measure in this situation. However, although it has a precise mathematical definition as a multiplier of a risk value, the meaning is difficult to explain. A more easily understood measure for the practical worker is the "percentage of improvement of the median survival time."

3. Wilcoxon–Mann–Whitney: There are two measures, the Hodges–Lehmann estimator, which is a robust measure of shift; it is usually calculated as the median of all possible differences between the observations of the two groups. There is also the Mann–Whitney superiority measure, which gives the number of winners of the test group if all values of the test group and control group are compared pairwise; this measure is made to vary from 0.0 to 1.0, with 0.5 denoting equality. It is of interest to note that this measure responds to differences of several different types: shift, stochastic superiority, and general superiority (more winners in the test group). We stress the point that the difference of medians is not a valid measure for the Wilcoxon–Mann–Whitney test. It should also be noted that for rating scales (with only a few categories) the Hodges–Lehmann shift measure is completely useless, whereas the Mann–Whitney statistic is still applicable.

4. Binary data, 2 × 2 tables: For this data type there are three well-known measures: the rate difference, the rate ratio, and the odds ratio. Until recently the rate difference was the most popular measure. The biological meaning of a difference, however, seems to change with the rate of the control group. This is why a department for anti-infective drugs in the FDA has defined varying differences for different rates of the control group as relevant (FDA, 1992). The odds ratio avoids this drawback and also has some theoretical advantages; thus it has been used more frequently in the last 5 years.

A new measure recently introduced is the "number needed to treat." (Laupacis *et al.*, 1988, Cook and Sackett, 1995). It is easily interpreted as the number of patients that must be treated in order to save one life or to have one patient cured. Computationally the measure is just the reciprocal of the risk difference.

(c) Two Correlated Samples. The following are the tests for two correlated samples:

1. t-test: Here again, the mean difference; others are possible.

2. Wilcoxon test for paired observations: Again the Hodges–Lehmann statistic and the Mann–Whitney statistic.

3. Binary data, 2 × 2 table: Some sort of odds ratio may again be used (see Noether, 1987), but also the difference of the two marginal rates from the table.

(d) Meaning of Certain Quantities. A very important question in this area is that of the meaning of certain quantities in the life sciences. When the meaning of a measurement scale is clear to the practical worker there is no problem: Most research workers would admit that a difference of 5 mm/Hg for the diastolic blood pressure is relevant. In other cases Cohen's empirical rule is useful. Cohen gave the following stipulations for the standardized difference (mean difference divided by the standard deviation): (a) 0.2 = small, (b) 0.5 = medium-sized or relevant, and (c) 0.8 = large (Cohen, 1969). Colditz *et al.* (1988), gave rules to translate this into analog boundaries of the Mann–Whitney statistic; Hasselblad and Hedges (1995) showed how the Cohen boundaries can be translated to those of the odds ratio for binary data. Another "translation" to the odds ratio was given by Tritchler (1995). Table 1 gives an overview of the boundaries of all three measures.

4.2. The Confidence Interval

Each measure of relevance can be associated with a confidence interval. This is defined as a random interval that captures the true parameter value of a measure of relevance with a predefined probability $(1 - \alpha)$. Another useful definition: The CI (confidence interval) contains all values which—if a hypothesis test were performed—would not give a significant result (all values outside the CI would give a significant result).

Based on these definitions the CI allows the practical worker to infer:

1. The possible location on the scale of measurement (or a derived scale) of the population parameter or measure of difference.

TABLE 1. Measures of Relevance with Relations

Interpretation according to Cohen's effect size, d	Measures[a]					
	SD $(x-y)/s$ Cohen's effect size d		Mann–Whitney statistics $P(X<Y)+0.5P(X=Y)$		Odds ratio	
	inferior	superior	inferior	superior	inferior	superior
No	0.0		0.5		1.0	
Small	−0.2	+0.2	0.44	0.56	0.696	1.438
Medium	−0.5	+0.5	0.36	0.64	0.404	2.477
Large	−0.8	+0.8	0.29	0.71	0.234	4.270

[a]SD is standardized difference.

2. The precision of the study: the wider the CI the less precise the study; within the framework of the t-test, e.g., the CI can be made smaller by making the standard deviation smaller and/or enhancing the sample size.
3. The null hypothesis decision can easily be derived from the CI: if CI covers the point of zero difference, a hypothesis test would give the decision "not significant," otherwise the decision would be "significant."

Thus, whereas a P-value would have given only the result "negative" (no significant difference), the CI also gives the reason: for instance, a small CI near but not including 0 indicates a trivial difference in a study with high precision, or a large interval including 0 indicates a potentially relevant difference in a study with very low precision (high variation and/or small sample size).

Table 2 gives some possible scenarios, explaining precision, significance, and relevance. Note that not only is scenario 5 a demonstration of equivalence, but—according to current test philosophy—so is scenario 3, although the difference is statistically significant.

5. Multiple Testing and the Control of α

Proof of efficacy (and also safety) can only be given with so-called confirmatory research. Remembering the fact that 1 in 20 of tested true null hypotheses will give a significant result, biometricians have recommended reducing the number of false positive results against the background of the "multiple testing problem" or "multiplicity of testing problem." The solution is "only one test per study" or the control of a global/multiple-level α for the entire study ($\alpha = 0.05$ per study, not per test!).

The German Guidelines of 1987 already contained the requirement for just one primary criterion (and implicitly: one point in time), so that from that time on there was little chance of a study being accepted by drug authorities if one presented

TABLE 2. Schematic Presentation of Potential Confidence Intervals and Their Interpretation

		Interpretation	
Scenario	Sign difference	Decisive for relevance	Relevance
(1) Zero difference ↓ Relevant difference ↓ 	No	No	Doubtdul
(2) Zero difference ↓ Relevant difference ↓ 	Yes	No	Doubtful
(3) Zero difference ↓ Relevant difference ↓ 	Yes	Yes	No
(4) Zero difference ↓ Relevant difference ↓ 	Yes	Yes	Yes
(5) Zero difference ↓ Relevant difference ↓ 	No	Yes	No

results with several primary criteria. The ICH Guideline now also requires a rigorous control of α in every study.

Control of a global α value is not new: The classical analysis of variance technique and also the well-known multiple comparison procedures are already global multiple-group tests with control of α. The new current guidelines require control of α for the entire study, i.e., for criteria, points in time, multiple-treatment groups, interim analyses, etc. The first tools to be used for this purpose were inefficient because many more patients had to be enrolled in a study than hitherto; otherwise the power was reduced to an unacceptably low level. Thus new methods had to be and were invented. In the following some general rules are given, starting with the all-purpose Bonferroni and Bonferroni–Holm rules.

The Bonferroni rule says that in order to control the multiple-level α one has to test a single hypothesis with an α divided by the number of tests. Thus with 20 tests one has a pertest-α of 0.0025 instead of 0.05. Of course, by this procedure some "significances" will vanish or else the sample size must be substantially increased. A small improvement of this rule was introduced by Holm, who recommended ordering the k P-values, and then testing the smallest by α/k, the second smallest by $\alpha/(k-1)$ and so on, so that the last could be tested with $\alpha/1 = \alpha$ (Holm, 1979). It is well known today that these "classic" procedures correctly control α but result in false-negative results: they have too little power.

Fortunately, three other simple procedures are available. The first is simply to select one confirmatory test and declare the others "exploratory" or "supportive" (and never say "significant" for those tests).

The second simple procedure is the principle of the *a priori* ordered hypotheses first discussed in some internal or informal papers in German-speaking countries (Maurer, 1987; Lehmacher, 1989) but now also explained in an English publication (Maurer *et al.*, 1995). According to this principle one orders the hypotheses according to medical relevance (and possibly the chance of success) and tests them all with full α. This can be done as long as the preceding hypothesis test in the *a priori* order is significant. Thus, e.g., for a chronic disease and a symptomatic treatment one starts with the last observation time and the most promising efficacy criterion and then turns to less promising times and/or criteria.

Yet another simple procedure is to condense *a priori*: One can summarize, e.g., locations, points in time, and similar facets of a syndrome to one single, powerful efficacy criterion: index value or composite index. In dermatological research the PASI index is a good example: It summarizes several aspects of psoriatic lesions with their severity, locations on the body, and the extent of the lesions. In general these composite index values are much more powerful than any of the single variables and thus should be used whenever this is medically feasible.

Serial observations over time under certain circumstances (so-called growth curves) can also be summarized to a single index value by several simple methods, e.g., sum, average, area under curve, response curve, linear trend component, or time to event. If the summarizing index is appropriate for the data situation, it is a powerful tool for avoiding the multiple-testing problem.

There are also statistically sophisticated tools now available for an efficient testing of several variables, locations, and time points, even if these all have different scales. These procedures give very powerful results because—as is done with the composite index values—all variables are pooled with their expected direction. Thus they test in principle as the composite index but in a much more flexible way. One well-known procedure for multiple-endpoints analysis is the O'Brien (1984) procedure. An even more flexible nonparametric directional multivariate test is known as the Wei–Lachin procedure (Wei and Lachin, 1984; Lachin, 1992), which is now commercially available as SmarTest Vers. 1.3.*

Some hints follow concerning different hypothesis situations:

1. Multiple criteria/endpoints: In some rare cases one criterion is sufficient, otherwise the principle of *a priori* ordered hypotheses or a composite index may be used. Finally all criteria can be tested together using the Wei–Lachin procedure,

*SmarTest, 1997, Software for efficient multiple endpoints analysis according to the nonparametric Wei–Lachin procedure, idv-Data Analysis and Study Planning, 82131 Gauting/Munich, Germany (Version 1.3).

either because the criteria are several aspects/locations of one syndrome or just because a powerful test is desirable (with some five criteria the sample size in general can be reduced by approximately 50% when using the Wei–Lachin procedure).

2. *Multiple points in time:* One tests only for one "endpoint" or summarizes several observations to one index value. Also the Wei–Lachin procedure may be used, perhaps even mixed, with several efficacy criteria.

3. *Multiple treatment groups:* The classical analysis of variance is a multigroup test with control of α and can still be used. However, if a reduction to two-group comparisons with an *a priori* order is feasible, this procedure is much more powerful (Maurer *et al.* 1995). The sample size may be reduced by at least 18% (Phillips, 1998).

4. *Interim analysis:* Naive interim analyses are no longer allowed as confirmatory research. The techniques employed for α adjustment are "group sequential design" or "two-stage procedure" (Bauer, 1989; Bauer and Köhne, 1994). The practical worker, intending to do a study with an interim analysis, is well advised to seek professional help in this very special and sophisticated area of biometrics.

5. *Multiple subgroups:* Subgroup analyses if done are usually performed in an exploratory manner. If the objective is a confirmatory statement, then α has to be controlled.

6. *Multiple tests and pretests:* Confirmatory testing means one preplanned test and no pretest. Thus for the same comparison of groups one should not calculate a *P*-value using the *t*-test *and* the Wilcoxon–Mann–Whitney. It is also no longer acceptable to do a pretest for a normal distribution and another for variance homogeneity and then decide which test is more appropriate. Thus the more robust Wilcoxon–Mann–Whitney will usually be the correct test to prespecify in the study plan.

6. *Tests for Difference and Equivalence*

Classical tests were all tests for differences: the null hypothesis was zero-difference, the alternative, a nonzero difference. For the well-known *t*-test the hypotheses are:

ONE-SIDED,

$$H_0: \mu_1 \leqslant \mu_2 \qquad H_A: \mu_1 > \mu_2$$

TWO-SIDED,

$$H_0: \mu_1 = \mu_2 \qquad H_A: \mu_1 \neq \mu_2$$

In the field of bioequivalence testing and also therapeutic trials it was found to be more useful to test for equivalence, with therapeutic equivalence being a one-sided test for noninferiority, and bioequivalence being a test for two-sided bioequivalence.

The hypothesis structure is now:

ONE-SIDED,

$$H_0: \mu_1 \leqslant \mu_2 - \Delta \qquad H_A: \mu_1 > \mu_2 - \Delta$$

TWO-SIDED (so-called composite hypothesis),

$$H_0: \mu_1 \leqslant \mu_2 - \Delta \quad \text{or} \quad H_0: \mu_1 \geqslant \mu_2 + \Delta$$

$$H_A: \mu_1 > \mu_2 - \Delta \quad \text{and} \quad H_A: \mu_1 < \mu_2 + \Delta$$

The problems with this definition are: (a) equivalence can only be defined with a statistic based on a scale (measure of relevance), and (b) it may be a matter of taste which Δ-value is relevant for defining either a relevant difference or a region of equivalence. For diastolic blood pressure, e.g., most researchers would agree that an amount of 5 mm/Hg is a relevant difference on the well-known scale of measurements. For mortality studies even 1 percentage point between the mortality rates of two groups might be regarded as relevant.

"One-sided or two-sided testing" has been debated for several decades in biometry. The US Food and Drug Administration has decided that the two-sided test with an $\alpha = 0.05$ is the rule. The ICH Guideline now allows one-sided directional tests but only if α is defined as 0.025 (although a one-sided $\alpha = 0.05$ could reduce the necessary sample size by more that 20%).

It should be noted that all test decisions could also be made using CIs. For a one-sided test for difference or equivalence this should be the 97.5% CI. For a two-sided test for difference it should be the 95% CI (as it was previously for several decades). For two-sided bioequivalence testing it must be—for some technical reasons—the 90% CI (!), in order to have a 5% test for equivalence. This is now the requirement of the current Bioequivalence Guidelines, either EU (1991) or US (1992).

7. Two Examples and Their Interpretation

The first example is a clinical study in dermatology, a comparison of two drugs—one test drug and one reference. The study objective may be operationalized as a statistical proof of noninferiority. Technically speaking, therapeutic equivalence results from a one-sided test for equivalence, with a definite lower bound of tolerable inferiority, Δ. The experimental design is a half-side study with symmetric lesions on both sides of the body, each side treated with one of the two drugs and a subsequent comparison within patients (each patient his own control). Primary criterion is defined as "overall assessment of efficacy," with scoring: 0 = none, 1 = slight, 2 = moderate, 3 = good.

Using the Mann–Whitney estimator we might formulate the hypothesis struc-

TABLE 3. Frequencies of Judgments[a]

	Test prepartion				
Reference	0	1	2	2	Sum
0	15	2	11	4	32
1	4	13	2	12	31
2	4	0	15	5	24
3	1	5	5	14	25
Sum	24	20	33	35	112

[a]Body of table: contingency table; marginals: judgments for the two treatments.

ture for noninferiority as follows:

$$H_0\text{: MW}_{TR} \leqslant 0.40 \qquad H_A\text{: MW}_{TR} > 0.40$$

where T is test preparation and R is reference or comparator. This means that an inferiority of the test drug condition as compared to the reference drug condition of 0.4 [between: 0.36 (=relevant) and 0.44 (trivial)] is accepted as negligable for practical purposes. The sample size of the study will then be planned in the standard way so that if the true difference between groups is zero (MW = 0.5) then the null hypothesis of inferiority can be rejected with high probability, say greater than 0.9.

An empirical study could, e.g., produce the observational frequencies for two test conditions on the same patient shown in Table 3. It should be noted that most people would prefer to compare and interpret the difference of the two marginal frequencies. It is, however, absolutely neccessary that we also have the frequencies of the cross classification for the situation of the within-patient comparison—whether we interpret these or not. The numbers of the cross classification are used for the complete calculation.

Table 4 shows the output of the module $k \times k$ table with ordered categories, a part of the program TESTIMATE. It can easily be seen that the test for linear trend is significant ($P = 0.0074$) so that there is an indication of a difference between treatment conditions. However, in order to test for noninferiority we need a measure of the treatment effect and its one-sided CI. Treatment effects of clinical relevance are often expressed by dominance of one cumulative distribution over the other, called stochastic superiority, for which the Mann–Whitney statistic is an appropriate measure. The Mann–Whitney statistic for the two marginal frequences is MW = 0.58 with the lower bound of the 97.5% CI being LB = 0.52 for $\alpha = 0.025$, one-sided. As the lower bound is larger than the predefined lower bound of the equivalence region MW = 0.4, we conclude that there is statistical proof for noninferiority at the preplanned $\alpha = 0.025$ (one-sided) level of significance. (In this case the lower bound is even larger than MW = 0.5, and we could also conclude that there is statistical proof of superiority; this, of course, is not always the case, but in some studies a lucky by-product).

TABLE 4. Overall Assessment of Efficacy Comparison of Test Drug and Reference Drug ($\alpha = 0.025$, One-Sided)

Contingency Table with Ordered Categories[a]

Reference / Test:	0	1	2	3	Total
0	15 13.4%	2 1.8%	11 9.8%	4 3.6%	32 28.6%
1	4 3.6%	13 11.6%	2 1.8%	12 10.7%	31 27.7%
2	4 3.6%	0 0.0%	15 13.4%	5 4.5%	24 21.4%
3	1 0.9%	5 4.5%	5 4.5%	14 12.5%	25 22.3%
Total	24 21.4%	20 17.9%	33 29.5%	35 31.3%	112 100.0%

[a]Note: 50.0% of (average of paired off-diagonal) cells have frequencies <5.

Marginal Homogeneity (Stuart–Maxwell test for trend)

Linear Contrasts

1:	−3	−1	1	3
2:	1	−1	−1	1
3:	−1	3	−3	1

Contrast	Chi^2	DF	P
1: Linear	7.1675	1	0.0074
2. Quadratic	.0833	1	0.7728
3: Cubic	3.7059	1	0.0542

Measure of Agreement/Association

Ordered Cat.: Kendall's Tau-b = 0.3091

Test and estimators (97.5% CI: LB, UB) one-sided; direction: Test > Reference

Isoasymmetry test (Marginal Homogeneity)

$P = 0.0029$ $(z = 2.7587)$

Mann–Whitney	0.5813	(0.5235, −.−)
Mann–Whitney difference (Goodman–Kruskal Tau)	0.1625	(0.0471, −.−)

Anisoasymmetry test (Generalized McNemar)

$P = 0.0109$ $(z = 2.2923)$

Mann–Whitney	0.5759	(0.5110, −.−)
Mann–Whitney difference	0.1518	(0.0220, −.−)

TABLE 5. *t*-Test Titel (Criterion: PASI)

Statistics	Group: Placebo		Group: Test
N	15		19
Valid number	15		19
Mean value	2.324		2.087
Standard deviation	0.5084		0.5691
Standard deviation pooled		0.5434	
SEM		0.1877	

Direction of location: (R): placebo > (T): test

Test for Difference (one-sided)
$P = 0.1079$ ($t = 1.2628$; DF = 32)

Estimators and confidence intervals (97.5% CI, two-sided)

—Mean Difference	0.237	(−0.145, -.-)
—Mean Quotient (Sasabuchi)	1.114	(0.935, -.-)
—Standard. Difference (Cohen's d)	0.4362	(−0.2524, -.-)
—Coefficient of Determination (R^2)	0.1598	(0.0000, -.-)
—Hedges's g (unbiased)	0.4259	

Test for homogeneity of variances
$P = 0.1050$ ($F = 1.2530$; DF = 18, 14)

Note: The F-test is a test for homogeneity of variances of both samples. If the F-test is significant, the following t-test for heterogeneous variances can be used.

t-Test for heterogeneous variances (Welch approximation)
$P = 0.1050$ ($t = 1.2801$; DF = 31)

A second example is a test for superiority that is planned based on a parallel-group design comparing a new test drug with a Placebo, with two randomized groups of $N_1 = 15$ and $N_2 = 19$ patients. The results of the PASI index at the end of the study will be analyzed using the well-known t-test, but interpreting measures of relevance and their CIs. Table 5 gives the output for the t-test module of the program TESTIMATE from idv. The upper part shows the statistics: the mean values indicate a superiority of the test group as lower values are preferable in a clinical sense. The different measures of treatment effects are now interpreted:

1. Mean difference: There is a superiority in mean values of 0.237. The lower bound is -0.145. As this lower bound does not exclude the value zero, there is no statistical proof for superiority or the null hypothesis of an inferiority cannot be rejected at the 2.5% level of significance. (This, of course, fits with the statement $P =$ 0.1079).

2. *Mean quotient:* The quotient is 1.11, i.e., the placebo mean value is 11% higher. The lower bound of the CI covers 1.0, so there is no statistical proof of superiority.

3. *Standardized difference:* The mean difference in standard units is 0.436, which is a relevant superiority as compared to Cohen's benchmark values (0.5: relevant or medium-sized superiority). The lower bound CI is −0.252. As the value 0.0 is not excluded there is no proof of superiority.

4. *Coefficient of determination:* R^2 is 0.16 which means that 16% of the total variance of the PASI index is explained by the group difference. The lower bound of the CI is 0.0 but only because negative values are not defined for this squared measure. Thus the bound of this coefficient is easily interpreted only if there is a positive value as a result.

Thus all four measures demonstrate a superiority for the test group. There is, however, no statistical proof for superiority as all lower bounds of CIs cover the value of a zero-difference. Had a test for noninferiority been planned we could have proven noninferiority as all lower bounds of CIs exclude large or relevant values of inferiority. It should be noted that switching hypotheses in this direction is not appropriate, but the other way around, first to test for noninferiority and—if significant—to test for superiority, is appropriate (see Röhmel, 1998).

8. Conclusions

This chapter gave an overview of the philosophy of statistics and also presented the ideas and ways of the new biometry, now required in clinical research by international guidelines and health authorities. It was shown how study objectives are now expected to be operationalized precisely as a test for superiority or as a test for noninferiority (therapeutic equivalence). Tests were set into perspective and it was described how the amount of a difference between treatments could be quantitated using one or more of several available measures of relevance. Study results should now always be presented as measures of relevance and their confidence intervals. The latter can be used for inferring either noninferiority, or superiority, or even relevant superiority. According to current guidelines only one statistical statement is accepted as confirmatory statement; other statements are either controlled concerning the so-called multiple-level α of the study or should be regarded as descriptive or supportive.

Two examples in dermatology, one from a half-side experiment (correlated observations) and rating scale values and the other from a standard parallel-group design with the t-test applied on PASI index values, were given in order to show how measures of relevance and their confidence intervals are used for the statistical statements concerning a test for noninferiority or a test for superiority. Thus it may be seen that—using the new toolbox of biometry—statements can be made that are statistical proofs and at the same time give a feeling for the medical relevance of the result.

References

Altman, D. G., 1996, Better reporting of randomised controlled trials: the CONSORT statement, *Brit. Med. J.* **313**:570–571.

Bauer, P., 1989, Multistage testing with adaptive designs, *Biom. Inf. Med. Biol.* **20**:130–148.

Bauer, P., and Köhne, K., 1994, Evaluation of experiments with adaptive interim analyses, *Biometrics* **50**:1929–1041.

Browner, W. S., and Newman, T. B., 1987, Are all significant *P* values created equal? The analogy between diagnostic tests and clinical research. *JAMA* **257**:2459–2463.

Cohen, J., 1969, *Statistical Power Analysis for the Behavioral Sciences*, Academic Press, New York (Rev. Ed. 1977).

Colditz G. A., Miller, J. N., and Mosteller, F., 1988, Measuring gain in the evaluation of medical technology. *Intern. J. Technol. Assess. Health Care* **4**:637–642.

Cook, R. J., and Sackett, D. L., 1995, The number needed to treat: A clinical useful measure of treatment effect, *Brit. J. Med.* **310**:452–454.

Diamond, G. A., and Forrester, J. S., 1983, Clinical trials and statistical verdicts: Probable grounds for appeal. *Ann. Int. Med.* **98**:385–394.

EU Bioequivalence Guideline, 1991, CPMP Working Party on Efficacy of Medicinal Products, Note for Guidance, Investigation of Bioavailability and Bioequivalence, Brussels, Belgium: CPMP (December 1991).

EU-Biostatistics Guideline, 1994, Commitee for the Proprietary Medicinal Products. A Note for Guidance on Biostatistical Methodology in Clinical Trials in Applications for Marketing Authorizations for Medicinal Products, Brussels, Belgium: CPMP (December 1994).

FDA Guideline 1988, Guideline for the Format and Content of Clinical and Statistical Sections of New Drug Applications. Rockville, MD, US Food and Drug Administration.

FDA, 1992, Points to consider, Division of Anti-Infective Drug Products.

Gardner, M. J., and Altman, D. G., 1986, Confidence intervals rather than *P* values: Estimation rather than hypothesis testing, *Brit. Med. J.* **292**:746–750.

Gardner, M. J., Machin, D., and Campbell, M. J., 1986, Use of check lists in assessing the statistical content of medical studies, *Brit. Med. J.* **292**:810–812.

German Guideline, 1987, Principles for the Proper Performance of a Clinical Study: Bekanntmachung von Grundsätzen für die ordnungsgemäße Durchführung der klinischen Prüfung, Bundesminister für Jugend, Familie, Frauen, Gesundheit (BMJFFG) (9.12.1987).

Graubard, B. I., and Korn, E. L., 1987, Choice of column scores for testing independence in ordered $2 \times k$ contingency tables, *Biometrics* **43**:471–476.

Hasselblad, V., and Hedges, L. V., 1995, Meta-analysis of screening and diagnostic tests, *Psychol. Bull.* **117**:167–178.

Holm, S., 1979, A simple sequentially rejective multiple test procedure, *Scand. J. Stat.* **6**:65–70.

ICH-Biostatistics Guideline, 1998, ICH Harmonized Tripartite Guideline, Note for Guidance on Statistical Principles for Clinical Trials, ICH Topic E9, Step 4, Consensus Guideline (February 5, 1998).

Johnson, A., 1985, Sample size: clues, hints or suggestions, *J. Chron. Dis.* **38**:721–725.

Lachin J. M., 1992, Some large-sample distribution-free estimators and tests for multivariate partially incomplete data from two populations. *Stat. Med.* **11**:1151–1170.

Laupacis, A., Sackett, D. L., and Roberts, R. S., 1988, An assessment of clinically useful measures of the consequences of treatment, *New Engl. J. Med.* **318**:1728–1733.

Lehmacher W., 1989, Schrittweises multiples Testen bei a-priori-geordneten Hypothesen, Int. Report, Gesellschaft für Strahlen- und Umweltforschung (GSF), MEDIS, Neuherberg/Munich, Germany.

Mantel, N., 1963, Chi-square tests with one degree of freedom: Extensions of the Mantel–Haenszel procedure, *J. Am. Stat. Assoc.* **58**:690–700.

Maurer, W., 1987, Statistische Besonderheiten bei klinischen Studien, Biometrisches Seminare, Locarno (September 21–25, 1987).

Maurer, W. Hothorn, L. A., and Lehmacher, W., 1995, Multiple comparisons in drug clinical trials and preclinical assays: A priori ordered hypotheses, in: *Testing Principles in Clinical and Preclinical Trials, Biometrie in der chemisch-pharmazeutischen Industrie* (J. Vollmar ed.) Gustav Fischer, Stuttgart, pp. 3–18.

Noether, G. E., 1987, Sample size determination for some common nonparametric tests, *J. Am. Stat. Assoc.* **82**:645–647.

O'Brien, P. C., 1984, Procedures for comparing samples with multiple endpoints, *Biometrics* **40**:1079–1087.

Phillips. A., 1998, Sample size estimation when comparing more than two treatment groups, *Drug Inf. J.* **32**:193–199.

Rahlfs, V. W., and Zimmermann, H., 1993, Scores: Ordinal data with few categories—how should they be analyzed? *Drug Inf. J.* **27**:1227–1240.

Röhmel, J., 1998, Therapeutic equivalence investigations: statistical considerations, *Statist. Med.* **17**: 1703–1714.

Rosenthal, R., 1991, *Meta-Analytic Procedures for Social Research*, Sage Publications, Newbury Park.

Streitberg, B., 1991, Review of: Daniel, W. W., *Applied Nonparametric Statistics*, 2nd Ed, PWS-Kent, Boston 1990, *Biometrics* **47**:1634–1635.

Tritchler, D., 1995, Interpreting the standardized difference, *Biometrics* **51**:351–353.

US Bioequivalence Guideline, 1992, Statistical Procedures for Bioequivalence Studies Using a Standard Two-Treatment Crossover Design, Division of Bioequivalence (HFD-650), Food and Drug Administration, Rockville, MD 20855.

Wei L. J., and Lachin J. M., 1984, Two-sample asymptotically distribution-free tests for incomplete multivariate observations. *J. Am. Stat. Assoc.* **79**:653–661.

Chapter 2

The Case Control Study

CORNELIS KENNEDY, HELEN J. HEACOCK AND JASON K. RIVERS

1. Introduction

Observations are fundamental to medical knowledge, but we must be able to describe our observations accurately so that correct information is transmitted to our peers. Observations may be communicated by means of case reports and these in turn may lead to the discovery of new diseases, such as Lyme disease, which was identified after the initial description of "epidemic arthritis" in Connecticut (Steere *et al.*, 1983). However, the drawback of case reports is that an observed association between an exposure and a disease cannot in itself prove causality nor can it provide a measure of the risk association between the observed exposure and the disease of interest.

By contrast, the case-control study design can provide a risk measure (the odds ratio, OR) for a particular exposure–disease association (Fig. 1). Examples where case-control studies would be appropriate include the assessment of sunburns as a risk for melanoma, where the number of blistering sunburns (the exposure or perceived risk factor) reported would be compared between a group of melanoma patients (cases) and a group of people without melanoma (controls). In an occupational setting, a case-control design could evaluate previous asbestos exposures in the ship-building industry among workers with mesothelioma (cases) and compare this to asbestos exposures among nondiseased workers (controls).

The purpose of this chapter is to provide an overview of the case-control study. Steps involved in its design and execution will be discussed so that the reader will feel competent both in critically appraising the published medical literature and in conducting such a study.

Cornelis Kennedy and Helen J. Heacock • Department of Health Care and Epidemiology, University of British Columbia, Vancouver, Canada. *Jason K. Rivers* • Division of Dermatology, Vancouver Hospital and Health Sciences Centre, University of British Columbia and British Columbia Cancer Agency, Vancouver, Canada.

Cutaneous Biometrics, edited by Doris A. Schwindt and Howard I. Maibach, Kluwer Academic/Plenum Publishers, New York, 2000.

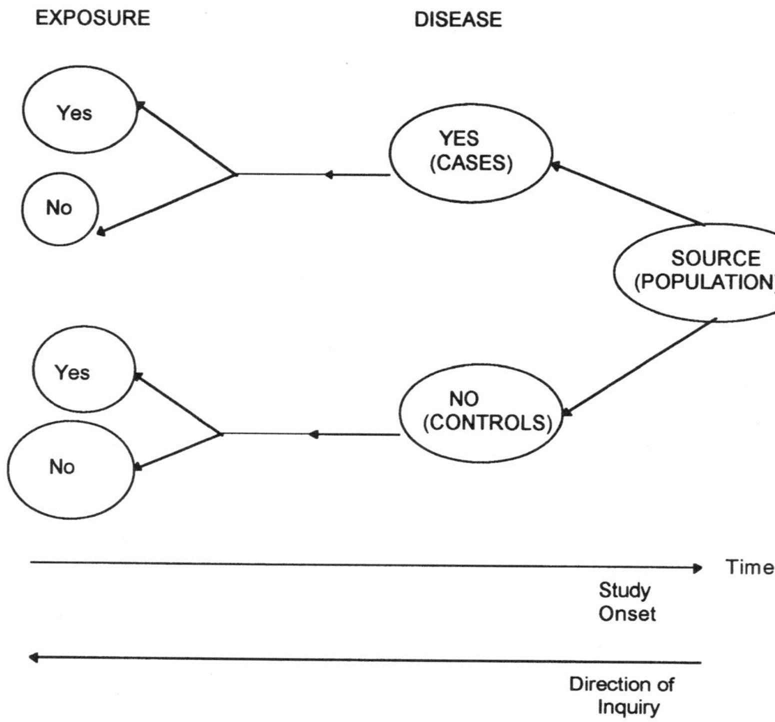

FIGURE 1. Schematic diagram of case-control study design (from Heacock and Rivers 1997*a*).

2. Reasons for Selecting a Case-Control Design

The case-control strategy begins by identifying persons with a particular disease (cases) and a sample of nondiseased controls, then looks back in time to elucidate risk factors or exposures that may be related to the disease/condition in question. Therefore, researchers do not have to wait for disease to develop as would be the situation for prospective designs such as cohort studies and clinical trials. Hence, this design saves time and money and is most suitable in the following circumstances:

1. When studying diseases with long latency periods: Skin cancers, both melanoma and nonmelanoma, generally develop more than 20 years after exposure to the suspected carcinogen (sunlight). Hence waiting for new cases to accrue while simultaneously assessing exposure is impractical. Therefore, skin cancer is ideally suited to study using the case-control design. Numerous case-control studies evaluating the association between sun exposure and skin cancer have been published. In one such study, the investigators identified skin cancer cases and nondiseased controls, then evaluated previous exposures using medical records and information provided directly by cases and/or controls (Healy *et al.*, 1995). In another study, investigators compared cases and controls by counting the number of nevi on different body sites (Rieger *et al.*, 1995).

2. In the evaluation of rare diseases: Temporal and geographical distributions of disease often make it difficult for an investigator to enroll sufficient cases of a rare disease from one center to achieve a meaningful exposure–disease association; therefore cases from multiple centers are often enlisted. The investigator can collect all identified rare cases (often as part of a multicenter study) and can then compare the exposure histories of these cases to exposure histories of a control group without the disease. For example, case-control studies have focused on rare diseases including toxic epidermal necrolysis (TEN) and Stevens–Johnson syndrome (Roujeau *et al.*, 1995), bacillary angiomatosis (Tappero *et al.*, 1993), and non-Hodgkin's lymphoma (Pearce *et al.*, 1986).

3. For the simultaneous evaluation of multiple exposures and for the evaluation of interrelationships among these different exposures: The case-control design is useful for diseases where the underlying cause(s) remains unknown. Using this research design, the investigator can collect information on many different exposures that may be considered risk factors for a particular disease with unknown etiology, and subsequently compare exposure status between a case and one or more controls. Case-control studies have been used to evaluate risk factors for melanoma (Garbe *et al.*, 1994; Carli *et al.*, 1995), for psoriasis (Naldi *et al.*, 1996), or occupational risk factors for various cancers (Siemiatycki *et al.*, 1986).

Many occupational studies of workplace hazards use a special type of case-control study known as industry-based or "nested." In this type of design, the study base includes all persons in a workplace or in a specific industry. Cases then are all persons in that industry or workplace who develop the disease of interest and controls are drawn from the nondiseased workforce. The groups are usually matched on gender, age, and workplace experience. For example, in a study of leukemia and organic solvent exposure among 10,000 chemical workers, a suitable design would include all cases of leukemia among the workers and a sample of other workers free of leukemia at the time the leukemia cases were diagnosed (controls). Readers interested in nested case-control studies are referred to Checkoway *et al.* (1989).

3. Planning and Carrying Out a Case-Control Study

1. *Develop a study protocol:* (a) state the study objectives and the research question; (b) identify the study base where the cases and controls will be selected; (c) define the disease under study and establish eligibility criteria for cases and controls; (d) determine the source of cases and controls; (e) describe the sampling and matching method(s); (f) describe the exposure assessment; (g) describe the analysis; (h) consider sample size and power; (i) consider consent and confidentiality; (j) prepare time-line and budget; (k) obtain ethics approval; (l) perform a pilot study.
2. *Data collection:* (a) identify and collect the cases and controls; (b) collect exposure data.
3. *Data analysis:* (a) analyze the data.
4. *Interpret the results:* consider factors such as: (a) flaws in the study design; (b) measurement error; (c) systematic errors in data collection and/or

interpretation (bias); (d) the influence of variables that may affect the observed association (confounding); and (e) the "luck of the draw" (chance).

3.1. Developing a Study Protocol

(a) Stating the Study Objectives and the Research Question. Much of today's medical research requires funding through peer-review committees at the state or national level, hence a well thought out and clearly described protocol is necessary before embarking on a study. The research protocol starts with a literature review that establishes the present state of knowledge in the area of interest and also indicates existing gaps in knowledge. Numerous computer data bases such as Medline and Cancerlit are available to access the published literature and help define the study's objective(s) and research question(s). The research question (or hypothesis) is a statement indicating that the study will examine an association between an independent (exposure) variable and a dependent (disease) variable. This statement often starts as a fairly broad question or statement about the exposure and the disease. An example of a research question is, "the purpose of this study is to examine the relationship between sunburn (exposure) and histologically confirmed melanoma (disease) in a Caucasian population in Hawaii." All research should include an objective or research question in the introductory section of the paper.

(b) Identifying the Study Base from Which the Cases and Controls Will Be Selected. Before selecting cases and controls, the researcher must define the population base from which cases will be selected and controls sampled. The population base is often defined using geographic and temporal boundaries, e.g., all males living in Washington State between 1980 and 1985.

(c) Defining the Disease under Study and Establishing Study Eligibility Criteria for Cases and Controls. It is imperative that strict diagnostic criteria as well as rules for inclusion and exclusion of cases and controls be established to ensure that only individuals with the disease of interest are included in the case group and individuals without the disease are included in the control group. Diagnostic criteria (such as histologic, radiographic, or laboratory confirmation) should clearly define the disease under study and a statement of how the diagnosis was reached should be provided. This will permit results to be generalized to similar population structures.

Most often, only incident (newly diagnosed) cases are recruited for case-control studies. Utilizing incident cases is advantageous because the investigator can be relatively sure that the case represents an individual who did not have time to change his/her behavior *because* of the disease, thereby invalidating the exposure history. For certain diseases (e.g., acute allergic contact dermatitis) the time of diagnosis is a reasonably good approximation to the time of onset, where as for many other diseases the actual time of onset is more difficult to ascertain (e.g., skin cancers). An example of a clear description of case selection that also includes a description of the study base is: "We studied consecutive male patients with psoriasis admitted to the outpatient clinics of the departments of dermatology of the university central hospitals in Helsinki, Oulu, and Tampere from September 1987 to August 1989... we restricted the study group to patients aged 19–50 whose onset of skin disease was 1976 or later... six patients refused." (Poikolainen *et al.*, 1990).

(d) Determining the Source of Cases and Controls. The objective of a case-control comparison is to discover differences in exposures between the cases and controls that can explain the occurrence of disease in the cases and absence of disease in the controls.

There are many sources available by which to obtain cases. These include hospital admission and discharge records, disease registries (e.g., cancer, congenital anomalies), medical records, workers' compensation, and other insurance claims, disability pension records, vital records, and death certificates.

The choice of the control group is an essential part of the case-control strategy. The purpose of the control group is to estimate the "expected" exposure history of the case population. Controls are drawn from the same study base that generated the cases and should represent a group of unaffected people who would have been considered "cases" had they developed the disease. The choice of control group is an important part of the case-control strategy because compatibility between the two groups is essential if the observed difference is to be attributed to the exposure of interest (Kelsey, 1986). Several sources are available from which to draw a control group:

1. Hospital controls are patients in the same hospital as the cases but admitted for conditions other than the disease of interest. They can be selected by using hospital admission/discharge summaries or disease registry data. The advantage of using hospital controls is that cases and controls both have had time to ponder their illness and previous exposures, so the study is less likely to be affected by differences between cases and controls with respect to remembered exposures (recall bias). In addition, the control patients are readily available and usually more motivated to participate, thereby helping to minimize differences in response rates between cases and controls (nonresponse bias). The disadvantage of using hospitalized controls is that like the cases the control patients are ill and therefore not representative of normal, healthy people.

2. General population controls represent a probability sample of the population from which the cases were recruited. In population-based studies, where cases are selected from the general population, e.g., all melanoma cases reported to the Cancer Registry in Massachusetts in 1995, the most appropriate control group would also be drawn from the general population. The advantage of population-based designs is that they are more representative of disease occurrence in the general population. The disadvantage is that this design is expensive, and it is time-consuming to collect the population controls from voters' lists, tax lists, telephone directories, census data, or health insurance plans.

3. Friends, neighbors, and relatives of the cases can serve as controls. By choosing controls who live in close proximity to the cases, differences owing to socioeconomic gradient, ethnicity, environmental exposures, income, and access to health care may be minimized. On the other hand, these types of controls may have experienced similar exposures to the cases thereby making it difficult to derive a true exposure difference.

There is no control group that is optimal for all situations. The selection of an appropriate control group depends on several factors, including the characteristics and source of the cases and issues of practicality and finance.

(e) Describing Sampling and Matching Methods. Once the population base from which cases are to be selected and controls sampled is determined, one must consider the optimal ratio of controls to cases, which factors ("confounders") will be matched for, and finally, the sampling method most appropriate for the study in question.

Deciding on the case:control ratio will depend on the cost of selecting and obtaining information from the controls in relation to the number of available cases. If the number of cases and controls is large and the cost of obtaining information is comparable between the two groups, then a 1:1 ratio will normally be sufficient. If the number of cases is small and the cost of obtaining information on controls is relatively low, then more controls may be matched to each case. By utilizing the latter strategy, one can increase the statistical power of a study. However, little benefit is achieved if the control-to-case ratio is increased above 4:1. (Hennekens and Buring, 1987).

Matching controls to cases is a process that draws on the similarities between them and allows the study to better investigate the risk of disease from exposures, thereby minimizing spurious associations based on confounding ("nuisance") variables that distort true exposure–disease associations. Factors that are often matched for in medical studies include age, gender, race, socioeconomic status, access to health care, and family history of a certain disease (Kelsey *et al.*, 1986). For instance, matching on age means that a case is comparable to a control based on age and if a difference occurs it can be attributed to a factor other than age. Matching on too many variables runs the risk of matching for variables that are related to the exposure of interest but not the disease. This is called overmatching and besides being costly may make it harder to detect an association between the exposure and the disease if one exists.

Once the type of control group has been selected and the number of controls per case has been decided, a method by which to sample them is needed. The most commonly used sampling techniques are:

1. Random sampling: This procedure is simple and utilizes random number tables to assign controls to the study. The disadvantage is that the investigator will need a complete listing of all eligible controls.

2. Systematic sampling: Eligible controls are selected by using a specified interval between each control selection. With this technique controls can be selected as the study progresses and the investigator does not need to have a complete listing of all eligible controls in advance. For instance, the investigator can select every third eligible control that is admitted to the hospital (given of course that the control does not have the same disease as a case).

3. Stratified sampling: Before the actual sampling occurs the target population is divided into strata or subgroups based on specific characteristics (e.g., race or socioeconomic status). The investigator subsequently takes a random sample from each stratum to ensure that each subgroup of the population is represented.

4. Matched sampling: By using this technique one or more controls are matched (paired) to a case. The advantage is that different controls can be compared to one selected case on the basis of different exposures; however, the process is more expensive. Other techniques, which are used less frequently, include cluster sampling, multistage sampling, area sampling, multiphase sampling, sequential sampling, and quota sampling. For a more in-depth overview of these sampling techniques the reader is referred to Schlesselman (1982).

(f) Describing the Exposure Assessment. In case-control studies, accurate exposure assessment is vital to achieving a valid exposure–disease association. Information on previous exposures of interest is usually obtained by directly questioning the cases and controls. Spouses, parents, friends, and relatives can be used as sources of information if direct questioning of the cases is not possible. A questionnaire administered either by interview (face to face or telephone) or by mail is the most commonly used instrument to obtain and quantify exposure information directly from the study subject. Choosing an open or closed question format depends on the kind of information to be collected. The information provided through open-ended formats is more detailed, harder to code, and more time-consuming than the structured answers of closed-format questionnaires.

Researchers need to ensure that their exposure assessment instrument is valid and reliable. If resources are available, validity of the questionnaire answers can be verified by comparing questionnaire responses with previously collected records such as medical, dental, employment, and death or with biomarkers such as urine, blood, and saliva. Reliability can be assessed by repeating the same question twice (once at the beginning of the questionnaire and once at the end) and by comparing responses to the same question by two or more respondents.

(g) Describing the Analysis. The analysis of a case-control study involves comparing the proportion of cases with an exposure of interest to the proportion of controls with the same exposure. This measure of association is called the odd's ratio (OR) which is the ratio of odds of exposure among cases to that of controls. Unlike prospective studies, case-control studies cannot estimate the rate of development of disease directly. However, for rare diseases, the odds ratio is a stable estimate of relative risk (Mantel and Haenszel, 1959). Data from case-control studies can conveniently be presented in a 2 × 2 table as depicted in Figure 2. Initial crude

		DISEASE	
		YES	NO
EXPOSURE	YES	a	b
	NO	c	d

FIGURE 2. Two-by-two contingency table used in case-control studies. (Analysis of case-control studies involves the calculation of an odds ratio: OR = ad/bc. An odds ratio of 3.0 indicates that cases had three times the risk of disease of that of controls for a given exposure. Conversely, an OR of 0.5 indicates that cases had half the risk of disease of controls).

estimates of the disease-exposure association can be followed by stratified analyses that can assess the impact of each separate confounding variable (each stratum representing a confounding variable). The Mantel–Haenszel chi-square test is the appropriate statistical test of association utilized in the case-control study. To assess the effects of multiple variables, logistic regression is used. By using a logistic regression model the investigator can analyze individual or joint effects of a set of variables on the risk of disease. An in depth description of these techniques is described by Schlesselman (1982, Ch. 7. Basic Methods of Analysis, and Ch. 8. Multivariate Analysis).

(h) Considering Sample Size and Power. Before embarking on a study, the researcher needs to consider the sample size (number of persons in each group) necessary to provide the study with sufficient power to detect differences between the two groups when differences truly exist. If the sample size is too small it is possible that the investigator will not detect a difference between two groups when a difference truly exists. This type of mistake is called a type II or beta error, while the power to detect this difference is defined as 1 – beta. Most textbooks of biostatistics and epidemiology will provide information on sample size and power calculations (Hennekens and Buring, 1987; Cohen, 1988; Dawson-Saunders and Trap, 1990; Schlesselman, 1982). Generally speaking, larger sample sizes produce more stable estimates of effect, hence more powerful results that are less likely to be chance findings.

(i) Considering Informed Consent and Confidentiality. All medical research involving humans requires informed consent from the study participants and an assurance of confidentiality to the study subjects. Case-control studies require the collection of personal and historical exposure information from people participating in a study. Hence, before participation occurs, the investigator should explain: (a) the study objectives; (b) the information to be collected; (c) the procedures to be performed during the study; (d) associated risks or benefits to the participants; (e) that the individual has the right to withdraw from the study and that in doing so this will not affect future medical care; (f) that all individual information will remain confidential; and (g) how the results will be disseminated.

(j) Preparing Time-Line and Budget. When preparing a research proposal, a realistic time line and budget are both necessary. Researchers often underestimate the time taken to successfully execute a study and the financial resources required to successfully complete a study.

(k) Obtaining Ethics Approval. All medical research needs to be submitted to an ethics committee for approval before a study can commence.

(l) Performing a Pilot Study. Time and money can be saved by evaluating the overall feasibility of the study by testing the clarity of the questionnaire and the procedures of the protocol on a small sample population. This is called a pilot study and is a useful strategy to allow for last minute changes if testing has shown problems in implementation procedures.

3.2. Data Collection

(a) Identifying and Collecting Cases and Controls. Once the cases and controls have been identified, permission to enter the study must be sought. Generally

speaking, for the cases, this involves contacting first the family physician and then the patient (case) for their consent. If the physician agrees to having his/her patient participate in the study, a letter describing the study is sent to each case inviting participation. Shortly thereafter, a follow-up telephone call is made to discuss participation and, if the person agrees, to set an appointment for an interview. Interviews are conducted either at the case's home or by telephone. Controls, too, are sent an introductory letter and then telephoned to determine whether they will participate. Often, researchers like to keep basic demographic information such as age, occupation, place of residence, and marital status, as well as the reason for nonparticipation on all individuals approached for study but who refuse. This information can then be compared to those who agreed to participate to see if there are any major differences between participants and nonparticipants (selection bias).

(b) Collecting Exposure Data. The exposure assessment instrument(s) should have been tested for validity and reliability prior to embarking on the full study. If questionnaires or interviews are to be used, they should be easy to administer in a short period of time (not more than 30 min). Interviewers should be trained in questionnaire administration. Procedures used to obtain information should be similar for both cases and controls. To prevent interviewer bias the interviewer should be blinded to the respondent's case or control status and to the hypothesis under investigation. Previously collected data, such as employment records, should be checked for completeness and biological, chemical, or hygiene sampling instruments should be calibrated before being used.

3.3. Data Analysis and Interpretation of Results

(a) Analyzing the Data. After the personal and exposure data have been collected, checks should be made to ensure there is no missing information, and that the data are correctly recorded and entered into the computer. The analyses start with the calculation of a crude odds ratio, followed by stratified and multivariate analyses.

(b) Interpreting the Results. In order to interpret the results of case-control studies correctly and assign causality to an exposure–disease association, one must be able to assess elements in the design and/or analysis that may invalidate the results of a case-control study. Such factors include: (1) flaws in the study design, (2) measurement error, (3) systematic errors in data collection and/or interpretation (bias), (4) the influence of variables that may influence the observed association (confounding), and (5) "the luck of the draw" (chance). Finally, when all the above factors have all been considered, the researcher must endeavor to assess causality, i.e., whether or not the results represent a true cause-and-effect relationship.

1. Flaws in the study design: The investigator should evaluate whether all the steps mentioned in the study protocol have been adhered to throughout the execution of the study. Flaws in the design phase are difficult to correct retroactively.

2. Measurement error: Errors in the measurement process can occur while the data are being collected and may lead to misclassification of disease and/or exposure. Disease misclassification can occur if diagnostic tests and criteria are not applied correctly and consistently. Exposure misclassification can be caused by errors in environmental measures or when the exposure is incorrectly remembered by the study subject (e.g., both cases and controls may incorrectly record the number of childhood sunburns they experienced). Nondifferential misclassification occurs when misclassification is equally distributed between the cases and controls, and will reduce observed differences between the two groups. Differential misclassification is more serious and occurs when one group is preferentially misclassified; this form of misclassification will result in an observed estimate that will either over- or underestimate the true association (Heacock and Rivers, 1997*b*).

3. Bias: Bias refers to a systematic error in the method of assessing exposure history and/or disease status and may result in spurious associations between exposure and disease. Bias can be prevented or minimized by careful study design but cannot be controlled for in the final analyses phase.

More than 56 different biases have been identified in analytic research (Sackett, 1979; Choi and Noseworthy, 1992). With respect to the case-control study, there are two main categories of bias, namely selection bias and information bias.

Selection bias refers to differences in the criteria used to select cases and controls and occurs in the design phase of the study. If this bias occurs, the measure of association between exposure and disease may invalidate the results. The most common forms of selection bias include: Berkson's bias, nonresponse bias, prevalence–incidence bias, and detection bias. Berkson's bias, also known as hospital admission bias, occurs when the risk of admission to hospital is higher for cases than controls (Last, 1995). Nonresponse bias occurs when people refusing to participate in a study differ systematically with respect to exposure or disease status from those who participate. Prevalence–incidence bias occurs when both incident and prevalent cases are studied simultaneously. Prevalent cases reflect factors that are related to survival whereas incident cases reflect factors that lead to the development of a disease. Detection bias occurs when a symptom or sign of disease initiates the search for that disease (Horwitz and Feinstein, 1977).

When a systematic difference between cases and controls exists in the way information was collected, recorded, or analyzed, information bias (also known as observation bias) may occur. The most common type of information bias is recall bias, which refers to a preferential recollection of previous exposures between cases and controls. Usually cases have had more time than controls to ponder their illness and may, therefore, recall more factors associated with their disease. For example, in one study, a questionnaire was administered in 1984 and again in 1986 to nurses who were surveyed about melanoma risk factors before and after diagnosis of melanoma. There was a significant shift toward reporting a reduced ability to tan (a considerable risk factor for melanoma) among nurses who were diagnosed with melanoma after the first questionnaire but before the second (Weinstock *et al.*, 1991). Recall bias can be minimized by reviewing existing records in order to verify exposure histories and by using hospital controls since they, too, have spent time thinking about their illness and possible earlier exposures. When the exposure history as obtained by an interviewer/questionnaire is extracted, recorded and/or interpreted differently between cases and controls, interviewer and questionnaire bias may occur. An interviewer may be more likely to probe for factors associated with melanoma (e.g., family history) among patients with melanoma than disease-free controls. This bias can be minimized by blinding the interviewer to the case/control status as well as to the hypothesis under investigation, by using structured questionnaires that are valid and reliable, and by training interviewers using strict protocols.

4. *Confounding:* A confounder is a variable that must be related to the exposure and at the same time be independently associated with the disease. Common confounders in medical research include demographic factors such as age, sex, ethnicity, socioeconomic status, income, profession, and behavioral factors such as smoking. In contrast to bias, which is usually caused by the investigator or study participants, confounding occurs because of interrelationships between exposures and disease. For instance, if age is related to melanoma then a study that had a case group with a greater proportion of older individuals than the control group may show statistically significant differences between the two groups based solely on the overrepresentation of older people in the case group. Methods to control confounding can be applied to both the study design and the data analysis. In the design phase, two strategies can be used: (1) restricting enrollment to individuals possessing particular characteristics, e.g., including only white females between the ages of 35 and 55 for a study of melanoma, and (2) matching subjects in such a way that potential confounders are equally distributed between study groups. For example, in a study examining environmental risk factors for bacillary angiomatosis, controls were matched to cases by HIV serological status and clinical institution (Tappero *et al.*, 1993).

Although matching may minimize confounding among certain factors, it also has some disadvantages: it can be expensive and time-consuming, and one loses the ability to assess the effect of a factor that has been matched for. In the analysis phase, confounding can be controlled by stratifying factors into homogeneous strata (subgroups), where each stratum represents a confounding variable. If the odds ratio for the stratified data (e.g., the data that are adjusted for the confounding variable) is different from the crude odds ratio, then one may conclude that confounding played a role in the exposure–disease association. Another method to assess confounding is the use of multivariable analyses. Unlike stratified analyses, which can only control for a single potentially confounding variable at a time, multivariate analyses can simultaneously assess effects of several potential confounders by means of a logistic regression model.

5. *Chance:* The investigator needs to be aware that the observed association between a disease and an exposure can occur by chance. To evaluate the role of chance, the *P*-values generated by the Chi-square test and the regression analyses are utilized. *P*-values represent the likelihood that the observed effect occurred by chance alone, given no association between exposure and disease. By convention, the *P*-value 0.05 has been set as the cutoff between "significant" and "nonsignificant" results. *P*-values larger than 0.05 suggest that chance is a likely explanation for the findings. In this case the researcher may not conclude that a statistically significant association between exposure and disease exists.

Confidence intervals (CIs) are also commonly reported in conjunction with *P*-values and represent a range of values within which the true magnitude of association lies with a certain (often 95%) degree of assurance. An odds ratio (OR) equal to 5.2 and a reported 95% CI 4.8–5.7 indicate that the cases were slightly more than five times as likely as the controls to have had the exposure in question and 95 times out of 100 the true OR lies between 4.8 and 5.7. Larger samples produce ORs with narrower CIs, which indicates that the interval surrounding the OR is a more precise measure of the association. It is helpful in the interpretation of results to report CIs since they indicate both the magnitude of effect and the precision of the OR.

Finally, when all the above factors have all been considered, the researcher must endeavor to assess causality, i.e., whether or not the results represent a true cause-and-effect relationship.

4. Criteria Used to Judge Causality

The earliest attempts to establish guidelines for causality were known as the Henle–Koch postulates (from Jacob Henle 1840 and his pupil Robert Koch in 1882) (Kelsey *et al.*, 1986). More recently, the British medical statistician, Sir Austin Bradford Hill developed a list of criteria for causation in epidemiological studies that continues to be used as a standard for the establishment of causality between an exposure and a disease (Hill, 1965; Schlesselman, 1982; Hennekens and Buring, 1987; Greenberg, 1993; Guyatt *et al.*, 1993). Some of these criteria include:

1. Strength of the observed association: The larger the risk (in case-control studies, this refers to the OR) the less likely the result can be explained by chance, bias, or confounding.
2. Consistency of results with other studies: If other researchers, using different methodologies, studying different populations in different settings find similar results, then a causal explanation is supported.
3. Biological plausibility: If the results are consistent with what is known about the biological mechanism(s) of the disease process, then the case for causality is strengthened.
4. Dose–response relationship: If increasing the level of exposure increases the risk of disease development, alternative explanations for the observed results are unlikely and causality is supported.
5. Temporal relationship: An exposure must always precede an outcome by a period of time that is compatible with known biological mechanism(s).
6. Specificity: For a factor to be considered causally related to a disease, its removal should prevent the disease from occurring (i.e., a specific exposure leads to a specific disease).

5. Conclusion

In summary, the case-control study is a powerful observational analytic research method used to investigate associations between exposures and disease. This chapter has briefly outlined issues surrounding the planning and execution of this type of design.

References

Carli, P., Biggeri, A., and Giannotti, B., 1995, Malignant melanoma in Italy: Risks associated with common and clinically atypical nevi, *J. Am. Acad. Dermatol.* **32**(5 pt. 1):734–739.

Checkoway, H., Pearce, N. E., and Crawford-Brown D. J., 1989, *Research Methods in Occupational Epidemiology*, Oxford University Press, New York, pp. 171–201.

Choi, B. C. K., and Noseworthy, A. L., 1992, Classification, direction, and prevention of bias in epidemiological research, *J. Occup. Med.* **34**(3):265–271.

Cohen, J., 1988, *Statistical Power for the Behavioral Sciences*, 2nd Ed., Lawrence Erlbaum Associates, Mahwah, New Jersey, pp. 52–66.

Dawson-Saunders, B., and Trapp, R., 1990, *Basic and Clinical Biostatistics*, Appleton and Lange, Norwalk, pp. 271–276.

Garbe, C., Buttner, P., and Weiss, J., 1994, Risk factors for developing cutaneous melanoma and criteria for identifying persons at risk: Multicenter case-control study of the Central Malignant Melanoma Registry of the German Dermatologic Society, *J. Invest. Dermatol.* **102**:695–699.

Greenberg, R. S., 1993, *Medical Epidemiology*, Appleton and Lange, Norwalk, pp. 125–126.

Guyatt, G. H., Sackett, D. L., and Cook D. J., 1993, User's guide to the medical literature. II: How to use an article about therapy or prevention. A. Are the results of the study valid? *JAMA* **270**(21):2598–2601.

Heacock, H. J., and Rivers, J. K., 1997*a*, Assessing scientific data: The case-control study as it applies to dermatology: Part 1. The case-control method, *J. Cut. Med. Surg.* **1**(3):151–154.

Heacock, H. J., and Rivers, J. K., 1997*b*, Assessing scientific data: The case-control study as it applies to dermatology: Part 2. Interpreting the results, *J. Cut. Med. Surg.* **2**(1):35–40.

Healy, E., Collins, P., and Barnes, L., 1995, Nonmelanoma skin cancer in an Irish population: An appraisal of risk factors, *Ir. Med. J.* **88**(2):58–59.

Hennekens, C. H. and Buring, J. E., 1987, *Epidemiology in Medicine*, 1st Ed., Little and Brown, Boston, pp. 39–43, 141–142, 258–264.

Hill, A. B., 1965, The environment and disease: Association or causation, *Proc. Roy. Soc. Med.* **58**:295–300.

Horwitz, R. I., and Feinstein, A. R., 1977, New methods of sampling and analysis to remove bias in case-control research, *Clin. Res.* **25**:459A.

Kelsey, J. L., Thompson, W. D., and Evans, A. S., 1986, *Methods in Observational Epidemiology*, Oxford University Press, New York, pp. 148, 168, 257, 31.

Last, J. M., 1995, *A Dictionary of Epidemiology*, Vol. 15, 3rd Ed., Oxford University Press, Toronto.

Mantel, N., and Haenszel, W., 1959, Statistical aspects of the analysis of data from retrospective studies of disease, *J. Natl. Cancer. Inst.* **22**:719-748.

Naldi, L., Parazinni, F., Peli, L., Chatenoud, L., and Cainelli, T., 1996, Dietary factors and the risk of psoriases: Results of an Italian case-control study, *Brit. J. Dermatol.* **134**(1):101–106.

Pearce, N. E., Smith, A. H., Howard, J. K. *et al.*, 1986, Non-Hodgkin's lymphoma and exposure to phenoxyherbicides, chlorophenols, fencing work, and meat works employment: A case-control study, *Br. J. Ind. Med.* **43**:75–83.

Poikolainen, K., Reunala, T., Karvonen, J., Lauharanta, J., and Karkkainen, P., 1990, Alcohol intake: A risk factor for psoriasis in young and middle aged men?, *Br. Med. J.* **300**(6727):780–783.

Rieger, E., Soyer, H. P., Garbe, C. *et al.*, 1995, Overall and site-specific risk of malignant melanoma associated with nevus counts at different body sites: A multi-center case-control study of the German Central Malignant-Melanoma Registry, *Int. J. Cancer.* **62**(4):393–397.

Roujeau, J. C., Kelly, J. P., Naldi, L. *et al.*, 1995, Medication use and the risk of Stevens–Johnson syndrome or toxic epidermal necrolysis, *N. Engl. J. Med.* **333**(24):1600–1607.

Sackett, D. L., 1979, Bias in analytic research, *J. Chron. Dis.* **32**:51–63.

Schlesselman, J. J., 1982, *Case-Control Studies: Design, Conduct and Analysis*, Oxford University Press, New York, pp. 20–25, 105–124, 144–290.

Siemiatycki, J., Richardson, L., Gerin, M. *et al.*, 1986, Associations between several sites of cancer and nine organic dusts: Results from an hypothesis-generating case-control study in Montreal, 1979–1983, *Am. J. Epidemiol.* **123**:235–249.

Steere, A. C., Grodzicki, R. L., Kornblatt, A. N. *et al.*, 1983, The spirochetal etiology of Lyme disease, *N. Engl. J. Med.* **308**:703–740.

Tappero, J. W., Mohle-Boetan, J., Koehler, J. E. *et al.*, 1993, The epidemiology of bacillary angiomatosis and bacillary pelliosis, *JAMA* **269**(6):770–775.

Weinstock, M. A., Colditz, G. A., Willett, W. C. *et al.*, 1991, Recall (report) bias and reliability in the retrospective assessment of melanoma risk, *Am. J. Epidemiol.* **133**(3):240-245.

Chapter 3

*Principles for the Conduct of Nondrug Clinical Studies**

GERHARD J. SCHMITT

1. Introduction

The world Medical Association's Declaration of Helsinki outlines the principles for the conduct of drug and nondrug clinical studies. The key difference between drug and nondrug biomedical research is to be seen in the fact that for nondrugs, the potential benefits, hazards and discomforts for the volunteer enrolled in a clinical study cannot be weighed against advantages of improving current diagnostic or therapeutic methods. The principles reflected in the Guidelines for Good Clinical Practice for Trials on Medicinal Products in the European Community, issued by the Commission of the European Communities in substantial part also apply for trials on nonmedicinal products. Specifically this is true with regard to the requirements for the privacy, integrity, and well-being of volunteers subjected to research, and the obligation to fully inform them about the risks and benefits potentially associated with the use of a test product. Good clinical practice as agreed under ICH[†] in May 1996 for pharmaceuticals for human use does not address any other product groups, but its spirit is equally applicable to nondrug clinical study. Skin care products are the most prominent group of nondrug products for which clinical studies are conducted. The key differences between executing the principles in a nondrug and a drug clinical study when using these products are highlighted.

*Based on a paper presented at the Regional Meeting of the International Society for Bioengineering and Skin, Lübeck, Germany, April 23–25, 1993.

†International Conference of Technical Requirements for Registration of Pharmaceuticals for Human Use.

Gerhard J. Schmitt • Beauty Care Europe, Procter & Gamble European Service GmbH, D-65823 Schwalbach/Taunus, Germany.

Cutaneous Biometrics, edited by Doris A. Schwindt and Howard I. Maibach, Kluwer Academic/Plenum Publishers, New York, 2000.

2. World Medical Association Declaration of Helsinki

... It is the mission of the physician to safeguard the health of people. His or her knowledge and conscience are dedicated to the fulfillment of this mission.... Biomedical research involving human subjects should be conducted only by scientifically qualified persons and under supervision of a clinically competent medical person. The responsibility for the human subject must always rest with the medically qualified person and never rest on the subject of the research, even though the subject has given his or her consent.

In the special case of nontherapeutic biomedical research involving human subjects... in the purely scientific application of medical research carried out on a human being, it is the duty of the physician to remain the protector of the life and health of that person on whom it is being carried out.... In research on man, the interest of science and society should never take presidence over considerations related to the well-being of the subject.

These excerpts taken from the World Medical Association's Declaration of Helsinki in its latest updated version (The World Medical Association, 1996) outline the key principles for the conduct of drug and nondrug clinical studies.

3. Nondrug vs. Drug Studies

A key difference between drug and nondrug biomedical research is that in case of nondrugs the potential benefits, hazards, and discomforts for the volunteer enrolled in a clinical study can not be weighed against advantages of improving current diagnostic or therapeutic methods. Nondrug clinical trials are conducted in healthy subjects with regard to the parameters under investigation. The biological endpoints chosen to determine safety and efficacy for all future consumers must not put a volunteer at an undue risk, specifically the risk of irreversible damage.

Because of the nonsystemic effects typically involved, biomedical research conducted with nondrug products is usually much less risky for subjects than that with therapeutic drugs. For application of the principles laid down in the Helsinki Declaration, national or supranational guidelines are in place that can provide adequate protection of the trial subject and the generation of scientifically valid data. Examples are the US FDA guidelines, or more recently, the Guidelines for Good Clinical Practice for Trials on Medicinal Products, issued by the Commission of the European Community (1990). While, for obvious reasons, there is a lot of similarity between those two guidelines, the more recently issued European ones are in part more stringent and the ones referred to here. But in either case, in substantial part, these GCPs apply equally for trials on medicinal or nonmedicinal products. Prior to their issue, important aspects that they addressed were considered when using human volunteers for testing nonmedicinal products such as cosmetics and toiletries (van Abbe, 1983, 1987).

The internationally harmonized ICH GCP Guidelines (EG GCPD, 1996) developed by the European Community, Japan, and the United States are to provide a uniform quality standard and thus facilitate the mutual acceptance of clinical data by the regulatory authorities in these jurisdictions. They are intended for pharmaceuticals for human use. However, the underlying spirit of respect and care for the subject involved in safety and efficacy tests are equally applicable for nondrug clinical trials, as explicitly stated in the introduction of the ICH-GCP Guidelines: "The principles established in this guideline may also be applied to the chemical investigations that may have an impact on the safety and well-being of human subjects."

4. Skin Care Products

Skin care products designed to maintain the physiological condition, prevent damage, or treat minor ailments are the most prominent group of nondrug products for which clinical studies are typically conducted, either to confirm their safety in use for consumers or to demonstrate their efficacy. Hence, these products have been chosen here to highlight the principles for the conduct of nondrug clinical trials.

5. Protection of Trial Subjects: Written Informed Consent

One prerequisite for the conduct of any valid, ethically justified clinical study is the selection of a suitably qualified responsible clinical investigator. Since it is the obligation of such an investigator to protect the integrity and welfare of the trial subjects, he/she must know about any potential risks. Contact irritant or allergic responses including their photodynamic variants are the risk commonly associated with the testing of topically applied skin care products. For allergic or photoallergic responses, both the risk of sensitization during extended use, and of elicitation of a preexisting sensitization have to be considered.

The responsible clinical investigator must obtain freely given informed consent from all subjects to be enrolled into a clinical study. In a nontherapeutic study, i.e., when there is no direct medical benefit to the trial subject, this consent must always be attested to by the signature of the subject. Before asking for written informed consent, the investigator has to ascertain that the test subject was given a full explanation of the risks associated with the trial, had the opportunity to ask questions about them, had those questions answered, and was informed about his/her right to discontinue participation in the trial at any time without giving a reason, without risk of personal or financial damage. Furthermore, there needs to be independent assurance provided by an independent ethics committee that subjects are adequately protected. As for drugs, health insurance coverage has to be provided and funded by the sponsor of the trial for all subjects.

6. Study Protocol

Both, the sponsor and the responsible clinical investigator must agree upon and sign the final protocol as an agreement of the details of the clinical trial and the means of data recording (e.g., case report forms). Any amendments to the procedures or the science in the protocol must have the agreement of both the sponsor and the investigator before being implemented, unless medically required to protect the subject. Any such agreement must be documented. There is no difference in this requirement between trials of drug and nondrug products. As with drug products, it is the obligation of the responsible clinical investigator to confirm in writing that he/she has read, understood, and will work according to the protocol and good clinical practice, accepting supervision of the study by a monitor and control procedures, and that he/she agrees with the sponsor on a publication policy. The investigator is required to provide informed consent to study subjects and to obey the notification and documentation mechanisms in case of serious adverse effects, and that appropriate measures are taken to safeguard trial subjects. Finally, the protocol must be followed throughout the trial.

7. Safety Documentation for Test Products

There is no difference between drug and nondrug products in the investigator's responsibility to be thoroughly familiar with the properties of the test products. In any case, no test product should be applied to a trial subject, for which the formulation, identity, and safety are not adequately known, including the description of the way the product was made to the extent that it is relevant to the safety pattern of the finished product.

For drug products, requirements for type and scope of safety data are defined by regulation and have to be met with complete documentation for all the raw materials used and the finished product. For nondrug products such as those for skin care, a detailed risk assessment by a competent toxicologist is needed and is typically provided by the sponsor. This may include historical data on individual raw materials or extrapolation to the state of art in toxicology from similar preparations onto the risk associated with the new product. This type of risk assessment is valid and justified by the kinds of chemicals typically used in skin care products. In part it is also a necessity as in recent years regulations in many countries have drastically restricted the conduct of animal testing for cosmetics. However, if this risk assessment is properly done by a responsible toxicologist, it is an acceptable situation for both the trial subject and responsible clinical investigator. Another safeguard comes from the fact that the effects permitted for skin care products are regulated under the EC Cosmetics Act resulting in formulations free of systemic effects, most constituents of which do not penetrate the barrier of the skin at all or to any relevant extent.

There are no legal requirements for the manufacture of nondrug products such as skin care products under GMP, but the industry generally adheres to the principles of these standards. In any case it should be required that a test material and its properties be known before trial subjects are exposed to it.

8. Data Handling and Documentation

The responsible clinical investigator undertakes to ensure that the observations and findings are recorded correctly and completely in ink on the case report forms, and that the forms are properly signed and dated. All corrections on a case report form and elsewhere in the hard copy raw data must be made in a way that does not obscure the original entry. Any corrected data must be inserted with the reason for the correction and the insert must be dated and initialed by the investigator. For electronic data processing, only authorized persons should be able to enter or modify data and there should be a record of changes and deletions. These quality standards apply equally to drug and nondrug trials.

For drug clinical trials the responsible clinical investigator must arrange for the retention of the patient identification codes for at least 15 years after the completion or discontinuation of the trial. For nondrugs such as skin care products, this obligation may be contracted out to an organization arranging for the clinical trial or to the sponsor. However, in cases where patients are eligible as trial subjects under the criteria of a nondrug clinical trial, the investigator must ensure that their files and other source data are kept for the maximum period of time permitted by the hospital, institution, or private practice they were recruited from, but not less than 15 years.

9. Statistics

Where and by whom the statistical work shall be carried out is subject to agreement between the responsible clinical investigator and the sponsor. However, what is agreed should be written in the protocol, and it has to address the validity of the experimental design, randomization and blinding of trial subjects and test products, and statistical analyses of the data generated.

10. Final Report

The final report has to be signed by the responsible clinical investigator. In view of the fact that for skin care products a treatment benefit free from health risk for the consumer is a legal requirement, the investigator has to place special emphasis on the interpretation of any side effects and any adverse events noted during a study. This consideration has to take into account that skin care products are intended to be used by consumers in an uncontrolled way and without medical surveillance. Also there will likely be fewer established feedback mechanisms in case of health problems. Hence, any indication of adverse effects observed during a clinical trial must be given careful attention and weighed against the clinical relevance of the benefits for the user.

11. Summary

The best possible protection of human health is the essential prerequisite for the conduct of both drug and nondrug clinical studies. This is reflected by the fact that key requirements for their proper conduct do not differ at all. Differences in the principles for the conduct of nondrug clinical studies occasionally requiring less rigorous executional detail are justified by and based on the fact that biomedical research conducted with them is typically much less risky to subjects than that with drugs. However, there is no excuse for nondrug clinical trials to be carried out with less quality or less good science. Obviously, the responsible clinical investigator has to be satisfied with the risk assessment for the test products provided by the sponsor, demonstrating that trial subjects are not at any undue health risk. Complete information about benefits and risks associated with the use of a nondrug product, the agreement to which is documented in a written informed consent, is a critical aspect, but it cannot substitute for the obligation of the investigator to safeguard the health and welfare of the trial participants.

In addition, assurance of proper protection of trial subjects by an independent ethics committee remains an indispensable requirement. While the requirements for proper statistical design, data collection, and analyses of data generated during a clinical trial do not differ between drug and nondrug clinical studies, the requirements for archiving of data are more flexible with nondrug products. In the evaluation of the results of a nondrug clinical study, any clinically relevant finding—positive or negative—has to be carefully assessed for its consumer relevance under consideration of the legal requirement of absence of any appreciable safety risk from a nondrug product under conditions of uncontrolled, recommended or reasonably foreseeable use.

References

Commission of the European Communities, Directorate General for Internal Market and Industrial Affairs Good Clinical Practice for Trials on Medicinal Products in the European Communities; III/3976/88-EN, Final. EC, Brussels, Belgium, 1990.

EG GCPD 12.WP6, 01 May 1996—ICH EWG EG: Good Clinical Practice.

The World Medical Association, Declaration of Helsinki, updated by the 48th World Medical General Assembly in South Africa, October 1996.

van Abbe, N. J., 1983, The use of human volunteers for testing cosmetics and toiletries, *Soap Perfum. Cosmet.* **56**:134–136.

van Abbe, N. J., 1987, Human testing, *Soap Perfum. Cosmet.* **60**:41–45.

Chapter 4

Questionnaire Design in Skin Disease Epidemiology

PÄIVIKKI SUSITAIVAL

1. Questionnaire Methods Used in Skin Disease Epidemiology

1.1. Some Epidemiological Concepts

Epidemiological studies are a means of getting information on the distribution and determinants (exposures, modifying, and confounding factors) of diseases in populations. Diseases are determined by outcomes; the signs or symptoms used as disease (case) criteria. When studying the effect of an exposure (e.g., detergent) on an outcome (e.g., eczema), modifying factors (e.g., the use of emollients) that can change the effect of the exposure on the outcome and risk factors for the outcome (e.g., atopy) which are confounding factors if distributed differently between the exposure groups, should be accounted for and included in the questionnaire.

Studies can be cross-sectional (data gathered from all participants at a specified point of time—prevalence studies) or longitudinal (data gathered over time from a population—incidence studies). Prevalence of a symptom, sign, or a disease is defined as the proportion of all individuals meeting the specified criteria (cases) in a population at a specified time (point prevalence) or time period (period prevalence). When dealing with diseases such as dermatitis, which recur but tend to affect the same individuals, the period prevalence of the disease (e.g., 1-year prevalence) is a more relevant estimate than the point prevalence. The incidence rate tells us how many new cases have occurred within a time period in a population unit (e.g., cases/1000 persons/year), while the incidence proportion (cumulative incidence) refers to the proportion of individuals within the population at risk developing the disease during a specific time period.

Päivikki Susitaival • Finnish Institute of Occupational Health, FIN-70701 Kuopio, Finland.

Cutaneous Biometrics, edited by Doris A. Schwindt and Howard I. Maibach, Kluwer Academic/Plenum Publishers, New York, 2000.

The study results depend on the definitions of cases and other determinants, and on the construction of the questions yielding the data. The questions, questionnaires, and definitions should be as reliable and valid as possible. Reliability refers to yielding the same results in repeated encounters. On the other hand, validity means that the question or the questionnaire is providing true correct information, e.g., doing the job it is intended to do. External validity refers to generalizability of the results, while internal validity refers to the lack of systematic error (bias). For example, misclassification, where study subjects are classified to wrong groups, e.g., noncases as cases, nonexposed as exposed, leads to information bias. This can arise from either erroneous or unfeasible definitions or questionnaire design failures. Sensitivity (the ability to detect the sick), specificity (the ability to detect the healthy), positive predictive value (the proportion of the detected cases that are sick), and negative predictive value (the proportion of those not detected as cases that are healthy) provide information on how well the method (e.g., questionnaire) differentiates between the sick and the healthy (Table 1).

There tends to be an inverse relation between sensitivity and specificity. When a method is very sensitive, it detects most people who have the disease, while the specificity tends to be lower, implying that a portion of the detected cases are false positives. On the other hand, when specificity is high, most of those cases detected do have the disease, but a portion of the diseased people are not found at all (false negatives). The positive predictive value gives the proportion for the sick among the cases detected. When a method's specificity is 100%, no false positives occur, leading to a positive predictive value of 100%. Similarly, when the sensitivity is 100%, no false negatives arise, making the negative predictive equal to 100%. Unlike sensitivity and specificity, which are properties of the method and hence theoretically invariant regardless of the population tested, the positive and negative predictive values are directly related to the prevalence of the disease in the screened population. Positive predictive value is positively associated and negative predictive value is negatively associated with an increasing frequency of disease in a studied population.

1.2. Hand Dermatitis and Other Dermatoses

An insight into the methods that have been used in skin disease epidemiology is very useful in both designing a questionnaire and in interpreting results from other questionnaire studies on skin diseases. At least six studies on the epidemiology of skin diseases or atopy have included some evaluation of the accuracy of the reporting (Agrup, 1969; Rea *et al.*, 1976; Bakke *et al.*, 1990; Meding, 1990; Susitaival *et al.*, 1995*a*), and three studies have been designed to evaluate specific methodology or questionnaires (Berg and Axelson, 1990; Smit *et al.*, 1992; Williams *et al.*, 1994*a*, *b*, *c*, 1996). In the study by Williams *et al.* (1994*a*, *b*, *c*, 1996), a UK working party systematically designed and validated the clinical criteria for atopic eczema. In several of the above publications, the clinical or the questionnaire case criteria are not clearly defined. Another problem in the interpretation and comparison of the studies is the variability in the vocabulary and meaning of dermatological terms in different countries. Both medical and lay terminology vary greatly in different languages and cultures (e.g., manifestations of atopic skin disease). English literature

TABLE 1. Sensitivity, Specificity, and Predictive Values

Disease		Reality		
		+	−	
Method	+	a	b	a + b
	−	c	d	c + d
		a + c	b + d	

Sensitivity = a/a + c   Positive predictive value = a/a + b
Specificity = d/b + d   Negative predictive value = d/c + d

uses terms like "childhood eczema" or simply "eczema" as a synonym for atopic eczema, which in German publications may be called "endogenous eczema" or "neuroderm(at)itis." The terms used and understood by lay people present an even more complex picture.

In a British survey concerning all skin diseases, 14% of those reporting "rash or other disorders of the skin, hair, or nails" were found to have no dermatosis, and in 13% of those reporting "no" a moderate to severe dermatosis was present (Rea *et al.*, 1976). As part of an epidemiological study of Finnish farmers (Susitaival *et al.*, 1995*a*), a question "Do you have a skin disease now?" was validated immediately after the subjects had completed a form including other questions as well. All the farmers answering the skin disease question affirmatively were diagnosed with a dermatosis. However, 22% of those answering "no" were also given a dermatological diagnosis. Thus, the specificity of the question, presented in this way, was high (100%) while the sensitivity was lower (60%). Most of the skin diseases not considered as dermatoses were toe-web maceration in feet while 2.5% of those with negative answers had eczema on their hands.

Berg and Axelson (1990) evaluated a questionnaire concerning skin symptoms and signs during the past two years. Their conclusion was that the questionnaire answers regarding clear-cut symptoms over a 2-year period were more "adequate" than those concerning the presence of skin signs at a particular point in time. The clinical diagnosis confirmed the self-reported answers in 92% of the cases of atopic eczema, in 86% of seborrhoeic dermatitis, and in 83% of hand dermatosis cases. The agreement between the questionnaire answers and the examination was poorer for other diagnoses.

Hand dermatitis is the most common skin condition surveyed in population studies. Two methods have been used to diagnose hand eczema in questionnaire studies: a self-report (self-diagnosis) of hand dermatosis or eczema ("Have you had..."), or a diagnosis based on a symptom list (symptom-based diagnosis). Some studies have included the forearms in hand eczema questions ("hands and/or forearms") (Coenraads *et al.*, 1983, 1984; Susitaival *et al.*, 1994), while some have only included recurring dermatoses or symptoms, or those lasting for at least 3 weeks (Coenraads *et al.*, 1983, 1984; Smit *et al.*, 1992). In Agrup's population-based study,

94% of tehose reporting any skin changes on their hands had a hand dermatosis confirmed by an examiner. Hand eczema, mostly mild, was also found in 0.9% of those reporting no dermatosis on hands (Agrup, 1969). A Swedish population-based study found that 89% of those affirmatively answering the question "Have you had hand eczema within the previous 12 months?" had hand eczema upon clinical examination (Meding and Swanbeck, 1987).

Smit *et al.* (1992), in their questionnaire evaluation study, compared a self-report of hand dermatitis (Dutch nurses) and a symptom-based hand eczema diagnosis with a dermatologist's diagnosis. The symptom-based diagnosis was derived from reported skin symptoms: (1) redness with edema, (2) redness with fissures, (3) vesicles, (4) scaling with fissures, or (5) itching with fissures. The criteria for the symptom-based diagnosis were: one or more of the above five symptom categories during the past 12 months, persisting for more than 3 weeks or recurring. The symptom-based diagnosis detected all hand eczema cases, but 62% of the detected cases were not diagnosed with the condition by the dermatologist. Thus the positive predictive value (Table 1) of the symptom-based diagnosis was 38%. The respondent's own opinion of having suffered from hand dermatitis in the past year was the self-reported diagnosis. Sixty eight percent of those reporting hand dermatitis (positive predictive value), and 8% of those not reporting it, had hand eczema diagnosed by a dermatologist. The sensitivity of the symptom-based and the self-reported diagnoses were 100 and 65%, respectively, while the specificity was 64 and 93%, respectively (Table 1). Thus, in this study, the symptom-based diagnosis gave a much higher hand eczema prevalence (48%) than the self-reported (17%) or medical diagnoses (18%). When the criteria of the symptom-based diagnosis included at least two, instead of one, of the above five symptom groups, the specificity (89%) and the positive predictive value (62%) increased and the sensitivity (80%) decreased.

1.3. Atopy

The history of atopic symptoms, skin prick-test reactivity, and the combination of both have been used for indicating atopic constitution. In a Danish study, two-thirds of those claiming to suffer from "allergic rhinitis" actually had the condition (Pedersen *et al.*, 1980). Bakke *et al.* (1990) evaluated a questionnaire on allergic symptoms and eczema by carrying out a clinical examination 2 years after the questionnaire survey. The questions used in this study were: "Have you ever had one of the diseases mentioned below?," and the list included eczema and hay fever. "Eczema" was defined in the clinical examination as atopic eczema using the Hanifin and Rajka criteria. Of those reporting hay fever, 89% were verified to be correct. The corresponding figure for eczema was 51%.

Meding (1990) has also evaluated questions on the personal history of atopy (childhood eczema, hay fever, bronchial asthma) by inquiring about these in both a survey and a clinical examination. There was a slight tendency to overreport childhood eczema in the questionnaire (4% of the questionnaire "yes" answers became "no" in the interview, and 2.5% vice versa), but for respiratory symptoms there were no significant differences. The Glostrup allergy study examined the associations between positive skin-prick tests and questionnaire answers on respiratory symptoms (Nielsen *et al.*, 1993). The conclusion was that reported symptoms (itchy or stuffy nose, sneezing, shortness of breath) on exposure to allergens (in the summer or near plants or animals) were highly associated with positive skin-test reactivity.

The criteria for atopic dermatitis have been designed and evaluated by a UK working party (Williams *et al.*, 1994*c*). The criteria are: (1) an itchy skin condition and (2) at least three of the following: (a) history of involvement of specific skin creases (at least one of the following: fronts of the elbows or behind the knees, around the neck, fronts of ankles, or around the eyes, eyelids), (b) history of allergic rhinitis or asthma, (c) history of dry skin, (d) dermatosis starting before 2 years of age. These criteria have been tested in a population of schoolchildren and gained a sensitivity of 80% and specificity of 97% when adjusted for the period of the past year (Williams *et al.*, 1996). The same group also evaluated the question "Have you had an itchy rash that has been coming and going for at least six months, which at some time has affected skin creases?," and found a sensitivity of 73% and a specificity of 87% (Williams *et al.*, 1994*a*).

1.4. Skin Cancer

Schrijvers *et al.* (1994) have validated the self-report of cancer ("Have you, within 5 years, suffered from any of the following," as one of 23 subcategories "any cancer or malignant condition") by comparing the answers with the cancer registers. Twenty-two percent of the registered nonmelanoma skin cancers and 50% of melanomas were reported. Underreporting of skin cancer was very common among men, those over 65 years of age, and those from rural communities or with low education level. Thus the presented way of detecting skin cancer prevalence had a very low sensitivity (22%). Out of 212 false-negative reports of any cancer, 107 were either nonmelanoma skin cancer or melanoma. If the prevalence of skin cancer is studied, a specific question should be used ("Have you ever been diagnosed as having skin cancer or melanoma?"). Even then, the self-report may tend to underestimate the prevalence, depending on the population or group studied because individuals may not perceive the conditions they have been diagnosed with as skin cancer.

1.5. Concluding Remarks

Several of the above studies have indicated that the specificity of reporting skin conditions was fairly high (i.e., above 90%), but the sensitivity was somewhat or much lower (i.e., less than 70%) (Smit *et al.*, 1992; Susitaival *et al.*, 1995*a*; Schrijvers *et al.*, 1994). Similar results have also been encountered in the reporting of many other health conditions (Bachus *et al.*, 1996). This indicates that false negatives more so than false positives may create bias in these studies. However, self-reporting of conditions, e.g., hand dermatosis, is probably an appropriate method for estimating the prevalence of *explicit* skin conditions (those of concern to the respondent) in a population. Differences in the perception of what the inquired skin condition (e.g., hand dermatitis) is will cause variation in this measure in different occupational groups and cultures. Questions on symptoms, duration, and treatment provide more depth to the self-diagnosis. Symptom-based hand eczema diagnosis, as used in the Smit *et al.* (1992) study, e.g., is a useful screening method when all cases are needed for further examination.

In the validation study of clinical atopic criteria for epidemiological research by the UK working party, the between-observer agreement was better with historical features than physical signs, thus implying that the prevalence of atopic dermatitis can be reasonably estimated on the basis of questions alone (Williams *et al.*, 1994*a*, *b*). The same group also found that the between-observer agreement was better among less experienced observers (Williams *et al.*, 1994*b*). Findings indicating that observers with high medical qualifications may not be ideal for epidemiological data collection have also been reported in other studies (Spiteri *et al.*, 1988). Some studies have found a good association between the self-reported aggravators of skin symptoms and positive skin test results to the reported agents (Susitaival *et al.*, 1995*b*; Johansen *et al.*, 1997). All the above results, together with the previously mentioned finding in the Glostrup allergy study (Nielsen *et al.*, 1993, see Section 1.3.) suggest that questionnaire data in skin disease and allergy studies can be valid and even superior to clinical data for epidemiological study purposes.

There is a definite demand for both clinical and questionnaire criteria for skin conditions such as hand eczema. The criteria must be agreed upon by experts and validated in different populations and countries before final acceptance for standard epidemiological use. The phrasing of the questions in each language, with the help of linguistic experts and lay people, is very important in providing consistent data on specific dermatoses and skin complaints in all target populations.

2. Designing a Questionnaire

2.1. Introduction

The questionnaire is a critical element in surveys that rely on information provided by respondents. Questionnaire data are based on human communication, which is highly prone to misunderstandings. Advice on questionnaire design can be found in most survey research textbooks (e.g., Sheatsley, 1983; Frey, 1989) and occasional journal articles (e.g., Stone, 1993). A helpful and also entertaining piece on the subject is Payne's old classic *The Art of Asking Questions* (Payne, 1980). The principles in designing a valid and reliable questionnaire are basically the same for skin diseases as for any other survey subject. The following discussion focuses mainly on issues that have not received much attention elsewhere.

The answers provided by a good questionnaire produce appropriate and valid data to meet the objectives relevant to the research. The questionnaire must be based on clearly specified objectives. The definitions for outcomes and other variables must be stated and based on a realistic scenario of expected answers, which requires a lot of information on the target population and pretesting the questions (Table 2). Pretesting, meaning simply administering the questionnaire to people prior to finalizing it, is the reality check of the questionnaire. For best results, the persons participating in the pretesting should be asked to comment on all aspects of the

TABLE 2. Minimal Requirements before Designing a Questionnaire

- Precise study objectives
- Definitions for outcomes, etc., that can be met with questioning
- Researcher must have working knowledge of the target population (language, work, culture, etc.)

questions and the questionnaire, and the researcher should interact closely with them. Doing this can also be very useful for the researcher in getting a better insight into the future data. The final pretesting should always be done with persons representative of the target population. This is often imperative before being able to choose the appropriate response categories. The mode of administration of the questionnaire (self-administered or telephone or face-to-face interview) has implications for the language and answer categories, as well as the question order. Documenting skin conditions with the help of photographs can also be used, especially in face-to-face interviews.

The questionnaire must keep the respondent interested in answering until the end. It is wise to begin with issue-oriented questions and place most of the basic demographic questions at the end of the questionnaire. The order of the items in the questionnaire should serve the purpose of being logical and pleasant to answer. The question items should always be from general to specific. The purpose of the questionnaire is to provide truthful answers to all questions. Thus neutrality of both questions and question sequences is imperative (Table 3).

2.2. The Art of Asking Questions

Designing questions seems simple but is actually very complex. Questions should be such that the respondent is able to read, understand, and provide answers quickly without any difficulty. One must ensure that the respondents are able (and willing) to produce reliable answers to the questions (pretesting). A common flaw is to ignore the fact that words may have dissimilar meanings, or no meanings, in different cultures and subgroups. There should be only one issue in a question; thus questions with the words "and" and "or" should be avoided. The more complex the questions, or the longer time frame they cover, the less valid the data. Questions should not be asked in the negative ("don't you...") or suggestive form ("has it often been impossible for you to..."), which is either confusing or implicative.

TABLE 3. A Good Questionnaire

- Introduces the goals of the study and the researchers
- Has ample "white space" and room for answers
- Flow of questions is logical and pleasant to the respondent
- Does not lead the respondents in choosing answers
- Captures respondents' motivation until the end
- Produces valid and reliable data on the surveyed issue

TABLE 4. Requirements for Good Questions

- Similarly understood by the researcher and all the respondents
- Short, neutral, and simple
- Specific—with only one possible interpretation
- Relevant and makes sense to the respondent

In research, all concepts have to be specific. In order for the questions to be specific, as much as possible should be defined. Time frames should always be set for events, symptoms, etc. (within the past week, 6 months, a year, ever, as long as you remember, etc.). Numerical quantifiers should be used whenever possible (times/day, week, etc.; percentage of time). They have been found to produce quite consistent results and should be used instead of verbal quantifiers (often, sometimes, etc.) (Hartley *et al.*, 1984). The units that are being asked for should always be defined (hours, days, pounds, etc.) (Table 4).

2.3. Answer Categories

It is wise not to assume specific answers before pretesting questions in a surrogate sample of the target population. The respondent may totally miss the point. In most cases it is not feasible to ask for more details than one needs for one's analysis. One should ask for best estimates, e.g., in time, amount, etc., to allow the respondent's memory more latitude. An exception to this is "easy" continuous data entries (age, height, number of children, etc.)—these should not be categorical. In large self-administered surveys, especially, one should avoid open-ended questions, where the respondents provide their own answers. Open-ended questions can be used in interviews when the interviewers are trained to code the answers to preselected categories. Closed questions (preselected answer categories) are much more easily processed and produce more uniform data. One should choose response categories relevant for one's research after pretesting the questionnaire. To make the categories all exhaustive, one should always add a category "other," which can be used by the respondent when feasible.

2.4. Final Tasks

A questionnaire must always have an introductory letter or page telling the respondents, in a language that they understand, about the goals of the study, the researchers, and their affiliations. It should always include a name and phone number of at least one of the researchers as a contact person. It should also include information on confidentiality and how the respondents can find information about the results of the study (specify media, approximate time). Last but not least, the potential (even marginal) benefits to the respondents from the study should be stated explicitly. There should always be instructions for the respondents, however trivial they may appear, at the beginning of the questionnaire, and possibly in connection with special questions as well.

TABLE 5. The (Order of) Steps in Designing a Questionnaire

- Specify the objectives of the study
- Define the outcomes and other variables within the limitations of the method
- Draw "dummy" tables with the defined variables to see what is needed
- Search for existing questions to yield data for the variables in the tables
- Draft or refine the questions that cannot be found or which do not fit the research
- Beware of vocabulary, wording, etc.
- Consult other researchers (with earlier experience)
- Pretest the questionnaire on colleagues and other individuals
- Rewrite
- Pretest the questionnaire on representatives of the target population
- Consider every comment. Rewrite
- Make data entry sheets
- Finalize the questionnaire and its layout

The format of the questionnaire can be just as important as the wording of the questions. The desirability of spreading out the questions cannot be overemphasized. Nothing puts a respondent off sooner than a busy form with tiny spaces for answers. If the questionnaire includes a lot of screening questions and sections that only some respondents are supposed to answer, it is a good idea to put a frame around the questions that pertain to all respondents, and explain this in the instructions (on "no" to a framed question skip over to the next framed question) (Table 5).

In questionnaire studies the results derive from answers that originate from questions, which thus constitute a part of the method of the study. The definitions for the outcomes (cases) and other important variables, and the questions used to generate these definitions, should always be stated in the publications, much like descriptions of laboratory methods are reported in experimental studies (Table 5).

3. *The Tuohilampi Questionnaire*

The comparison of survey data on the prevalence of skin diseases is difficult because of the lack of uniform methods and criteria for questionnaire diagnoses of skin diseases. The Tuohilampi questionnaire was the first one published for epidemiological research of contact dermatitis (Susitaival *et al.*, 1996). It includes a group of questions and question sets for epidemiological research, with instructions to the researcher and a literature review. In addition to English, the questionnaire has been translated into Swedish, German, and French. In a validation study of detecting hand dermatitis, the questionnaire had a specificity of 99% and sensitivity of 73% (Yngveson *et al.*, 1997). The Tuohilampi questions are recommended for epidemiological studies on contact dermatitis or environmental exposure and skin symptoms. Other studies with the Tuohilampi questionnaire have shown that the questions worked adequately. In supplying the past history, the respondents seemed to go into more detail with the questionnaire when compared to the interview by the dermatologist. Location and symptoms of the dermatitis were accurately described in the questionnaire by all cases. Future objectives of the Tuohilampi project include validation of the questions. The long-term objective of this dynamic project is to obtain epidemiological tools that are operational and yield pertinent data.

The Scientific Committee on Occupational and Environmental Dermatoses of the International Commission on Occupational Health (ICOH), and the Working Group on Occupational Dermatoses of the European Society of Contact Dermatitis (ESCD) recommend the use of the Tuohilampi questionnaire for epidemiological studies of occupational contact dermatitis.

ACKNOWLEDGMENTS: The author wishes to thank Anssi Auvinen, MD, Matti Hannuksela, MD, Philip H. Kass, DVM, PhD, Kevin M. Kelly, PhD, and Antti Lauerma, MD for valuable comments in revising the manuscript.

References

Agrup, G., 1969, Hand eczema and other hand dermatoses in South Sweden, *Acta Derm. Venereol. (Stockh.)* Suppl **61**:1–91.

Bachus, J., Williams, M., Chan, P., Smith, G. D., Grainge, M., Donovan, J., and Frankel, S., 1996, Deprivation and cause specific morbidity: Evidence from the Somerset and Avon survey of health, *Br. Med. J.* **312**:287–292.

Bakke, P., Gulsvik, A., and Eide, G. E., 1990, Hay fever, eczema and urticaria in southwest Norway, *Allergy* **45**:515–522.

Berg, M., and Axelson, O., 1990, Evaluation of a questionnaire for facial skin complaints related to work at visual display units, *Contact Dermatitis* **22**:71–77.

Coenraads, P. J., Nater, J. P., and van der Lende, R., 1983, Prevalence of eczema and other dermatoses of the hand and arms in the Netherlands: Association with age and occupation, *Clin. Exp. Dermatol.* **8**:495–503.

Coenraads, P. J., Nater, J. P., Jansen, H. A., and Lantinga, H., 1984, Prevalence of eczema and other dermatoses of the hands and forearms in construction workers in the Netherlands, *Clin. Exp. Dermatol.* **9**:149–158.

Frey, J. H., 1989, *Survey Research by Telephone*, 2nd Ed., Sage, Newbury Park, pp. 1–289.

Hartley, J., Trueman, M., and Rodgers, A., 1984, The effects of verbal and numerical quantifiers on questionnaire responses, *Appl. Ergon.* **15**(2):149–155.

Johansen, J. D., Andersen, T. F., Veien, N., Avnstorp, C., Andersen, K. E., and Menné, T., 1997, Patch testing with markers of fragrance contact allergy: Do clinical tests correspond to patients' self-reported problems? *Acta Derm. Venereol* **77**(2):149–153.

Meding, B., 1990, Epidemiology of hand eczema in an industrial city. *Acta Derm. Venereol. (Stockh.)*, Suppl **153**.

Meding, B., and Swanbeck, G., 1987, Prevalence of hand eczema in an industrial city, *Br. J. Dermatol.* **116**:627–634.

Nielsen, N. H., Dirksen, A., and Madsen F., 1993, Can subjects with a positive allergen skin test be selected by a short questionnaire?, *Allergy* **48**:319–326.

Payne, S. L., 1980, *The Art of Asking Questions.* 1st paperback ed., Princeton University Press, Princeton.

Pedersen, P. A., Kamper-Jørgensen, F., and Weeke, E. R., 1980, Forekomst af høfeber i den voksne danske befolkning, *Ugeskr Lég* **49**:3269–3271.

Rea, J. N., Newhouse, M. L., and Halil, T., 1976, Skin diseases in Lambeth, *Br. J. Prev. Soc. Med.* **30**:107–114.

Schrijvers, C. M. T., Stronks, K., van de Mheen, D. H., Coeberg, J.-W. W., and Mackenbach, J. P., 1994, Validation of cancer prevalence data from a postal survey by comparison with cancer registry records. *Am. J. Epidemiol.* **139**(4):408–414.

Sheatsley, P. B., 1983, Questionnaire construction and item writing, in: *Handbook of Survey Research* (P. H. Rossi, J. D. Wright, and A. B. Anderson, eds.). Academic Press, Orlando, pp. 195–230.

Smit, H. A., Coenraads, P. J., Lavrijsen, A. P. M., and Nater, J. P., 1992, Evaluation of a self-administered questionnaire on hand dermatitis, *Contact Dermatitis* **26**:11–16.

Spiteri, M. A., Cook, D. G., and Clarke, S. W., 1988, Reliability of eliciting physical signs in examination of the chest, *Lancet* **1**(8590):873–875.

Stone, D. H., 1993, How to do it—Design a questionnaire, *Br. Med. J.* **307**:1264–1266.

Susitaival, P., Husman, L., Horsmanheimo, M., Notkola, V., and Husman, K., 1994, Prevalence of hand dermatoses in Finnish farmers, *Scan. J. Work. Environ. Health.* **20**:206–212.

Susitaival, P., Husman, L., Hollmén, A., and Horsmanheimo, M., 1995*a*, Dermatoses determined in a population of farmers in a questionnaire based clinical study including methodology validation, *Scan. J. Work. Environ. Health.* **21**:30–35.

Susitaival, P., Husman, L., Hollmén, A., Horsmanheimo, M., Husman, K., and Hannuksela, M., 1995*b*, Hand eczema in Finnish farmers: A questionnaire based clinical study, *Contact Dermatitis* **32**:150–155.

Susitaival, P., Kanerva, L., Hannuksela, M., Jolanki, R., and Estlander, T., 1996, Tuohilampi questionnaire for epidemiological studies of contact dermatitis and atopy, People and Work, Research Reports 10, Finnish Institute of Occupational Health, Helsinki.

Williams, H. C., Burney, P. G. J., Hay, R. J., Archer, C. B., Shipley, M. J., Hunter, J. J. A., Bingham, E. A., Finlay, A. Y., Pembroke, A. C., Graham-Brown, R. A. C., Atherton, D. A., Lewis-Jones, M. S., Holden, C. A., Harper, J. I., Champion, R. H., Poyner, T. F., Launer, J., and David, T. J., 1994*a*, The U.K. Working Party's diagnostic criteria for atopic dermatitis: I. Derivation of a minimum set of discriminators for atopic dermatitis, *Br. J. Dermatol.* **131**:383–396.

Williams, H. C., Burney, P. G. J., Strachan, D., and Hay, R. J., 1994*b*, The U.K. Working Party's diagnostic criteria for atopic dermatitis: II. Observer variation of clinical diagnosis and signs of atopic dermatitis, *Br. J. Dermatol.* **131**:397–405.

Williams, H. C., Burney, P. G. J., Pembroke, A. C., and Hay, R. J., 1994*c*, The U.K. Working Party's diagnostic criteria for atopic dermatitis: III. Independent hospital validation, *Br. J. Dermatol.* **131**:406–416.

Williams, H. C., Burney, P. G. J., Pembroke, A. C., and Hay, R. J., 1996, Validation of the U.K. diagnostic criteria for atopic dermatitis in a population setting, *Br. J. Dermatol.* **135**:12–17.

Yngveson, M., Svensson, Å., and Isacsson, Å., 1997, Evaluation of a self-reported questionnaire on hand dermatosis in secondary school children, *Acta. Derm. Venereol.* **77**:455–457.

Part II

Noninvasive Evaluation of Skin Properties

Chapter 5

Scale Development for Clinical Assessment

GERHARD J. SCHMITT

1. Assessment Criteria and Scales

Critical elements of all clinical investigations are the parameters and clinical scales employed for the assessment of relevant parameters. Their suitability to correctly reflect tactile or visual findings about skin condition collected by experienced investigators as morphological parameters or as overall clinical diagnoses determine the relevance and effectiveness of an investigation. Hence, when first beginning to design a clinical investigation into skin physiology/morphology and their pathologic variants, the foremost relevant question is whether the parameters to describe endpoints or surrogate endpoints of the effect are covered by existing clinical grading scales for proper, valid documentation of changes. These scales must be suitable to generate data addressing the study objectives, i.e., they have to be adequately sensitive. If they are close to, but not fully in line with the respective needs, the obvious question is whether they can be adjusted to meet the needs or whether they have to be replaced by new grading scales to be developed by the investigator, often in close cooperation with the study sponsor, using the same, similar, or new parameters for assessment of changes in skin physiology/morphology.

When defining grading scales for visual changes, it has to be determined that they are detailed enough to document the initial (patho-)morphological findings on the one hand and deviations potentially developing over time as a result of application of the test products on the other. In defining scale grades, the clinical relevance of the individual steps in the scale should be given priority. Otherwise, there is the risk that by overemphasizing the technically possible numerical discrimination between findings, the sensitivity for clinical developments is lost. Suitability of

---

Gerhard J. Schmitt • Beauty Care Europe, Procter & Gamble European Service GmbH, D-65823 Schwalbach/Taunus, Germany.

Cutaneous Biometrics, edited by Doris A. Schwindt and Howard I. Maibach, Kluwer Academic/Plenum Publishers, New York, 2000.

0 No findings
1 Faint erythema on small area(s)
2 Large area(s) with mild erythema; (few) very small area(s) with severe erythema, yet without oedema
3 Large area(s) with severe erythema—without oedema—(few) very small area(s) with severe erythema, yet with oedema
4 Large area(s) with severe erythema—with oedema

FIGURE 1. Severity of erythema.

a visual grading scale should be gauged against the possibility of documenting (patho-)morphological properties, e.g., roughness, color, surface structure, with sufficient precision and good reproducibility. The method of examination, whether the eye, a magnifying optical instrument, or the sensitivity of the palpating finger and hand, respectively, both as individual parameters and as an overall finding based on these must be taken into consideration.

Historically, grading scales were based on a series of continuous numbers or letters following each other in discrete steps. Each step described an increase or decrease in an effect relative to the previous one over a range between two antagonistic endpoints, e.g., "no finding" and "severe damage" (Fig. 1). Their use suggested quasi-linear, continuous progression of a certain effect, which in a biological environment very often is only true within narrow ranges. Statistical analyses of these numerical severity ratings were typically conducted using analysis of variance procedures (ANOVA) with more or less sophistication to compensate for the nonlinearity of the scales. More modern ranking approaches take more account of the biological environment. They often use continuous-mark scales to define the endpoints, e.g., "no finding"/"severe damage" rather than discrete grading steps to describe the increase or decrease of a parameter. Typically, a description of the various transient stages in progress/development of a finding is attached to the scales to guide the investigator in his/her considerations for marking his finding on the grading line (Fig. 2). An in-between possibility is to give ranges of changes documenting a coherent series of steps resulting in a ranking scale rather than a direct grading scale (Fig. 3). For statistical analysis of nonlinear, biological ranking scales, state-of-the-art methods for nonparametric data (e.g., the Mann–Wittney U-Test) are used.

A self-assessment of the respective skin condition or change in skin condition by the user of the test product can be useful for determining the clinical relevance of (patho)morphological/(patho)physiological changes. Use of a scale formulated in lay

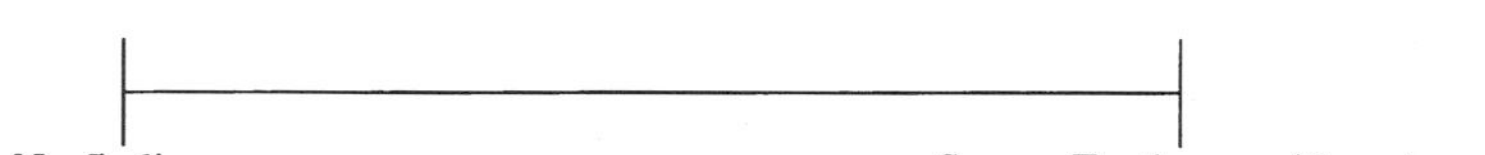

FIGURE 2. Severity of erythema. Erythema increasing in severity and spreading over the test area or even beyond, with edema of increasing severity and spread over an increasingly large area.

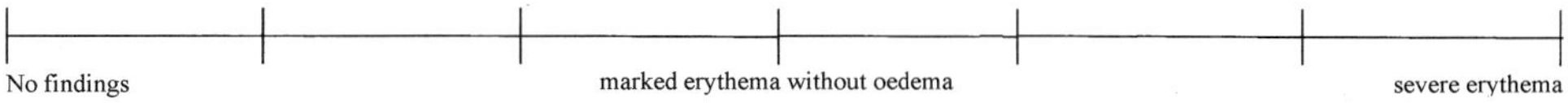

FIGURE 3. Severity of erythema. Erythema increasing in severity and spread over test area or even beyond and with edema.

terms, ideally with the same number of grading steps as on the clinical grading scale can be of scientific as well as practical interest. To achieve unbiased results, the self-assessment must take place at the same location at the same time under the same light and in the same temperature and relative humidity conditions, but prior to the clinical assessment, to avoid bias from the latter. This way, it can be determined whether the user of the test product can recognize seemingly irrelevant changes in skin condition that might be early indications for later-developing clinically relevant pathological changes. This can be useful for an early tentative diagnosis, and, if necessary, for therapy by the dermatologist. Furthermore, self-assessment can also give valuable information as to whether statistically significant and clinically relevant changes (benefits or impairments) resulting from the use of the test product that are detected by an expert grader are also consumer-recognizable. It will also help determine to what extent these changes, if at all recognized by the consumer, are considered by him/her to be relevant.

A scale that properly documents findings must also allow for the generation of reproducible findings by different but equally well-trained investigators. Specifically for multicenter studies and relevant meta-analysis of data generated in them, the robustness of a scale to allow comparable assessments by different investigators is of critical importance. A scale that is overengineered and geared to the level of sophistication of one particular experienced investigator, is critically limited in terms of its use for reapplication by others in a multicenter study or in trials to generate additional data in independent studies. Furthermore, in case of challenge and legal action, there may be questions as to the validity of the data if other recognized experts using the same scale are not able to reproduce the findings.

2. Case Report Forms

Records for documenting findings should not only provide the possibility for unequivocal data documentation as a basis for statistical analysis, but should also provide space for documentation of clinically irrelevant findings in the test area, e.g., artefacts, or other clinical observations that do not recognizably influence the investigation, e.g., the general health of the test subject. Such information may be of importance in the interpretation of unexpected side effects. This also holds true with regard to pathological changes in skin condition outside of the test area.

For listing of individual test parameters on a case report form, suitability for simple and fast data recording and error-free documentation is particularly important in cases of serial examinations. The structure of the form should be based on typical clinical routine and practice for examinations, with a logical sequencing of

Case Report Form for Diaper Rash Grading

001
Serial number)

Visit No.

	0	1	2	3
Grader Initials . . .	0	0	0	0
. . .	1	1	1	1
. . .	2	2	2	2

FINDINGS (Physician)

LOCALISATION	GENITAL	INGUINAL	PERIANAL	GLUTEAL	SYSTEMATIZED	VENTRAL	DORSAL	GENERALIZED
Visit No.	0 1 2 3	0 1 2 3	0 1 2 3	0 1 2 3	0 1 2 3	0 1 2 3	0 1 2 3	0 1 2 3
Erythema w oedema { {	0 0 0 0 1 1 1 1 2 2 2 2 3 3 3 3 4 4 4 4	0 0 0 0 1 1 1 1 2 2 2 2 3 3 3 3 4 4 4 4	0 0 0 0 1 1 1 1 2 2 2 2 3 3 3 3 4 4 4 4	0 0 0 0 1 1 1 1 2 2 2 2 3 3 3 3 4 4 4 4	0 0 0 0 1 1 1 1 2 2 2 2 3 3 3 3 4 4 4 4	0 0 0 0 1 1 1 1 2 2 2 2 3 3 3 3 4 4 4 4	0 0 0 0 1 1 1 1 2 2 2 2 3 3 3 3 4 4 4 4	0 0 0 0 1 1 1 1 2 2 2 2 3 3 3 3 4 4 4 4
Papules yes	+ + + +	+ + + +	+ + + +	+ + + +	+ + + +	+ + + +	+ + + +	+ + + +
Vesicles yes	+ + + +	+ + + +	+ + + +	+ + + +	+ + + +	+ + + +	+ + + +	+ + + +
Loss of epithelium yes	+ + + +	+ + + +	+ + + +	+ + + +	+ + + +	+ + + +	+ + + +	+ + + +
Desquamation yes	+ + + +	+ + + +	+ + + +	+ + + +	+ + + +	+ + + +	+ + + +	+ + + +

Visit No.	0 1 2 3
Overall Skin Rash Grade	0 0 0 0
	1 1 1 1
	2 2 2 2
	3 3 3 3
	4 4 4 4

CLINICAL DIAGNOSIS

Visit No. / Diaper Area	0	1	2	3
w/o Finding	0	0	0	0
Erythema	1	1	1	1
Diaper Rash	2	2	2	2
Candidosis (clinically)	3	3	3	3
Oral Candidosis	4	4	4	4

Grading Scales

I. Erythema Grades

0 No findings
1 Faint erythema on small area(s)
2 Large area(s) with mild erythema; (few) very small area(s) with severe erythema, yet without oedema
3 Large area(s) with severe erythema — without oedema —; (few) very small area(s) yet with severe erythema — with oedema
4 Large area(s) with severe erythema — with oedema

II. Rash Grades

0 Complete absence of rash, might include redness so slight as to be indistinguishable from affect of pressure of clothing
1 One or more of the following: small areas of "pinpoint" rash, mild erythema over large area
2 One or more of the following: moderate erythema — without oedema — limited to labia/scrotum or perianal area, moderate erythema over larger area, some desquamation, few papules
3 One or more of the following: large area of severe erythema — without oedema —, very small area excoriated, papules, few vesicles
4 One or more of the following: large area of severe erythema — with oedema —, excoriation on large areas, papules, vesicles, pustules; candidosis (clinically)

FIGURE 4. Case report form for diaper rash grading.

test parameters for the respective skin area. This in turn should be in logical correlation with the preceding and subsequent area of assessment without regard to its contribution to the overall assessment. Accordingly, clinical diagnosis and overall finding, respectively, should be taken and documented after recording of the individual clinical parameters. Because of the relevance of these clinical aspects in data generation, they have to be given priority over considerations about ease of subsequent data entry into electronic data carrier systems for statistical analysis. Figure 4 shows an example of a case report form including scale description (attached to case report form) as used for the clinical assessment of the skin care efficacy of panty diapers.

3. Validation and Calibration of Investigators

The reproducible evaluation of (patho)morphological/(patho)physiological skin conditions and their corresponding representation on a grading scale is highly dependent on the clinical experience of the investigator. Based on differences in weighing individual aspects, deviations in the overall assessment of a finding can occur even with equally experienced investigators. Hence, in case of multicenter studies, a calibration of the individual investigators relative to one another is especially desirable and highly recommended for achievement of comparable, well-discriminated results. The same holds true for several investigators sharing a single-center investigation. In this case, if at all possible, each investigator should see a randomized, roughly equally large subgroup of the total test population for assessment. These subgroups should not change during the course of the study. Particularly for the initial assessment (at test product placement) and for the final one (end of test product use), changes of test subjects among groups have to be avoided. Further, investigators should be calibrated in comparison with one another.

To collect valid findings of changes in individual morphological/physiological aspects of skin condition and for their diagnostic association with descriptive scales, visual documentation of the corresponding skin conditions in their full scope of variation within the respective scale grade are of decisive assistance. The use of photodocumentation of typical cases for each grade on the scale will enable the investigators to better understand the skin structures and symptoms described as grading criteria. It will also contribute to a more uniform grading of cases, in which overemphasis of individual symptoms to be integrated into the overall diagnostic assessment grade could lead to undue variation, e.g., giving too much or too little weight to the severity of a lesion relative to its size. Obviously, the photodocumentation has to be via a professionally executed photograph. Photodocumentation of a biological scale should cover as complete a range of variation as possible to enable standardization of findings to be collected in line with scale descriptions. It also significantly facilitates the calibration of the investigators.

In our own research, the procedure typically chosen involved giving the individual investigators comprehensive photodocumentation for the individual scale units of a skin condition grading scale, initially with open identification of the corresponding assessment values. With this, they could familiarize themselves with the scale descriptions and the correlating clinical pictures. Subsequently, they were shown pictures of the same or similar findings without scale grade identification and in random order for assessment based on the description of the grading scale. This process of examining pictures of skin conditions with and without identification of the scale grade was continued until there was agreement between expected and associated gradings in at least 8 out of 10 cases. Another joint open assessment of skin conditions among subjects not part of the clinical investigation followed to establish that the precision and power of discrimination in assessment of investigators trained by means of photodocumentation is readily translated into a practical framework. Following such a procedure, among clinically experienced investigators, reliability of 0.8–0.9 can be achieved among them or in a reference to an investigator specifically experienced with use of that particular scale. Such a high level of accuracy in assessment of individual skin conditions is always a specific advantage, especially for large-scale epidemiologic investigations using broad cross sections of the population across country borders. It is equally important if there has to be a statistically valid correlation among individual investigations conducted using the same test methodology and following the same test protocol (meta-analysis). This is even more important if such a clinical investigation is conducted as a multicenter study or a single-center multiinvestigator study.

To validate the findings of different investigators with regard to true test-product-specific or, seemingly, investigator-specific changes in skin condition, the inclusion of an experienced investigator assessing important individual parameters of skin condition as well as the overall findings and diagnoses of all test subjects concurrently, but independently from the respective individual investigator, can be advantageous.

4. Relevance of Biophysical and Biochemical Parameters

To the extent that biophysical and biochemical measurements are taken as additional means of describing changes in skin condition, they can be used in support of the interpretation of the clinical findings. Depending on the nature of the (patho)physiological/(patho)morphological parameters, they may offer early indications of preclinical processes in responding to a test product. Such biophysical or biochemical measurement parameters, independent of their statistical significance, must be critically evaluated with regard to clinical relevance and relative importance. As those measurements are typically surrogate endpoints of physiological regulation processes, whose relevance to a change in clinical condition may not be certain in each case, overassessment and overinterpretation should be avoided. On the other hand, the intelligent combination of clinical assessment and diagnosis of (patho)-morphological/(patho)physiological changes assessed by the experienced investigator with objective biochemical and biophysical measurements of surrogate endpoints can add significantly to the understanding of findings, the mode of their development, and the discrimination of early indications for change not yet accessible to clinical diagnosis. Advances in diagnostic and experimental dermatology in the recent past were only possible because of close cooperation between clinical dermatologists and researchers using bioengineering methods who targeted the same goal, namely a better understanding of human skin physiology and its interactive regulation mechanisms.

Chapter 6

Evaluation of Cosmetics by Sensory Assessment

PETER BUSCH AND THOMAS GASSENMEIER

1. Introduction

Cosmetics are used to provide external body care or to change the external appearance of the body in a decorative way in order to achieve a positive impact on our senses. This "impact on our senses" is a highly complex matter. It encompasses a variety of processes — physicochemical, chemical, physical, and/or physiological. It is also linked to a spectrum of phenomena that can best be subsumed under the heading "psychological" or "mental." This means that sensory assessment is an aspect of objective natural science on the one hand, but on the other also has to do with subjective perception and thus with the humanities. This gives it a dual assignment, a twofold goal, and that is what we want to examine more closely here, with examples taken from the field of cosmetics.

2. The Relationship between Effect and Perception

As a general statement it is surely true to say: The world is not the way it is because this is how it (really) is; it is like this because this is how we perceive it with our sensory organs. Expressed differently: The world can be regarded as the sum total of diverse impulses formed into an "image of the world" by and via a perceiver. For this reason, it basically makes no difference whether the impulses are emitted by some actual reality or by a "false reality" that conjures up certain illusions in us. After all, illusions are an aspect of all types of sensory perception. Moreover, illusions play a considerable role in the field of cosmetics, e.g., when makeup is used to make certain "undesirable" facial features less conspicuous by darkening them, i.e., making them disappear from our field of perception, while

---

Peter Busch and Thomas Gassenmeier • Henkel KGaA, D-40191 Düsseldorf, Germany.

Cutaneous Biometrics, edited by Doris A. Schwindt and Howard I. Maibach, Kluwer Academic/Plenum Publishers, New York, 2000.

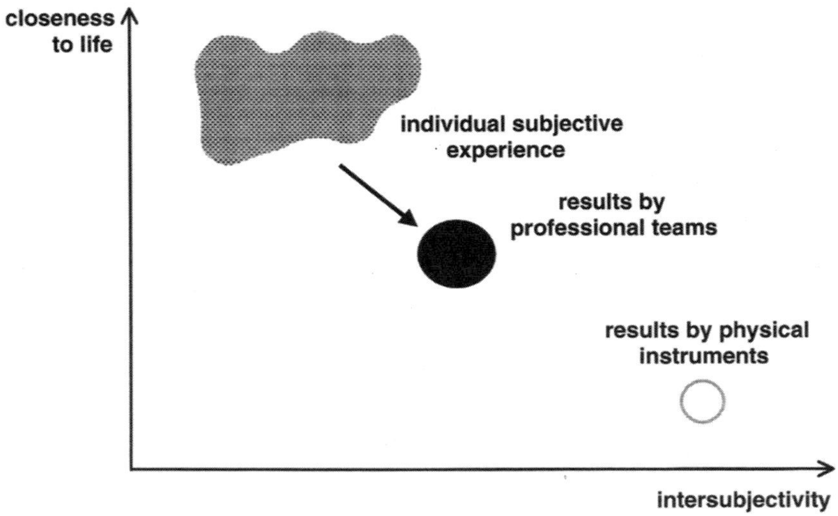

FIGURE 1. Intersubjectivity box.

favored features are made to impinge more strongly on our conscious perception by means of highlighting.

What all this implies is that it is not so much the interplay between the physical or chemical impulse and the relevant sensory receptor that plays the decisive role in registering effects; what really matters is the way in which the effects are interpreted, in a mental process that takes place more or less simultaneously. However, this process of interpretation is itself influenced by an extremely large number of factors, which explains why the subjective findings typical of sensory assessment tend to seem diffuse rather than clearcut or unambiguous. A perceptual paradox applies: That which can be registered very precisely with the help of objective measuring instruments is often not very "lifelike," while something experienced as "realistic" or "natural" is almost always a bit vague and difficult to sum up. The so-called intersubjectivity box illustrates this dilemma (Fig. 1). Thus it would be wrong to expect sensory assessment to deliver the kind of precision that measurement by means of instruments provides.

3. *Reasons for Sensory Vagueness in Assessing Cosmetic Effects*

The range of circumstances and factors responsible for the "fuzziness" of sensory experiences is actually quite stunning. There are anatomical differences among individuals that affect the receptive capacity of the sensory organs, and this capacity also changes quite considerably during the life of any one person. The substrate that is the object of a cosmetic treatment, e.g., the hair or the skin, varies in each individual—between one part of the body and another, for instance—but also between one person and the next, depending, say, on race, standard of cosmetic care, climate, and age.

These individual variations, which are by themselves virtually impossible to systematize, are then multiplied to an unmanageable extent by the different mental or psychological characteristics of any group of people. These mental/psychological characteristics are in turn influenced by factors such as cultural background, religion, level of education, membership in a particular social group, or age—to name but a few. The overall result is a very wide spectrum of different interpretations of sensory experiences. Beauty lies in the eye of the beholder—but it is not just the appreciation of beauty that is extremely subjective. So, too, is our perception of seemingly straightforward factors, such as skin smoothness or regularity. Even here, our senses come up with different interpretations.

4. The Significance of Language

4.1. Correlation of Experience and Language

Differences in interpretation are reflected especially clearly in the way in which we express what we experience. Language is the indicator of experience, and it is so twice over: on the one hand, it is language that grants us access to experience of all kinds; on the other, our linguistic ability deepens our ability to experience. That is to say, the more linguistic resources we have at our disposal, the more we experience. This can be shown by means of "open interviews" of the kind that provide an introduction into every sensory assessment session.

We selected five panels with four people of differing knowledge of cosmetics and asked them to depict their experience in using a face cream. Panels with middle-aged women who work in a skin cosmetics laboratory used about 110 words for this, while younger women from the same laboratory used an average of about 86 words, showing that they still lack experience—and thus the vocabulary that reflects experience—in this sector. Middle-aged and older women without any training in skin cosmetics used about the same number of linguistic terms, presumably because through lengthy used such products they had gathered a lot of experience. Men, on the other hand, and extremely young interviewees had not yet generated any vocabulary in this field and thus had only a limited horizon in this sector (Fig. 2).

Since the reality around us is always formed by our subjective interpretation, an infinite number of modes of comprehension are possible. A good example is the culture-dependent comprehension of fabrics by the hand (i.e., tactile sensation). It was found that summer cloth was judged very differently in Japan, India, New Zealand, and Australia. Associated with the concept of hand are quality words such as "soft," "warm," "thin," etc., which are common ideas created by communication among people (Mahar *et al.*, 1982). Evidently the meaning content of these words is specific to the different cultural circles. There was a satisfactory correlation between people in Australia, New Zealand, and India, while the Japanese perceived the cloth in a completely different way. They used verbal terms which implied almost an antagonistic meaning (e.g., instead of "stiff" they said "coarse," instead of "crisp" they said "smooth") (Table 1).

The generation and the subsequent use of appropriate verbal terms is the key to any successful subjective assessment, especially when our taste, smell, or tactile

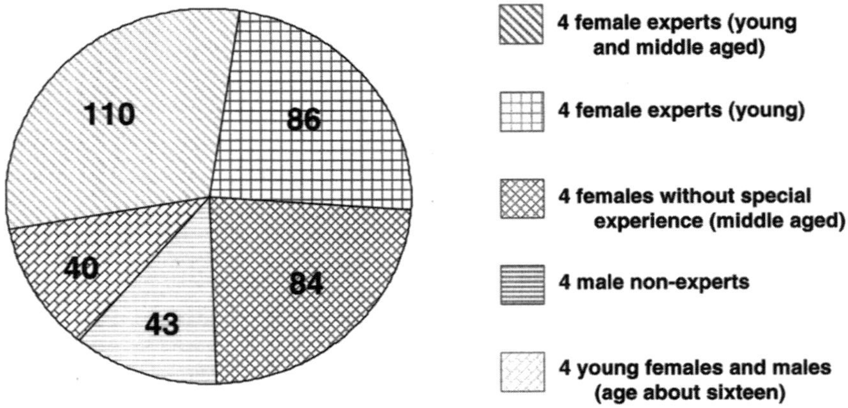

FIGURE 2. Panel test for generation of words.

organs are addressed. Therefore the technique of subjective evaluation is widely applied for judging food, beverages, and perfumes. While literature dealing with these substrates is abundant, only very few papers exist describing the systematic subjective assessment of cosmetics (e.g., Civille and Dus, 1991).

4.2. Generation of Verbal Terms

Each product leads to individual sensory impressions. The more associated words that are available, the better the product performance can be understood, communicated, and shaped. Therefore the generation of words plays an important role for the subjective assessment. It can be achieved by two main routes:

1. Asking test panels to find words while applying the product (in our case the cosmetic product) on different body sites.
2. Finding words via linguistic methods.

(a) Use of Test Panels. In a first run, panels of not more than four or five people should be asked to participate for this sort of word collection. Needless to say, the atmosphere should be relaxed and friendly. Most important is to have somebody who carefully writes down all spoken terms. The terms can be nouns, verbs, adjectives, exclamations, half and whole sentences, proverbs, idioms—"anything

TABLE 1. Correlations of Handle

Japanese/Indians	−0.40
Japanese/Australians	−0.35
Japanese/New Zealanders	−0.26
Indians/Australians	+0.78
Australians/New Zealanders	+0.84

goes." Any evaluating comments such as "your statement is false, exaggerated," etc., are ruled out. Depending on the vividness and language capabilities of the team, more or fewer words will be found; e.g., compare the terms generated in connection with a face cream: reminds me of Jell-O • I like it best • fragrance is light • very light cream • very nice • absorbs fast • you use more of it because of quick absorption • fragrance is almost unnoticeable • no residue afterward • light and fluffy cream • fragrance is very faint • absorbs quickly into the skin • leaves a soft afterfeel • cooling effect • not greasy on skin • it is smoother on my hands after two washings • can mix with perfume and wouldn't conflict • cream is not heavy • afterfeel is greasy • don't have to wait for absorption • spreads easily and quickly • soothing • I see a film but it doesn't have a soft or silky feel • not so crazy about it • somewhat greasy afterfeel but not entirely • very light to the touch • much nicer feel • I like it for an everyday lotion, not good for night use. It is recommended to have this word finding procedure repeated by a second group or even by several other teams.

(b) Processing the Terms. The terms are divided into two categories: hedonic and diagnostic. Hedonic words are terms of like and dislike. Most of the time, the individual esthetic appreciation plays a decisive role. In our case, statements such as: "I like it best," "very nice," "not so crazy about it" are typical hedonisms. *Diagnostic judgments* can—basically speaking—be substituted by objective methods, e.g., the coloring effect of a lipstick can be measured by means of a colorimeter, and greasy skin caused by an overproduction of sebum can be evaluated by use of a sebumeter.

All words are grouped for positive and antagonistic meanings (e.g., skin can be smooth or not rough; it can be dull or not shiny). Only one type of word is used. The words are organized by use of superordinating terms, e.g.:

General statements: very nice • I like it best • not so crazy about it • I love it.

Fragrance: fragrance is light • fragrance is almost unnoticeable • fragrance is very faint.

Texture of cream: very light cream • light and fluffy cream.

Absorption: absorbs fast • absorbs quickly into the skin • you use more of it because of quick absorption.

Residue: no residue afterward.

Cooling: cooling effect.

Greasiness: reminds me of Jell-O • not greasy on skin.

Rinsability: it is smoother on my hands after two washings.

Compatibility: can mix with perfume and wouldn't conflict.

Spreadibility: spreads easily and quickly.

Care effect: soothing • I see a film but it doesn't have a soft or silky feel • afterfeel is greasy • somewhat greasy afterfeel but not entirely • leaves a soft afterfeel • very light to the touch • much nicer feel.

Recommendation for applying: I like it for an everyday lotion, not good for night use.

The word clustering, of course, is in some respect arbitrary, since each word can be subordinated under different generic terms; e.g., "no residue afterward" can be a special generic term or be subordinated under "grease." The term "it is smoother on my hands after two washings" may also belong to "care effects."

(c) Support of Word Finding by Preformed Verbal Terms. Since each cosmetic formulation produces individual sensory impressions, the found vocabulary will be unique. Nevertheless, a guide to expected word terms is helpful. The first impression is the visual aspect of the product (in its package, jar): color, gloss, pearlescence.

At a very early stage, the notion of the fragrance/odor is decisive for the positive or negative acceptance of the product. Most appropriate would be to use descriptive perfumery terms such as green, flowery, woody, jasmine. However, as most people are not educated to distinguish among different odors, they use either very simple terms such as "light," "faint," "unnoticeable," "strong," or only hedonistic terms ("like it," "smells terrible," etc.). Very striking for the evaluation of a product is its texture, such as pasty, liquid, solid, gellike, fluffy, etc. Most of the time, texture is experienced by special movements of fingers with the product adhering or by dipping them into it; visual observation extends the information concerning the product (e.g., it is stringy, there are particles between the fingers, the product surface is even and uniform).

Evaluation of the product is one aspect, and the interaction with the skin is another. Since the product performance is correlated with special movements of one or more fingers, the palm, or the back of the hand on different body sites, it is necessary to supply the test person with a procedure guide outlining the testing actions as precisely as possible. An overview of evaluation terms is given below (Table 2).

For a comprehensive assessment, temporal effects must also be taken into account, e.g.,

- Initial feel
- Afterfeel
- Feel after a certain period of time

(d) Linguistic Methods. People who are trained in linguistics can use semantic fields in thesaurus collections to enlarge the vocabulary of special properties. The problem is not finding the words but limiting oneself to the relevant ones. This is because semantic fields overlap, i.e., it is easy to jump from a word to another

TABLE 2. Sensory Parameters Concerning Skin Care Products

⇒ Pick up	⇒ Absorption
⇒ Consistency	⇒ Smoothness
⇒ Peaking	⇒ Stickiness
⇒ Cushion	⇒ Oiliness (residues)
⇒ Distribution	⇒ Waxiness (residues)

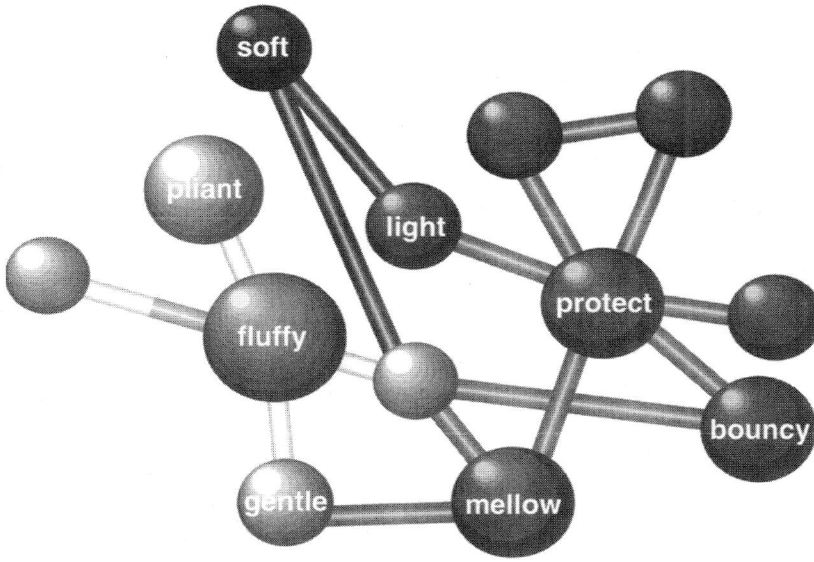

FIGURE 3. Interconnection of semantic fields dealing with the word "cushion."

similiar word without coming to an end. In Fig. 3, the interconnection of semantic fields around relevant words deals with the word "cushion" (American term for soft, smooth body of cream as it can be felt between rubbing fingers). An example of systematic checkup is listed in Table 3. The quoted numbers help to approach the semantic fields in *Roget's Thesaurus* (1984) systematically.

5. Quantitative Testing

5.1. Principles of Quantification of Subjective Tests

In contrast to objective tests by mechanical or chemical devices, subjective assessment does not produce absolute values. Most commonly used are scaling methods. The panelists are asked, e.g., to rate the magnitude of a certain parameter by assigning it a value on a numerical intensity scale or by using ratio scales or unstructured hedonic scales (ISO, 1987).

It is common experience that the results of such tests are highly variable. This can be explained by the individual disposition of each test person. Furthermore, one should bear in mind that we lack the reference by which we can rank the subjective findings. Thorough training of the panelists should alleviate this situation. An analysis, however, showed that panelists need at least 30 hr of training to acquire the knowledge needed to handle ten test parameters, and this is not economical. A better approach is to use comparison tests. Based on this knowledge, we developed a new method for sensory assessment in cosmetics, which is described below.

TABLE 3. Cushion

- ❖ alleviate 163.3
  - ➢ calming
  - ➢ palliative
  - ➢ pacifier
  - ➢ soother
- ❖ bedding 216.20
- ❖ partion 237.5
- ❖ silence 451.4
- ❖ softness 357.4
- ❖ muffle3 451.9
- ❖ protect 699.18
  - ➢ guard
    - ■ shelter
  - ➢ shield
  - ➢ cover
  - ➢ compass around
  - ➢ fence
- ❖ relieve 886.5
- ❖ soften 357.6
  - ➢ ease
  - ➢ gentle
  - ➢ mollify
  - ➢ milden
  - ➢ mellow
- ❖ soften 357.6 (cont.)
  - ➢ supple
    - ■ pliant 357.9
      - ● flexible
        - ◆ resilient
          - ➢ recuperative 694.23
          - ➢ elastic 358.7
            - ■ springy
            - ■ bouncy
            - ■ adaptable
            - ■ lively
            - ■ buoyant
        - ◆ docile
      - ● yielding
      - ● moldable
      - ● shapable
      - ● willowy
      - ● springy
  - ➢ knead
  - ➢ fluff
    - ■ fine texture 351.3
    - ■ lightness 353.2
      - ● air
      - ● either
      - ● feather
        - ◆ lightness
        - ◆ softness
        - ◆ plumage 230.18
    - ■ make pliant 357.5
    - ■ softness 357.4
- ❖ support 216.21

5.2. Comparison Test vs. Defined References

Although sensory assessment is widely conducted in R&D laboratories of the cosmetic industry, no common method is applied. The methods vary more or less among laboratories of different companies, even within one company or among individuals in one laboratory. In the following, a commonly applicable, reproducible test methodology for the sensory assessment of cosmetics for skin is presented. The method was designed in order to gather information about consumer-relevant sensory properties of cosmetic products as simply and easily as possible.

The two key factors are:

1. Very precise outlining of the test procedures.
2. Use of a special test kit with appropriate formulations for determining the desired parameters.

The application of the method enables meaningful and precise sensory profiles to be structured.

TABLE 4. Sensory Parameters Concerning Surfactant-Based Products

⇒ Distribution	⇒ Creaminess
⇒ Slipperiness	⇒ Foam Stability
⇒ Flash foam	⇒ Foam cushion
⇒ Amount of foam	⇒ Smoothness of the skin
⇒ Foam bubble size	⇒ Skin dryness

(a) Procedures for the Evaluation of Sensory Properties. For each product class in the cosmetic area, the parameters and the testing procedures are conceived from a practical point of view, i.e., how consumers would proceed with their testing. For skin care products, the first parameter is the pickup and in the end possible residues on the skin are evaluated. For surfactant-based cleansing products, other parameters such as the foaming behavior are evaluated (Table 4). The test procedures for each parameter were developed with the objective of obtaining information as praxis-relevant as possible. Therefore we started with procedures close to the usual application process, which were slightly modified with the aim of achieving a more precise evaluation allowing us to distinguish different products significantly.

The developed test procedures comprise exact instructions, i.e., the panelists are clearly informed about manual actions and the product quantities they have to apply. In Figs. 4 and 5 the exact definiton of the parameters "Consistency" (for skin care products) and "flash foam" (for shower gels) are given. Figures 6 and 7 show the testing procedure.

(b) Test Controls. General guidelines on the design of test rooms are described in the literature (ISO, 1988; ASTM, 1986). Specifically our experiments were conducted under defined conditions (temperature 23°C, 40% relative humidity) in a neutral, odor- and noise-free test room with standardized illumination. Prior to the start of the test, the panelists rest for 10 min in the climate-controlled test room.

We place two jars containing the test product and the standard product in front of us. We dip the index fingers of both hands synchronously in an angle of approx. 60 ° and a depth of one finger- tip into the products. (The elbows should not rest on the table but should be slightly lifted). Immediately after the dipping procedure we remove the fingers out of the jars. The consistency correlates with the pressure we experience during the dip in and the dragging force while we pull out the fingers. This procedure can be repeated one to three times. (Don't repeat this too often otherwise the emulsion gets liquified to a certain amount by shear forces). We compare with several standards.

FIGURE 4. Consistency.

Both hands are wetted with warm water (38° C) before the test. Then 0.5 g of the product is placed in the inner side of the hand together with 2.5 g water. The panelist is asked to start the washing procedure by using the second hand, which is moved with slight pressure and a frequency of two movements/sec. for 10 sec. over the area where the product has been applied. After that, again 2.5 g of water are applied and the procedure is repeated for another 10 sec.

The speed of foam formation is evaluated in comparison to the reference product.

FIGURE 5. Flash foam.

(c) Seven-Point Comparison Test for Quantification of Test Results. As already mentioned, the use of intensity scales tends to deliver highly variable results unless the panelists have been given adequate training.

For an easily applicable sensory assessment, reference samples for the desired test parameter had to be developed. This was achieved by starting with "root formulations." For skin care products, the root formulation chosen was the "Cremor Basalis DAC" (Fig. 8), which represents a pharmaceutical O/W cream based on PEG-20 glyceryl stearate combined with glyceryl stearate and cetearyl alcohol as an emulsifier/coemulsifier system. This nonionic cream is very robust against changes of the oil phase and water phase and the addition of other ingredients with completely different chemical structures. By systematic modifications of this formula,

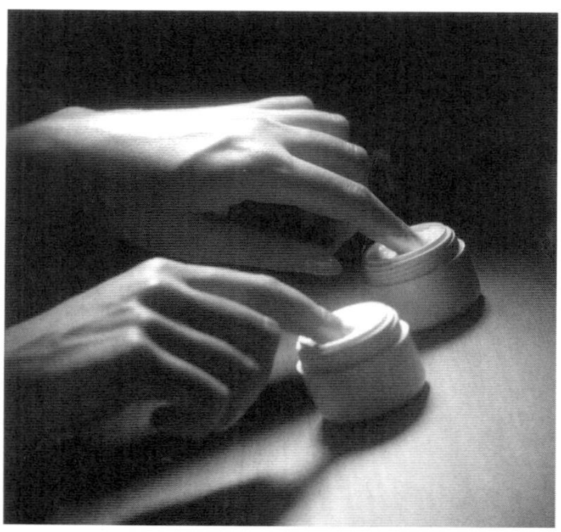

FIGURE 6. Test procedure for the evaluation of consistency.

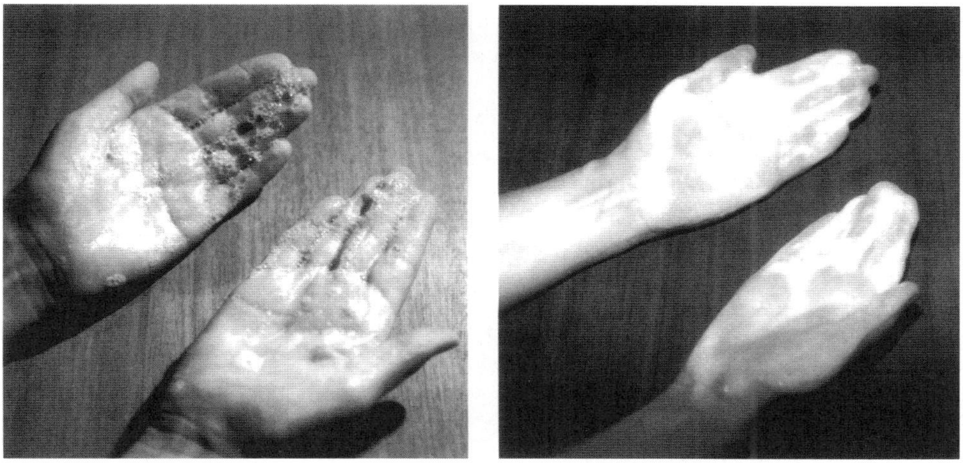

FIGURE 7. Test procedure for the evaluation of flash foam.

reference products with desired properties were developed (Fig. 9). In case of the sensory assessment for surfactant-based products, the root formulation is simply a solution of 10% sodium laureth sulfate, which was modified by changes in the surfactants used and the addition of special ingredients.

For each sensory parameter, three reference products are available, which represent a “medium,” “high,” or “low” comparison standard. The resulting set of formulations is like a fractal web that can be refined more and more, and thus produce any number of detailed reference standards.

The test evokes three possibilities: (1) the test product and reference product 2 are equal, (2) the test product is different in the one or (iii) other direction. In cases (ii) or (iii), the test product is tested again vs. the reference product 1 or 3. This procedure is repeated for each of the sensory parameters. In the end, one number on

Ingredient	Amount [% (w/w)]
Glyceryl Stearate	4.0
Cetearyl Alcohol	6.0
Caprylic/Capric Triglyceride	7.5
Petrolatum	25.5
PEG-20 Glyceryl Stearate	7.0
Propylene Glycol	10.0
Aqua	40.0
	100.0

FIGURE 8. Cremor basalis DAC.

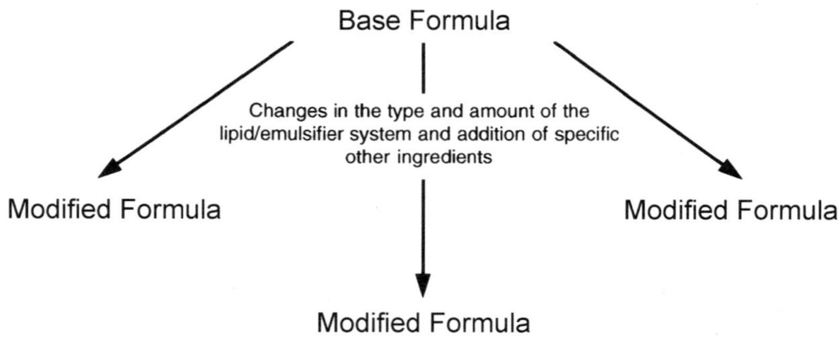

FIGURE 9. Strategy for generation of reference products.

a seven-point scale is obtained for each parameter and noted on a special sensory profile sheet. The principle of ranking and the sensory profile sheet used for skin care products are shown in Fig. 10.

This test methodology can be described as a special descriptive sensory assessment conducted as a comparison test vs. defined reference products. By using reference products, it is possible to achieve reproducible results with only a few panelists. Normally, a panel of ten members is sufficient. A further advantage is that the training time required for this panel is brief. The panelists need only be given a demonstration of the test procedure and shown some practical exercises for each sensory parameter. In contrast to other test methods, the influence of panelists' different skin types on the ranking is minimized. For example, a panelist with very dry skin will exhibit different absolute times for complete absorption compared to a panelist with greasy skin; the difference between the test product and the reference product, however, will usually be in the same direction.

Sensory Parameters			**Ref. 1**		**Ref. 2**		**Ref.3**		
Evaluation*		**1**	**2**	**3**	**4**	**5**	**6**	**7**	
Pick-up	good								less good
Consistency	low								high
Peaking	little								high
Cushion	little								much
Distribution	easy								difficult
Absorption	fast								slow
Smoothness	little								much
Stickiness	little								much
Oiliness	little								much
Waxiness	little								much

FIGURE 10. Sensory profile sheet.

5.3. Mathematical Aspects

For the seven-point comparison test, the normal situation is as follows: ten panelists judge n products, leading to n score values for each sensory parameter evaluated. An example is given in Fig. 11, which shows the score values that were produced for the test parameter "cushion." In this case, ten panelists judged three test products. By applying the Friedman test, a P-level of 0.0014 is computed, i.e., the three test products are significantly different at the 95% level for the test parameter cushion.

The subsequent multiple comparisons (Nemenyi) show the following results:

Comparison	*P-level*
1 vs. 2	0.8321
1 vs. 3	0.0205
2 vs. 3	0.0031

It may also happen that 30 panelists judge three products in the following way: Panelists 1–10 judge product 1, panelists 11–20 and 21–30 judge products 2 or 3. In this case the resulting scores are independent, and the Kruskal–Wallis test (as a global test), followed by muliple comparisons, has to be used. If the sensory assessment is conducted as a paired comparison test, the test according to Wilcoxon–Mann–Whitney has to be applied (independent scores), or alternatively the Wilcoxon test for dependent scores.

To summarize: with a given database, the appropriate statistical method has to be chosen very carefully in line with the given test design.

Panelist	Product 1	Product 2	Product 3
1	4	4	7
2	6	3	7
3	4	3	6
4	4	4	6
5	4	4	6
6	4	4	6
7	4	4	6
8	4	1	5
9	4	5	4
10	6	1	7

FIGURE 11. Score values for the parameter "cushion."

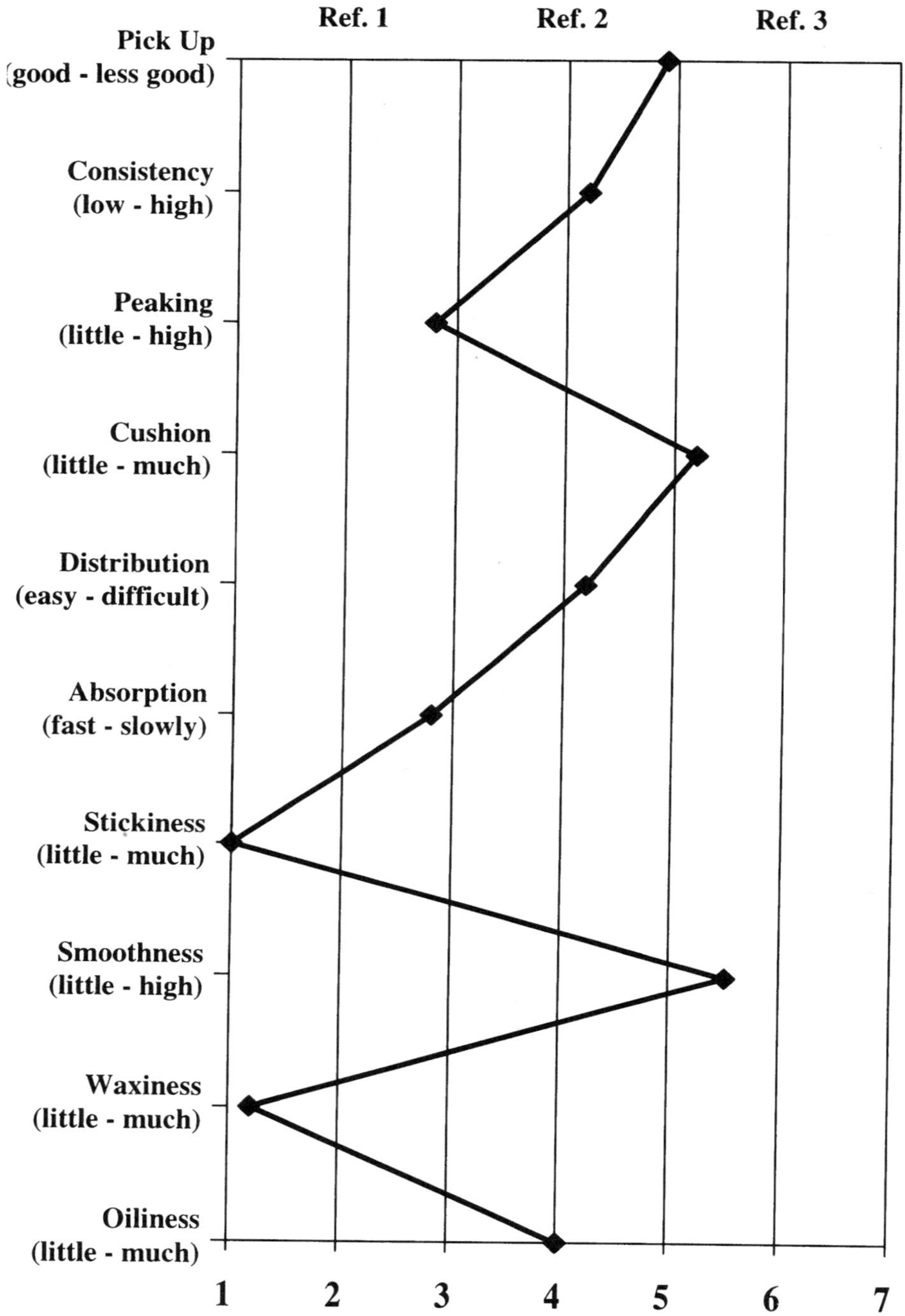

FIGURE 12. Sensory profile of an all-purpose moisturizing cream.

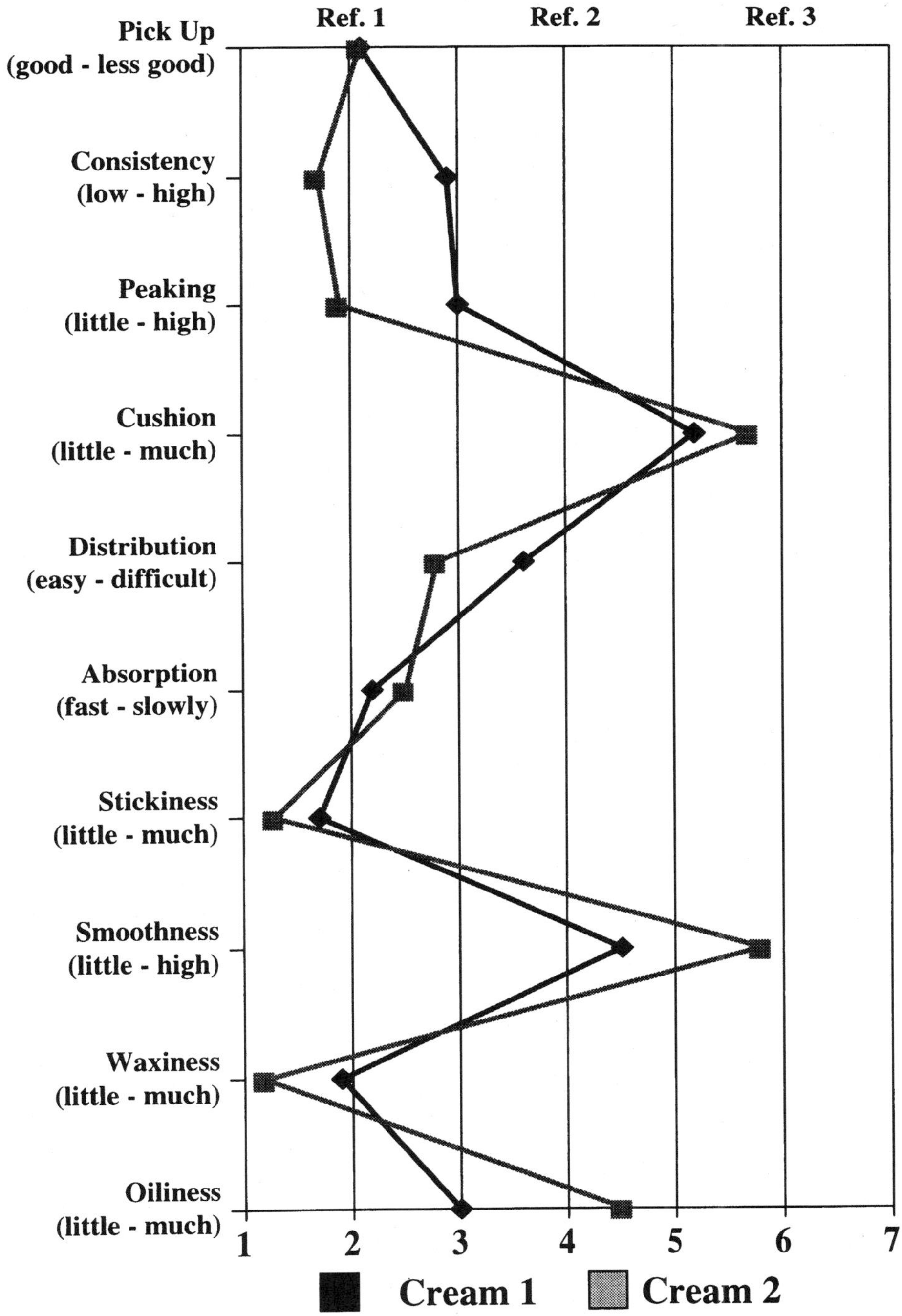

FIGURE 13. Sensory profile of two market products vs. reference.

5.4. Structuring of Sensory Profiles

Using the results generated with the above-described procedure, it is possible to draw sensory profiles. For that purpose, the numerical values for each parameter are transferred to sensory profile sheets and marked points are connected. In Figs. 12 and 13 examples of the sensory profiles of skin care products and shower gels are given.

It should be noted that the sensory profiles show only the diagnostic parameters (i.e., values that are "objective" and can — basically speaking — be determined by an appropriate physical device). Hedonic judgments can be depicted on a separate sheet.

6. Areas of Application for Sensory Assessment in the Cosmetic Field

Sensory assessment is a useful tool for product and concept development in the cosmetic industry. The method can be utilized, e.g., for

- Quick and cost-saving measurement of consumer-relevant effects.
- Claim substantiation.
- Providing advice regarding product optimization.
- Checking of own vs. competitive products by sensory profiles.
- Selection of the best candidate for a consumer test from among several development products.
- Examining factors affecting shelflife.
- Influence of process or formula changes.
- Development of low-cost formulations with high performance.

References

ASTM: Society for Testing and Materials, 1986, *Physical Requirement Guidelines for Sensory Evaluation Laboratories* (J. Eggert and K. Zook, eds.), ASTM Special Technical Publication 913, Philadelphia.

Civille, G. V., and Dus, C. A., 1991, Evaluating tactile properties of skincare products, *Cosmet. Toilet.* **106**:83–88.

ISO 4121, 1987, (E) Sensory analysis — Methodology: Evaluation of food products by methods using scales.

ISO 8589, 1988, (E) Sensory analysis — General guidance for the design of test rooms.

Mahar, T. J., Dhingra, R. C., and Postle, R., 1982, Comparison of fabric handle assessment in Japan, Australia, New Zealand and India, in: *Objective Specification of Fabric Quality, Mechanical Properties and Performance* (Kawabata, Postle, and Niwa, eds.), Textile Machinery Society of Japan, Osaka, pp. 149–159.

Roget's International Thesaurus, 1984, Harper & Row, New York.

Chapter 7

Cutaneous Bioengineering

DIMITER ILIEV, PETER ELSNER, AND HOWARD I. MAIBACH

1. Introduction

Bioengineering methods are becoming more and more useful tools not only for research but also for the objectivation of clinical findings and even for making clinical diagnoses. In this chapter we describe the most important noninvasive methods for characterization of skin properties. The application of irritancy tests for the assessment of individual irritant reactivity is discussed. The lack of reproducibility of single irritancy tests makes it necessary to apply a series of tests in order to assess proclivity to irritant contact dermatitis. The potential of bioengineering methods for the prevention of irritant contact dermatitis is discussed, and important influencing factors of the measurements are noted.

2. Barrier Function of Stratum Corneum as Measured by Transepidermal Water Loss

Cutaneous irritation is a major dermatological problem especially in occupational dermatology and accounts for more than 50% of all cases of occupational contact dermatitis (Fregert, 1975). Transepidermal water loss (TEWL) as an indicator for stratum corneum barrier function has been used to assess the irritancy potential of chemicals and to quantify the skin water barrier function in healthy skin as well as in clinical contact dermatitis. Clinically it may be especially useful to diagnose a tendency to cumulative irritant contact dermatitis before multiple subthreshold damages to the skin lead to clinically visible damage (Iliev and Elsner, 1997).

Dimiter Iliev • Department of Dermatology, University of Zurich, CH-8091 Zurich, Switzerland. *Peter Elsner* • Department of Dermatology, University of Jena, Jena, Germany. *Howard I. Maibach* • Department of Dermatology, School of Medicine, University of California, San Francisco, California 94143-0989.

Cutaneous Biometrics, edited by Doris A. Schwindt and Howard I. Maibach, Kluwer Academic/Plenum Publishers, New York, 2000.

There are currently two companies producing commercial equipment for the measurement of the TEWL. These instruments are the Tewameter (Courage and Khazaka, Cologne, Germany) and the Evaporimeter (ServoMed, Stockholm, Sweden). The physical principle of measurement is based on the evaluation of the water vapor pressure gradient just above the surface of the skin, which is surrounded by a water vapor boundary layer. Apart from skin disorders there are other important factors influencing the measured values of TEWL: individual factors such as anatomical sites (Rougier *et al.*, 1988), age (Leveque *et al.*, 1984), race (Berardesca and Maibach, 1988*b*), sex (Lammintausta *et al.*, 1987), menstrual cycle (Harvell *et al.*, 1992), skin surface temperature (Mathias *et al.*, 1981) and related vascular effects and environment-related factors such as air temperature, humidity, convention, season, and geographical variation. To avoid misleading results the Standardization Group of the European Society of Contact Dermatitis (SGESCD) published guidelines for TEWL measurement (Pinagoda *et al.*, 1990). These guidelines also state that if it is impossibile to perform measurements at the same air temperature (T) and humidity, the TEWL-values may be converted to a reference temperature of 30°C using a formula developed by Mathias *et al.* (1981) as follows: $\log \text{TEWL}_{30} = \log \text{TEWL}_{\text{T}} + 0.035\,(30 - T)$.

TEWL-measurements can also be used to quantify irritant reactions with sodium lauryl sulfate (SLS), sodium hydroxide (NaOH), dimethylsulfoxide (DMSO), and other irritating substances. Irritancy can be categorized into the aim of provocative testing and susceptibility evaluation. When applying SLS the guidelines of the SGESCD should be followed (Tupker *et al.*, 1997). Another practical application of irritation tests are bioassays for the purpose of identifying sensitive skin. A 24-hr patch test with SLS and a repetitive patch test, such as the 21-day cumulative irritation assay, the chamber scarification test, and the soap chamber test have been utilized (Lee and Maibach, 1994).

3. Skin Moisture

The moisture of the skin can be measured by two different physical principles: conductivity and capacity. Despite the number of specially designed and commercially available instruments, the basic physical principle is that electrical current is transported easily in hydrated or damaged skin (Berardesca and Maibach, 1988*a*). This method was first applied by Tagami *et al.* (1980) for evaluation of skin surface hydration. The currently available devices include Skicon (IBS Skin Hygrometer, Hamamtsu, Japan) and the corneometers CM 420 and CM 820 (Courage and Khazaka, Cologne, Germany). A third measurement device (DPM 9003, Nova Technologies) delivers impedance-based capacitance readings by integrating selected measurements at varying frequencies of the applied alternating current. It appeared to be less sensitive for grading the dry stratum corneum state but very suitable for measurements of the hydrated state (Gabard and Treffel, 1994).

This assessment can be used in a similar way as TEWL for the prediction of eczema risk, detection of a subclinical eczema, and therapy control. However, comparative studies of skin moisture measurements should not be performed with different devices without prior calibration (Van Neste, 1991). Standards for evaluating interinstrumental variation are currently unavailable. It is important to know that extended application of moisturizers can influence capacitance values and the interpretation of electrical measurements with respect to skin moisture should be made with caution (Loden and Lindberg, 1991). Another problem may be the presence of body hair, which has an important influence and needs to be considered when the hydration status of the skin is examined with the use of a corneometer (Loden *et al.*, 1995). The measurement of TEWL, conductance, and capacitance is complementary, and their combined use is recommended (Blichmann and Serup, 1988).

4. Acid Mantle of the Skin

The pH value of the skin can be measured with a flat surface glass electrode for pH measurement. One of the commercially available devices is the pH-meter PH 900 PC (Courage and Khazaka, Cologne, Germany). A skin pH between 5.4 and 5.9 can be considered as normal in adults (Braun-Falco and Korting, 1986). Lower normal values were reported for babies (Iliev *et al.*, 1997*a*). In monitoring eczematous skin conditions significant skin pH changes have been shown (Rovenský *et al.*, 1968). However, detergents applied to the skin may also alter its pH. In contrast to other biophysical properties of the skin it has been shown that there is no difference between skin pH of males and females (Wilhelm *et al.*, 1991).

Skin pH is an important indicator for the microbiological flora that can be expected at the appropriate skin site. For instance Proprionibacterium acnes grows well at a pH between 6.0 and 6.5, but not at a pH of 5.5 (Korting and Braun-Falco, 1996).

5. Skin Color

Most of the clinical symptoms in dermatology are associated with changes in the color of the skin. Although it is possible to describe these alterations very well, it was only in 1984 that objective measurement devices were used in dermatology (Diffey *et al.*, 1984). When talking about color we should be aware that it consists of its hue, lightness, and saturation (Elsner, 1995). In 1976 the Commission Internationale de l'Eclairage (CIE) defined the $L^*a^*b^*$ color space (CIELAB), which is defined by three axes parallel to each other: a^* is the red–green axis, b^* is the yellow–blue axis, and the third axis L^* is for lightness. There is a wide variety of instruments available to measure skin color such as Dermaspectrometer (Cortex Technology, Hadsund, Denmark), Erythema Meter (Dia-Stron, Andover, United Kingdom), Chromameter CR 300 (Minolta, Osaka, Japan) and Micro Color (Dr.

TABLE 1. Important Noninvasive Bioengineering Methods for Characterization of Skin Properties (Iliev *et al.*, 1997*d*)

Skin property	Possible significances in dermatology	Measured parameter	Principle of measurement	Device
Barrier function of stratum corneum	Prediction of eczema risk, detection of a subclinical eczema, therapy control	Transepidermal waterloss	Measurement of air moisture gradient in defined distance to the skin	Evaporimeter TEWA meter
Skin moisture	Prediction of eczema risk, detection of a subclinical eczema, therapy control	Conductivity	Measurement of conductance between two electrodes	Skicon
Skin moisture	Prediction of eczema risk, detection of a subclinical eczema, therapy control	Capacity	Measurement of capacity between two electrodes	Corneometer
Activity of sebaceous glands	Unknown	Lipids of the skin surface	Photometrically determined transparency of a plastic film	Sebumeter
Acid mantle of the skin	Buffer capacitance against alkali substances (prediction of eczema risk), detection of a subclinical eczema, therapy control	Skin surface pH	Hydrogen ion activity at a pH electrode	pH meter
Skin blood flow	Tolerance to temperature, detection of a subclinical eczema	Capillary erythrocyte flow	Wavelength shift of reflected light because of the Doppler effect	Laser Doppler velocimeter
Skin elasticity	Quantification of UV-induced elastosis, scleroderma	Viscoelasticity ratio, biological elasticity	Measurement of skin deformation under the influence of power	Torsionmeter, customer, etc.
Skin roughness	Prediction of eczema risk, detection of a subclinical eczema, therapy control	Adhesional and glide friction	Measurement of the power necessary to move an object on the skin surface	Frictionmeter
Skin roughness	Prediction of eczema risk, detection of a subclinical eczema, therapy control	Deviation of elevations and grooves from the mean	Mechanically or laser controlled profilometry of a skin surface replica	Profilometer, visiometer
Stratum corneum cohesion	Mechanical resistance	Corneocyte cohesion	Measurement of the power	Cohesiometer
Skin color	Objectivation of erythema and pigmentation as proof of eczema and light-induced changes	Light reflection	Measurement of reflection of a defined light flash with a photoelement	Tristimulus colorimeter spectrometer
Skin thickness	Quantification of eczema-induced lichenifications	Distance between skin surface and corium/subcutis border	Reflection of a high-frequency ultrasound signal at the skin surface and at the acoustic border between corium and subcutis	*a*-scan or *c*-scan ultrasound

Environmental influences	Individual influences	Room requirements	Patient preparation	Duration of one measurement
Temperature, air moisture, wind	Physical activity, sweating	High (air-conditioning recommended)	Physical rest for at least 15 min	5 min
Temperature, air moisture	Sweating	Medium	None	1 min
Temperature, air moisture	Sweating	Medium	None	1 min
None	Recently applied creams, drugs	Low	None	1 min
None	Sweating	Low	None	1 min
Temperature	Physical activity, temperature adaptation, drugs, coffee nicotine	Medium	Physical rest for at least 15 min	5 min
None	None	Low	None	3 min
None	None	Low	None	3 min
None	None	Low	None	5 min
None	None	Low	None	5 min
None	None	Low	None	1 min
None	None	Low	None	2–5 min

Bruno Lange, Düsseldorf, Germany). Whereas the first two devices are based on the emission of red and green light, the latter two emit an intensive white light covering the whole visible spectrum. Several studies have shown that color measured with a colorimeter is proportional to visual scoring (Serup and Agner, 1990) and (Wilhelm *et al.*, 1989). Therefore in skin bioengineering it has been used for the quantification of erythema in the study of irritant dermatitis (Wilhelm and Maibach, 1989), in dermatopharmacology (Wilhelm and Maibach, 1989), and in photodermatology (Seitz and Whitmore, 1988).

6. Skin Blood Flow

Skin blood flow can be measured as capillary erythrocyte flow by laser Doppler velicometry. This is an optical technique first used by Nilsson *et al.* (1982) in dermatology and now widely utilized. The method is based on a wavelength of 632 to 785 nm applied to the skin via a small probe. The effect of light on moving erythrocytes and nonmoving components of a limited volume of tissue is collected by an optical fiber that carries light back to a photodetector. At this point the signal is converted to an electrical signal (Bernardi and Berardesca, 1995). A selection of the commercially available devices includes Laserflo BPM (Vasamedics, Yorktown, USA), Perimed 4000 (Perimed, Jarfalla, Sweden), and Moor MBF3D (Moore Instruments, Axminster, UK).

This methodology has been applied to characterize the effect of cutaneous irritants (Dykes *et al.*, 1991), but it can also be used to evaluate diseases affecting the superficial dermal plexus and capillary loops as eczema (Staberg *et al.*, 1984) and psoriasis. Staberg and colleagues applied a standard patch test of 31 substances and monitored the response by visual scoring and laser Doppler flowmetry after 48 hr. A significant correlation between visual scores and laser Doppler flow values was found. Another group reported a good correlation between laser blood flow and visible erythema but found the naked human eye more sensitive (Blanken *et al.*, 1986).

7. Skin Roughness as Measured by Visiometry

The roughness of the skin can be measured as the deviation of elevations and grooves from the mean. One of the commercially available devices is the Skin-Visiometer SV 400 (Courage and Khazaka, Cologne, Germany), which uses silicon rubber replicas that have to be obtained from the skin side of interest. These replicas are processed by the visiometer, which consists of a bright light source and a CCD camera. Images can be digitized and analyzed by a computer program. Other possible methods to measure skin roughness include skin profilometry and frictionmetry (Grove and Grove, 1989).

It has been shown that skin roughness can be changed by lowering humidity in a climate chamber (Eberlein-König *et al.*, 1996). An increased roughness can be detected at a subclinical level from the age of 30 (Berardesca *et al.*, 1991). Therefore the method has been used to show treatment effects of different local therapeutics (Leyden *et al.*, 1989).

8. Skin Thickness

Skin thickness measurement may be useful to determine the distance between the skin surface and the corium/subcutis border in cases of eczema-induced lichenifications, tumors (Hoffmann *et al.*, 1992), or other skin conditions where the corium thickness is affected. Currently two different methods—high-frequency ultrasound and laser Doppler—imaging are available. In high-frequency ultrasound imaging the A- and B-mode scan can be distinguished. The A-scan shows a line representing echos from different layers, whereas in the B-mode scan the transducer is automatically moved tangentially over the object and a number of A-scans are depicted showing a cross-sectional image of the object in two dimensions (Serup, 1995). The viewing depth depends of the frequency used and is about 2 mm with 100 MHz up to 8 mm with 20 MHz (el Gammal *et al.*, 1997). However, a lower frequency goes with a lower resolution, making images with a frequency of much less than 20 MHz unsuitable for showing detail structures of the skin. Analyzing skin structures by ultrasound needs experience and the echogenity of the different structures cannot be discussed here. In experienced hands, this device may be a useful tool for noninvasive diagnosis of dermal tumors and infiltrates as well as for therapy control.

9. Other Methods and Relations among Different Measurements

A huge variety of different bioengineering devices has been developed in the last decades. Table 1 provides a summary of currently available methods with a short description of the technical principle. A newer technique is magnetic resonance microscopy, but that has to be improved for *in vivo* application (el Gammal *et al.*, 1996).

When choosing a bioengineering technique to monitor diagnosis or therapy or for clinical investigation, it is important to know the basic relations among the different measurements as well as between clinical findings and cutaneous bioengineering. For instance, we could show that there is no correlation between clinical atopy score and TEWL-measurement in a cohort of metal worker trainees (Iliev *et al.*, 1997*b*). Other important aspects are the reproducibility (Iliev *et al.*, 1997*c*), sensitivity, and specificity of results obtained with bioengineering devices. Therefore some of the methods described still need to be evaluated before they find their way into routine diagnostic and therapy control.

ACKNOWLEDGMENT: Dr. Dimiter Iliev is a staff member of the Zurich Prospective Metal Worker's Eczema Study (PROMETES) supported by the Swiss National Fund and the Swiss National Accident Insurance.

References

Bernardi, L., and Berardesca, E. , 1995, Measurement of skin blood flow by laser-Doppler flowmetry, in: *Bioengineering of the Skin: Methods and Instrumentation*, CRC Press, Boca Raton, pp. 13–28.

Berardesca, E., and Maibach, H. I., 1988*a*, Bioengineering and the patch test, *Contact Dermatitis* **18**:3–9.

Berardesca, E., and Maibach, H. I., 1988*b*, Racial differences in sodium lauryl sulphate induced cutaneous irritation: Black and white, *Contact Dermatitis* **18**:65–70.

Berardesca, E., Farinelli, N., Rabbiosi, G., and Maibach, H. I., 1991, Skin bioengineering in the noninvasive assessment of cutaneous aging, *Dermatologica* **182**:1–6.

Blanken, R., Van der Valk, P. G. M., and Nater, J. P., 1986, Laser Doppler flowmetry in the investigation of irritant compounds on human skin, *Dermatosen* **34**:5–9.

Blichmann, C. W., and Serup, J., 1988, Assessment of skin moisture. *Acta. Derm. Venereol.*, **68**:284–290.

Braun-Falco, O., and Korting, H. C., 1986, Der normale pH-Wert der menschlichen Haut, *Hautarzt* **37**:126–129.

Diffey, B. L., Oliver, R. J., and Farr, P. M., 1984, A portable instrument for quantifying erythema induced by ultraviolet radiation. *Br. J. Dermatol.* **111**:663.

Dykes, P. J., Hill, S., and Marks, R., 1991, The effect of area of application on the intensity of response to a cutaneous irritant, *Br. J. Dermatol.* **125**:330–334.

Eberlein-König, B., Spiegl, A., and Przybilla, B., 1996, Change of skin roughness due to lowering air humidity in climate chamber. *Acta. Derm. Venereol. (Stockh.)* **76**:447–449.

el Gammal, S., Hartwig, R., Aygen, S., Bauermann, T., el Gammal, C., and Altmeyer, P., 1996, Improved resolution of magnetic resonance microscopy in examination of skin tumors, *J. Invest. Dermatol.* **106**:1287–1292.

el Gammal, S., Hoffmann, K., Stücker, M., and Altmeyer, P., 1997, Bildgebende Verfahren in der Dermatologie, *Hautarzt* **48**:432–450.

Elsner, P., 1995, Skin color, in: *Bioengineering of the Skin: Methods and Instrumentation* (E. Berardesca, P. Elsner, K. P. Wilhelm, and H. I. Maibacch, eds.), CRC Press, Boca Raton, pp. 29–40.

Fregert, S., 1975, Occupational dermatitis in a 10-year material, *Contact Dermatitis* **1**:96–107.

Gabard, B., and Treffel, P., 1994, Hardware and measuring principle: The Nova DPM 9003, in: *Bioengineering of the Skin: Water and the Stratum Corneum* (P. Elsner, E. Berardesca, and H. I. Maibach, eds.), CRC Press, Boca Raton, pp. 177–195.

Grove, G. L., and Grove, M. J., 1989, Objective methods for assessing skin surface topography noninvasively, in: *Cutaneous Investigation in Health and Disease* (J.-L. Leveque, ed.), Marcel Dekker, New York, pp. 1–32.

Harvell, J., Hussona-Saeed, I., and Maibach, H. I., 1992, Changes in transepidermal water loss and cutaneous blood flow during the menstrual cycle, *Contact Dermatitis* **27**:294–301.

Hoffmann, K., Jung, J., el Gammal, S., and Altmeyer, P., 1992, Malignant melanoma in 20 MHz B-scan sonography, *Dermatology* **185**:49–55.

Iliev, D., and Elsner, P., 1997, Clinical irritant contact dermatitis syndromes, *Immunol. Allergy Clin. North. Am.* (in print).

Iliev, D., Bangha, E., and Elsner, P., 1998, Einfluss eines Baby-Molke-Bades auf haut-physiologische Parameter beim Kleinkind, *Schweiz Rundsch Praxis* **87**:617–621.

Iliev, D., Hinnen, U., and Elsner, P., 1997*b*, Clinical atopy score and TEWL are not correlated in a cohort of metal workers, *Contact Dermatitis* **37**:235–236.

Iliev, D., Hinnen, U., and Elsner, P., 1997*c*, Reproducibility of a non-invasive skin irritancy test in a cohort of metal worker trainees, *Contact Dermatitis* **36**:101–103.

Iliev, D., Hinnen, U., and Elsner, P., 1997*d*, Skin bioengineering methods in occupational dermatology, in: *Skin Bioengineering, Techniques and Applications in Dermatology and Cosmetology* (P. Elsner, A. O. Barel, E. Berardesca, B. Gabard, and J. Serup, eds.), Vol. 26, *Current Problems in Dermatology*, S. Karger AG, Basel.

Korting, H. C., and Braun-Falco, O., 1996, The effect of detergents on skin pH and its consequences, *Clin. in Dermatol.* **14**:23–27.

Lammintausta, K., Maibach, H. I., and Wilson, D., 1987, Irritant reactivity in males and females, *Contact Dermatitis* **17**:276–280.

Lee, C. H., and Maibach, H. I., 1994, Study of cumulative irritant contact dermatitis in man utilizing open application on subclinically irritated skin, *Contact Dermatitis* **5**:271–275.

Leveque, J.-L., Corcuff, P., de Rigal, J., and Agache, P., 1984, *In vivo* studies of the evaluation of physical properties of the human skin with age, *Int. J. Dermatol.* **23**:322–329.

Leyden, J. J., Grove, G. L., Grove, M. J., Thorne, E. G., and Lufrano, L., 1989, Treatment of photodamaged facial skin with topical tretinoin, *J. Am. Acad. Dermatol.* **21**:638–644.

Loden, M., and Lindberg, M., 1991, The influence of a single application of different moisturizers on the skin capacitance, *Acta Derm. Venereol. (Stockh.)* **71**:79–82.

Loden, M., Hagforsen, E., and Lindberg, M., 1995, The presence of body hair influences the measurement of skin hydration with the corneometer, *Acta Derm. Venereol. (Stockh.)* **75**:449–450.

Mathias, C. G. T., Wilson, D. M., and Maibach, H. I., 1981, Transepidermal water loss as a function of skin surface temperature, *J. Invest. Derm.* **77**:219–220.

Nilsson, G. E., Otto, U., and Wahlberg, J. E., 1982, Assessment of skin irritancy in man by laser Doppler flowmetry, *Contact Dermatitis* **8**:401–406.

Pinagoda, J., Tupker, R. A., Agner, T., and Serup, J., 1990, Guidelines for transepidermal water loss (TEWL) measurement. A report from the standardization group of the European Society of Contact Dermatitis, *Contact Dermatitis* **22**:164–168.

Rougier, A., Lotte, C., Corcuff, P., and Maibach, H. I., 1988, Relationship between skin permeability and corneozyte size according to anatomical site, age and sex in man, *J. Soc. Cosmet. Chem.* **39**:15–26.

Rovenský, J., Micánek, Z., and Najbrt, V., 1968, Importance of actual pH of skin in children, *Cs. Derm.* **4**:224–226.

Seitz, J. C., and Whitmore, C. G., 1988, Measurement of erythema and tanning responses in human skin using a tri-stimulus colorimeter, *Dermatologica* **177**:70–75.

Serup, J., 1995, Skin imaging techniques, in: *Bioengineering of the Skin* (E. Berardesca, P. Elsner, K. P. Wilhelm, and H. I. Maibach, eds.), CRC Press, Boca Raton, pp. 65–79.

Serup, J., and Agner, T., 1990, Colorimetric quantification of erythema—a comparison of two colorimeters (Lange Micro Color and Minolta Chroma Meter CR-200) with a clinical scoring scheme and laser-Doppler flowmetry. *Clin. Exp. Dermatol.* **15**:267.

Staberg, B., Klemp, P., and Serup, J., 1984, Patch test responses evaluated by cutaneous blood flow measurements, *Arch. Dermatol.* **120**:741–743.

Tagami, H., Ohi, M., Iwatsuki, K., Kanamaru, Y., Yamada, M., and Ichijo, B., 1980, Evaluation of the skin surface hydration *in vivo* by electrical measurements, *J. Invest. Dermatol.* **75**:500–507.

Tupker, R. A., Willis, C., Berardesca, E., Lee, C. H., Fartasch, M., Agner, T., and Serup, J., 1997, Guidelines on sodium lauryl sulfate (SLS) exposure tests. A report from the Standardisation Group of the European Society of Contact Dermatitis, *Contact Dermatitis.* **37**:53–69.

Van Neste, D., 1991, Comparative study of normal and rough human skin hydration *in vivo:* Evaluation with four different instruments, *J. Derm. Sci.* **2**:119–124.

Wilhelm, K. P., and Maibach, H. I., 1989, Skin color reflectance measurements for objective quantification of erythema in human beings, *J. Am. Acad. Dermatol.* **21**:1306–1307.

Wilhelm, K. P., Surber, C., and Maibach, H. I., 1989, Quantification of sodium lauryl sulfate irritant dermatitis in man. Comparison of four techniques: skin color reflectance, transepidermal water loss, laser Doppler flow measurement and visual scores, *Arch. Dermatol. Res.* **281**:293–295.

Wilhelm, K. P., Cua, A. B., and Maibach, H. I., 1991, Skin aging, affect on transepidermal water loss, stratum corneum hydration, skin surface pH and casual sebum content, *Arch. Dermatol.* **127**:1806–1809.

Part III

Skin Diseases and Therapy

Chapter 8

Atopic Dermatitis: Severity Scoring

ALAIN TAÏEB AND JEAN FRANÇOIS STALDER

1. Introduction

Atopic dermatitis (AD) is generally diagnosed without difficulty in most patients because of pattern and chronicity of pruritic lesions, but establishing consensus criteria to include patients in clinical trials has been more challenging than expected, owing to the absence of specific features (Hanifin and Rajka, 1980). Only recently have clinical diagnostic criteria, which are largely history-based, been validated (Williams *et al.*, 1994*a, b, c*). Scoring severity has proven even less amenable to standardization, and at the time of this writing, despite numerous efforts, no international consensus exists similar to PASI score for clinical trials in psoriasis (Frederiksson and Petterson, 1978). The reasons are multiple. First, in terms of severity-oriented description of disease only pruritus, the major burden for the patients, is shared at least intermittently by all sufferers. However, pruritus, if correlated with excoriations and lichenification, remains a subjective assessment of disease and, thus, not a first choice for hard-core cutaneous biometrics. Second, AD lesions are pleomorphic, i.e., made up of a mixture of several individual signs that are difficult to describe accurately and to choose/stage clinically in terms of severity. Individual signs linked to inflammation such as erythema, edema, and oozing can change rapidly according to site and time in an individual patient. Third, extent of lesional skin is far more difficult to delineate than in a monomorphic disorder such as psoriasis. Fourth, there are marked differences in clinical presentation according to age and ethnic group, making it difficult to design a universal scoring system. Fifth, but this is not specific for AD, as it concerns all chronic and relapsing diseases, instant (snapshot) determinations of severity may be misleading when they do not refer to the overall course of the disease.

Alain Taïeb • Pediatric Dermatology Unit, Hôpital Pellegrin-Enfants, 33076 Bordeaux, France.
Jean François Stalder • Department of Dermatology, CHU Hôtel Dieu, 44035 Nantes, France.

Cutaneous Biometrics, edited by Doris A. Schwindt and Howard I. Maibach, Kluwer Academic/Plenum Publishers, New York, 2000.

In this chapter, we review the field critically and take advantage of our experience within the European Task Force on Atopic Dermatitis (ETFAD), a group of dermatologists who have worked specifically on scoring severity in AD since 1990 and have proposed a scoring system, the SCORAD (for Scoring AD) index (ETFAD, 1993; Kunz *et al.*, 1997).

2. Historical Perspective

In a review article Finlay (1996) noted how difficult it is to trace the origin of a given scoring system in AD. Clendenning *et al.* (1973) did pioneering work in this field. Ring *et al.* (1979), and Zachary *et al.* (1983) made valuable contributions to developing variants. Special mention should probably be made of Saurat and coworkers Queille Roussel and Costa, who first tried to compare scoring systems based on their own experience, but eventually indicated their preference for the simpler technique (Queille-Roussel *et al.*, 1985; Costa *et al.*, 1989). A workshop on severity grading was organized at the 3rd International Symposium on AD in Oslo (Hanifin, 1989; Abrams, 1989). Bahmer *et al.* (1991) proposed a system derived from the PASI named ADASI (Atopic Dermatitis Area and Severity Index), which introduces a grid to estimate the body surface involved more easily. Another variant has been used in Japan (Kezawa *et al.*, 1991). The ETFAD introduced the SCORAD index, based on a consensus of definition of intensity items associated with an atlas and sophisticated statistical treatment of clinical data to define the weight most appropriate for extent, intensity signs, and daytime pruritus/sleep loss. Validation of this system included several clinical trials (Kowalzick *et al.*, 1995; Ruzicka *et al.*, 1997) and interinvestigator assessments of variability, with new recommendations for users and education tools (Kunz *et al.*, 1997; Oranje *et al.*, 1997; Stalder, 1995). The SASSAD (six area, six sign atopic dermatitis) system was proposed by Berth-Jones (1996) as a variant of the Leicester and Costa scores, on the basis of several studies performed with this system, but without formal validation. The EASI (Eczema Area and Severity Index) system, which was proposed by Hanifin (1999), corresponds to another adaptation of the PASI score to AD (ongoing FK506 trial). A Japanese symposium devoted to severity assessment in AD has also been published (Aoki, 1998).

3. Defining Extent

Everyone agrees that accurately delineating the body surface involved in active inflammatory changes is very difficult in AD. To circumvent this problem, distribution of signs has been considered as an alternative, since cutaneous involvement is grossly symmetrical, as in Broberg *et al.* (1994) system. In such a system distribution varies from 0 for no affected site to 4 for four or more affected symmetrical sites. Another alternative is to grade individual items of intensity on several sites that cover the body surface (multiple area assessment), completely or not, without taking the exact area into account (Stalder, 1995; Lever *et al.*, 1988). However, surface assessment of lesions on several segments (face and neck, trunk, limbs), which allows use of the rule of nine and thus necessitates an adjustment for age especially in infants (less than 2 years of age), remains the major standard (ETFAD, 1993). This method avoids the emphasis on some areas given in other systems, and allows scoring of all patients regardless of the distribution pattern of the lesions.

Some shortcuts may be used to compute the total area segment by segment, such as the patients's palm, which is around 1% of the body surface. To maximize the reproducibility of extent assessment, the gross delineation of lesions on a front–back drawing is mandatory. This method is judged cumbersome by some authors, and has also been criticized in cases of psoriasis (Ramsay and Lawrence, 1991) because it can overestimate surface area, but it is the only one that allows a careful follow-up of the patient based on previous assessments on a standardized sheet (Kunz *et al.*, 1997). The grid system advocated by Bahmer (1991) may represent an improvement in computing the total score.

4. Choosing/Defining Severity/Intensity Items

Given their weight in outcome measurement, the selection of relevant descriptive items to assess disease severity in AD is probably the most important consideration in designing a tool that can be used in clinical trials. The items should describe the disease as comprehensively as possible, but without unnecessary overlap and redundancy; be accurate in reflecting the true clinical status of the patient; and provide enough sensitivity to detect small clinical modifications.

It is interesting to compare the descriptive severity items found in the literature as was done by Finlay (1996). Erythema and lichenification are used by nearly all authors. Other items with variants in definitions used so far have been: edema, papules-vesicles-pustules/purulence, induration, excoriations, exudation/weeping/crusts, scaling, xerosis/dryness, cracking/fissuring, and pigmentation. Among members of the ETFAD, it took several hours to obtain a consensus on which severity items to select for preliminary studies. It was agreed that items related to inflammation were sensitive markers of disease activity: erythema, edema (also papulation) and weeping/oozing/crusts. Markers of pruritus such as excoriations and lichenification should also be included. There was a debate as to whether or not xerosis/dryness should be included, as it is considered by most investigators to be a diagnostic sign rather than a marker of disease activity. Dryness was also judged to be difficult to assess clinically on inflamed skin. Eventually, this item was included for the sake of comprehensiveness and graded on noninflamed skin. Appendix 1 includes the consensus definitions and grading of the ETFAD. An atlas has been published together with the definitions (ETFAD, 1993) and is currently electronically available at the following web site: http://ADserver.sante.univ-nantes.fr.

Another difficulty is how to grade these items—globally or in multiple areas. The multiple area system has advantages, but the final weight for each individual item may be highly dependent on the choice of areas for assessment. The item that has an uneven/small distribution is not covered any better in such a system than it is in a global one. For global evaluation of severity items, the worst affected area has been chosen by Costa's simple system but this may lead to less sensitivity of the score if those sites are refractory to treatment. The ETFAD has made the choice of a representative area for each item, except for dryness (See Appendix 1).

In some patients, severity of skin involvement varies dramatically in different anatomical sites. This feature was noted especially on the face and hands in a series of patients. In our initial consensus report (ETFAD, 1993), we indicated that "the area chosen for grading must be representative (average intensity) for each (intensity) item in a given patient, thus excluding one target area or the worst affected site." Based on the review of the slides of the patients examined compared to clinical scores, we have proposed the following change (Kunz *et al.*, 1997): the "most representative area" for a given intensity item should be present on at least two distinct sites accounting for at least 3% of total extent (e.g., in a patient with mostly facial involvement, forehead and the two cheeks represent three distinct sites and more than 3% of the total extent; on the other hand, ear fissures even very inflamed ones cannot reach "representativity").

5. Interobserver Variability

The reasons for interobserver variation not related to the patient's skin status [AD may change rapidly over time and made it necessary to exclude a patient from a formal scoring session (Kunz *et al.*, 1997)] may be listed as follows:

5.1. Physician Profile

We (Kunz *et al.*, 1997) determined the ranking for the 12 physicians who had attended at least two formal scoring sessions concerning objective data for each patient. Except for a minority of those present, there was a consistency indicating idiosyncratic low or high scoring profiles among individuals on the scoring panel. This tendency to score on a high/low basis was the same for both extent and intensity (Fig. 1). There was no evidence of a special educational influence correlated with their country of origin.

5.2. Intensity Items

When formal sessions were organized to validate the SCORAD system (Kunz *et al.*, 1997), intensity items were scored with overall consistency but there were variations for lichenification (Fig. 2). An adult with low-grade erythroderma was graded with high variations for edema and lichenifications. The possible reasons for variation were analyzed as follows: for edema, the upgrading by some observers was assumed to be related to the presumption of edema in an erythrodermic patient; for lichenification, only extremities were involved and were disregarded as unimportant by some panelists.

Some differences among observers are related to a different interpretation of the item oozing/crusts. Linear crusts resulting from excoriations should be included within the excoriation category. Only crusts following exudation should be included under the heading oozing/crusts.

With regard to lichenification, we found at a formal validation session a surprising difference in assessment for an item that was one of those with minimal interobserver variation based on photographs examined by members of the same

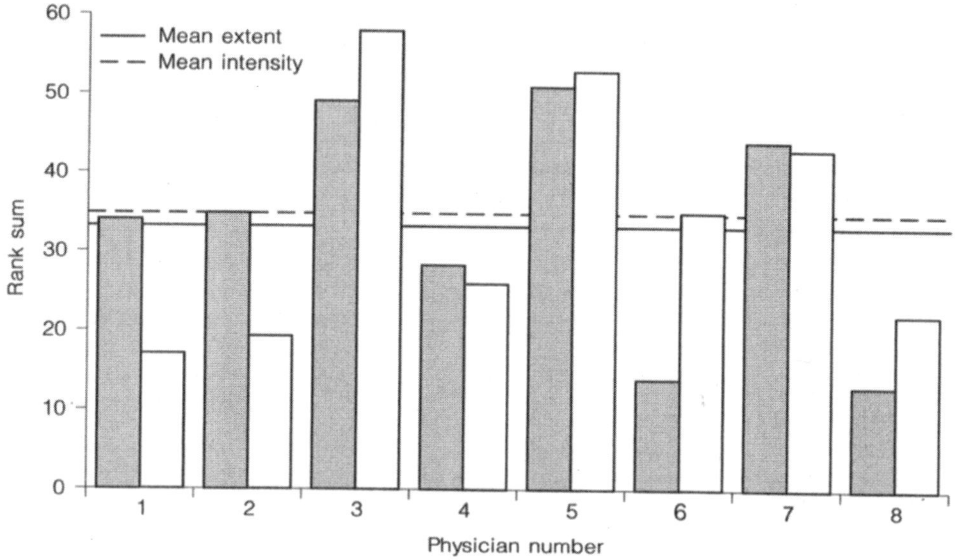

FIGURE 1. Ranks sums and means: ranking of physicians for extent and intensity at a formal scoring session concerning nine children with AD (from Kunz *et al.*, 1997). White bars: extent; gray bars: intensity. Note a general consistency for low scorers and high scorers for both types of scores (extent and intensity), with exceptions like physician 6 (high scorer for extent, low scorer for intensity).

group (ETFAD, 1993). In general, when looking at intensity items, those with the lowest variability such as erythema and dryness had a more even distribution over the body surface. On the other hand, more localized types of lesions such as those of lichenification are less consistently scored probably owing to differences in the choice of "representative" areas.

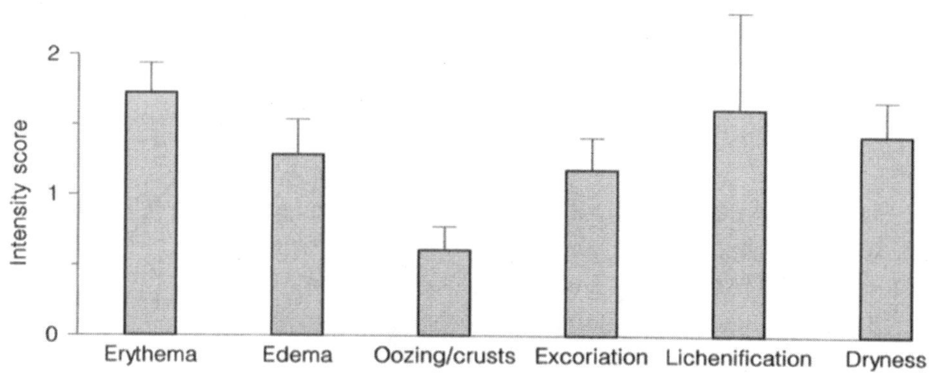

FIGURE 2. Intensity item means and standard deviations as scored by a group of 12 physicians in 19 patients. Lichenification is scored inconsistently. See interpretation in text. (from Kunz *et al.*, 1997).

5.3. Extent

The definition of areas involved in inflammation in patients with low-intensity eczema but significant body surface involvement, say, more than 10–20%, may pose a problem because of the ill-defined borders of lesions even when conditions are optimized, e.g., light, time... The extreme patients near-erythrodermic or with limited disease are more easily scored for extent.

We analyzed the reasons for variation among observers when variations were greater than usual for extent. An example is given in Fig. 3. In this child, mean score for extent for a panel of observers was 40.8 ± 19% with extreme values ranging from 20 to 80%. When the photographs were reviewed, two reasons were found for the variations: (1) the areas involved were easy to delineate on the face but not on the rest of the body, and (2) the intensity of facial lesions distracted the attention of the observer from the less obvious lesions on other sites. Overall, the extent of lesions according to the rule of nine induces interobserver variability mostly for patients with lesions of moderate intensity involving 20–60% of the body surface.

6. Subjective Items and Quality of Life

Clinical experience with AD clearly indicates that pruritus and resultant loss of sleep are the usual symptoms of flare-ups of the disease. How much of the overall severity is represented by those symptoms? Pruritus was given a weight varying from 7 to 33% in a series of scores compiled by Finlay (1996), with various methods of

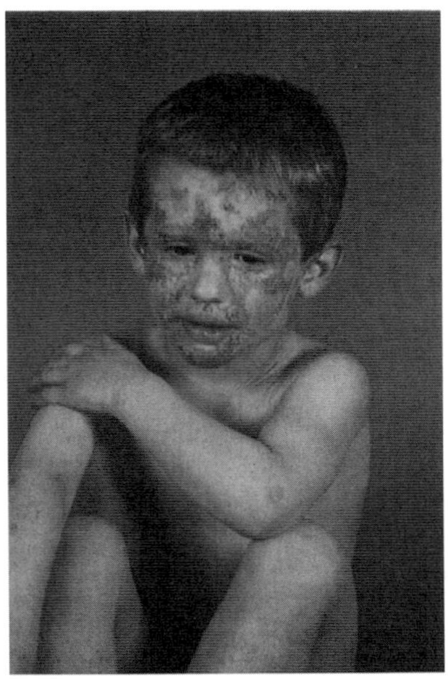

FIGURE 3. Patient with the greatest interobserver variability for extent at a formal scoring session (out of a total of 19 patients) (Kunz *et al.*, 1997). See text for comments.

assessment. Based on statistical evidence, subjective symptoms correspond to 17% of a global cumulative index combining extent, intensity items, and subjective symptoms, the SCORAD index (ETFAD, 1993). How should they be recorded? Based on comparative experiences, the ETFAD recommends the use of a 10-cm visual analog score. The patients or their parents answer the question concerning the three previous days/nights for pruritus and sleep loss. Visual analog scales are difficult to use in children under 7 years of age and it should be remembered that parental global judgment on health status is subject to systematic error.

Quality of life measures are a useful adjunct to subjective symptom assessment in AD. Standardized questionnaires are generally difficult to handle in a busy clinic and global assessment on a visual analog scale is redundant with pruritus and sleep loss evaluation (ETFAD, 1993). Recenly developed questionnaires for dermatologic disorders in adults and children are currently being evaluated in AD (Finlay and Khan, 1994; Lewis-Jones and Finlay, 1995; Herd *et al.*, 1997).

7. Assessment of Outcome with a Cumulative Compound Index: The SCORAD Index

The idea of trying to develop a universal scoring system might be misleading, since different studies may have to use different evaluation criteria depending on their goals. However, there is also an immediate advantage to combining assessment data in a single number to follow patients as an outcome measurement. Two methods are generally proposed (Bombardier and Tugwell, 1982) for combining individual items into an index: (1) experts may provide a global appreciation of various combinations of individual measures, and the weight given to each item in the global assessment is determined statistically in a second step; (2) experts assign an *a priori* arbitrary weight to each item. We chose the first alternative to design the SCORAD index, and it is interesting to note that the extent value derived from statistical calculations (20%) is even lower than the 30% given by the *a priori* method (Costa *et al.*, 1998; Hanifin, 1989).

The statistical study undertaken to design the SCORAD index has shown that apart from a severity component, a profile component was evidenced explaining that patients with an identical SCORAD may differ greatly in the measures of individual items (ETFAD, 1993; Schäfer *et al.*, 1997). Defining this profile further may allow for the inclusion of more homogeneous populations in future trials and take account of the fact that drugs affecting separate manifestations of the disease may have an apparently similar impact on the cumulative index. Figure 4 shows the SCORAD evaluation sheet in its 1997 version (Kunz *et al.*, 1997). It takes a trained individual about 3 min to fill in the evaluation sheet, which records the type and amount of steroids used, as well as number of flare-ups in the previous month, but these data are not included in the final score.

The validation of the SCORAD index is progressing. Some of the recommended criteria for building a cumulative index (Bombardier and Tugwell, 1982) have been met (1) comprehensiveness, (2) credibility including feasibility, sensitivity, and usefulness, as judged by various investigators. Other features tested in past and

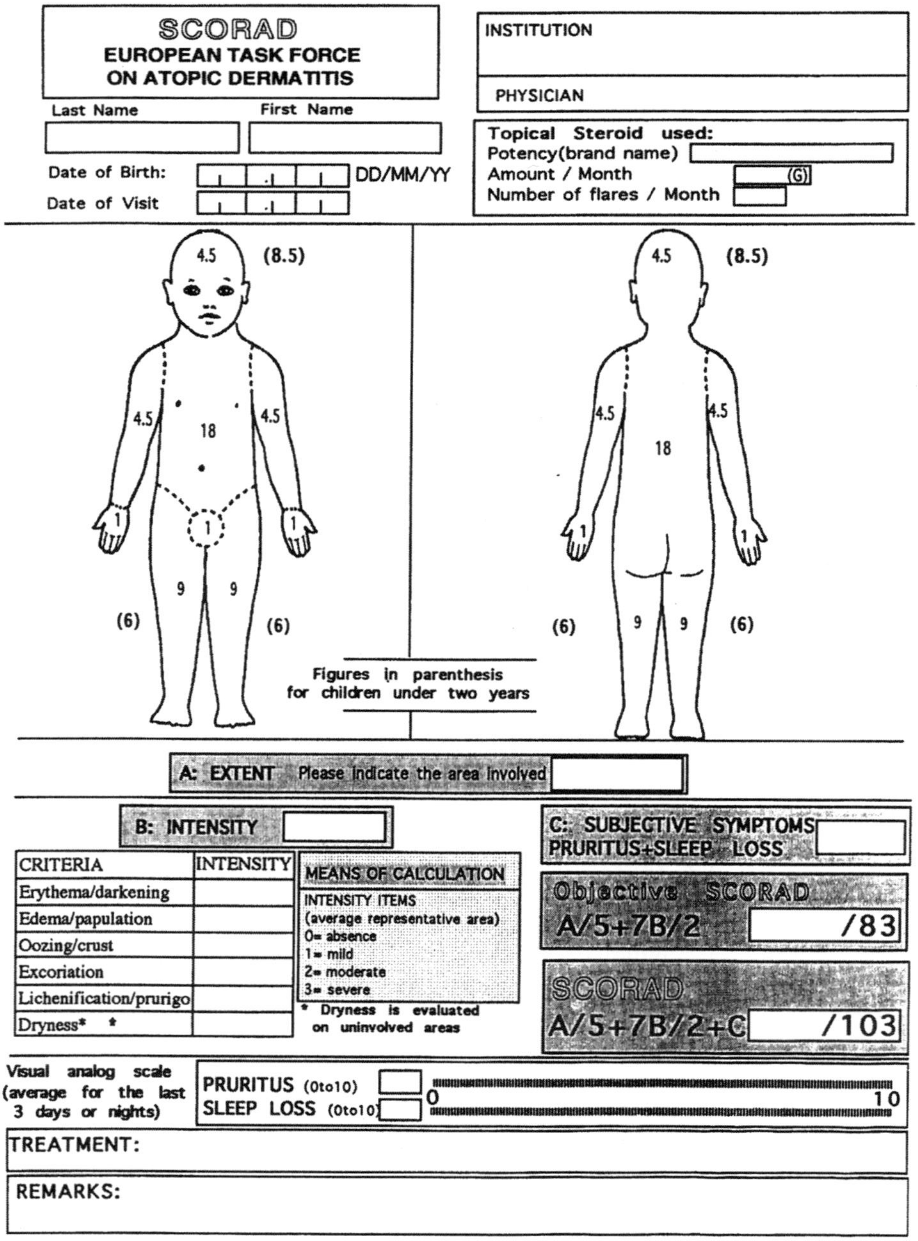

CRITERIA	INTENSITY
Erythema/darkening	
Edema/papulation	
Oozing/crust	
Excoriation	
Lichenification/prurigo	
Dryness* *	

FIGURE 4. Updated SCORAD sheet (Kunz *et al.*, 1997). Erythema/darkening can be scored as the same item, as well as lichenification/prurigo. The objective SCORAD is mentioned.

ongoing clinical studies have validated accuracy, or reflection of the true clinical state of the patient; and sensitivity, allowing detection of the smallest possible clinical modification.

What remains now is to test biological relevance: does SCORAD agree with expected results based on current knowledge of the disease? Current sudies such as the 2-year ETAC (Early Treatment of the Atopic Child, UCB pharma) trial, which will monitor serum markers together with clinical scoring with SCORAD, may provide the needed answers (Oranje *et al.*, 1997). As it happens that social and cultural factors have much influence on the grading of subjective symptoms, these remain important to assess the quality of life of the patient and should serve as a follow-up measurement tool. A cumulative index such as SCORAD can be used to evaluate patients once the objective criteria for severity have been included in clinical studies and could become the major endpoint study parameter.

8. Severity and Time Course of the Disease

Owing to the erratic course of the disease in any individual patient, long-term assessment of severity is sometimes needed, especially for an epidemiological determination of the disease as a public health problem or before the patient is included in clinical trials. Methods in this particular area have never been validated, and suffer from being history-based. The most commonly used techique was proposed by Langeland (1983) and modified by Rajka and Langeland (1989), (Table 1), and there other variants (Hanifin, 1989). Finlay (1996) suggested adding future parameters such as: number of years affected, frequency of hospital admission, frequency of visits to the doctor, volume of topicals, and days lost from school or work. For long-term prospective studies, diary cards grading eczema 0–3 have proven useful for many investigators.

9. Quality Control and Routine Scoring of Atopic Dermatitis

Quality assurance and quality control are key issues in the development of public health systems at the turn of the third millennium. The application of those concepts to dermatology depends on the implementation of scoring systems to assess the quality of care given to patients and the relative costs (Penneys, 1997; Jemec and Wulf, 1997*a*). There is a need to control the treatment outcome of common skin disorders in the setting of hospital outpatient clinics or in private offices to make appropriate decisions concerning treatment. This is most important when costly systemic treatments are envisaged, in parallel with patient information concerning the risks and benefits of such treatments. Simple systems usable in routine settings are the most valuable in this context. The implementation of a global scoring system has been advocated for inflammatory skin diseases, but significant disease-specific differences exist (Jemec and Wulf, 1997*b*). Since the routine use of SCORAD in our departments was introduced, it has been noted that regular scoring of patients is helpful in making such decisions. Scoring during routine visits obliges physicans to perfom a complete assessment of disease and not to overlook details as is more likely

TABLE 1. Grading (Severity) of Atopic Dermatitis [according to Rajka and Langeland (1989)]

Extent	
(*a*) Childhood and adult phase	
Less than ~9% of the body area	1[a]
Involvement evaluated to be more than score 1, less than score 3	2
More than ~36% of the body area involved	3
(*b*) Infantile phase	
Less than ~18% of the skin involved	1
Involvement evaluated to be more than score 1, less than score 3	2
More than 54% of the skin involved	3
Course	
More than 3 months of remission during a year[b]	1
Less than 3 months remission during a year[a]	2
Continuous course	3
Intensity	
Mild itch, only exceptionally disturbing night's sleep	1
Itch, evaluated to be more than score 1, less than score 3	2
Severe itch, usually disturbing night's sleep	3
Score summation	
3–4 = mild	
4.5–7.5 = moderate	
8–9 = severe	

[a]When doubt, score 1.5 or 2.5, may also be used.
[b]May be adjusted in infants or if onset was less than 1 year before grading.

when common global judgment (better, worse, stable) is used. It may also help to determine the optimum frequency of visits. Furthermore, patients' compliance is improved when they can see that the scoring figure follows a certain trend.

10. Scoring with SCORAD for Special Purposes

10.1. Localized Assessment

A variant of the SCORAD system can be used, e.g., in case of hand or head and neck dermatitis, as was already done by Broberg and Faergeman (1995), or for a given extent of affected skin (Ruzicka *et al.*, 1997).

10.2. Inclusion Criteria for Clinical Trials

Based on our previous observations and on the patients examined by an international group of physicians, the severity of subjective symptoms is a cause of large variations, which may be useful during follow-up of individual patients but which may cause confusion in multicenter trials for inclusion criteria. The ETFAD proposes to use a severity scale based on objective criteria only to define patients as having mild, moderate, or severe AD. The criteria are different from those proposed by Rajka and Langeland (1989), but on the basis of experience using the SCORAD system at different institutions for 4 years or more, the overlap between the two is important. The importance of functional impairment has also been included in the definition (see Appendix 2).

10.3. Epidemiologic Studies

Only recently prospective studies using a severity measurement tool have shown regional differences in populations for the severity of atopic dermatitis. Such indicators may be useful to trace environmental factors at the population level (Schäfer *et al.*, 1997).

11. Biophysical and Biochemical Evaluation of Severity

When functional aspects are evaluated, intervals between severity grades within a given item are not equal either quantitatively or in terms of skin function changes (Van Neste, 1992). Thus clinical grading for specific purposes may be completed with noninvasive techniques such as ultrasound, transepidermal water loss, erythem-ameter, blood flow, and roughness measurements. Photography of AD needs highly professional skills and cannot be used routinely. Accurate measurement methods have been applied to itching (Wahlgren *et al.*, 1990; Endo *et al.*, 1997). Invasive methods, e.g., serum markers of disease activity (ECP, inflammatory cytokines), have not yet been found clinically reliable for staging patients independently of clincal scores. They need to be further tested against validated clinical scoring systems, and recent papers have used the SCORAD index (Thurau *et al.*, 1996; Frezzolin *et al.*, 1997; Wolkerstorfer *et al.*, in press).

12. Training

Based on the experience of the ETFAD, the homogeneity of results of the formal scoring sessions improve with time, especially as concerns intensity scoring, owing to a better identification of difficulties in using the SCORAD system and to the progressive adoption of more accurate guidelines. Thus the need for training individuals to perform adequately must be emphasized. Educational packages in the form CD-Roms, software for calculation, or internet sites featuring the basic features of scoring system with exercises and a basic review of AD are available for this purpose (Stalder, 1995).

Appendix 1. ETFAD Recommendations and Definitions for Grading Intensity Items*

Each intensity item should be graded 0–3 (0 = absent; 1 = mild; 2 = moderate; 3 = severe) according to reference photographs. Half-point scoring is not recom-

*Adapted from ETFAD (1993) and Kunz *et al.* (1997) (corresponding atlas available at http://AD server.sante. univ-nantes.fr).

mended. The area chosen for grading must be representative (average intensity) for each item in a given patient as determined by the investigator; thus one "target" area or the worst affected site is excluded. However, the same site may be chosen for two or more items. For example, one site may serve for evaluation of both excoriations and erythema. On the other hand, dryness must be noted on areas that do not have acute lesions or lichenifications. In areas with localized disease, where involvement may cause severe functional/relational impairment such as face, hands, or genitals, it may be difficult to score patients. When functional impairment associated with limited disease occurs (i.e., perlèche, hand fissures, genital involvement with problems of micturition, this should be noted separately and not included in the items being scored.

- *Erythema or Redness/Darkening:* There is no problem in defining this item on light-colored skin. In pigmented patients it is still possible to grade skin darkening instead of erythema, on the same basis. If grading is totally impossible (black skin), it should be noted on the evaluation sheet under remarks.

- *Edema/Papulation:* Edema/papulation stands for palpable infiltration of skin that may occur in acute erythematous and excoriated lesions as well as in chronic lesions during flare-ups. Palpation of the lesion should be taken into account in assessing this item.

- *Oozing/Crusts:* This item applies to exudative lesions resulting from epidermal edema and vesiculation.

- *Excoriations:* This item is *per se* an objective marker for pruritus, more visible in nonlichenified lesions. Number and intensity of scratching marks for each grade are illustrated in the atlas.

- *Lichenification/Prurigo:* This item is synonymous with epidermal thickening in chronic lesions. Grossly accentuated skin furrows separate shiny rhomboid areas, and the color is grayish or brownish. Prurigo lesions and common large-fold involvement undergo lichenification. This is generally noted in patients over 2 years of age. Prurigo lesions are frequently mixed with other types especially on the legs of school children, adolescents, and adults. Based on the analysis of cases with such lesions in patients examined collegially we propose to grade prurigo lesions within the lichenification category.

- *Dryness:* Whenever possible, this item should be noted at best at a distance from inflammatory lesions, and without prior application of emollients/moisturizers. Scales from healing inflammatory lesions should not be taken into account. Palpation is also important to assess the roughness of the skin. It is important to note if there is an appearance of ichthyosis vulgaris (under Remarks in the evaluation sheet). The presence of fissures is usually associated with severe dryness on the extremities.

Appendix 2. A Proposal for Severity Grading of Atopic Dermatitis

- *SCORAD* (0–103) = Extent (0–100)/5 + intensity (0–18) × 7/2 + Subjective items (0–20). On the SCORAD grading scale, *objective items only* should be used, i.e., extent (20-point scale) and intensity (63-point scale). The total objective SCORAD score has a point scale of 0–83.

- *MILD AD* = mean objective score < 15 on two baseline measurements at a minimum interval of 2 weeks.

- *Moderate AD* = 15 < mean objective score < 40 on two baseline measurements at a minimum interval of 2 weeks.

- *Severe AD* mean objective score > 40 on two baseline measurements at a minimum interval of 2 weeks.

- *Functional Impairment:* 10 points could be added on the initial score for perlèche, hand or foot fissures, or genital involvement, when lesions interfere with normal activities (physician's judgment).

References

Abrams, B. B., 1989, Atopic dermatitis: Elements in clinical study design and analysis, *Acta Derm. Venereol. (Stockh.) Suppl* **144**:15–19.

Aoki, T. (ed.), 1998, Evaluation of treatment effect on atopic dermatitis, Proc. 1st Symp., *Skin Res.* **38**(Suppl 18):1–146.

Bahmer, F. A., Schafer, J., and Schubert, H. J., 1991, Quantification of the extent and the severity of atopic dermatitis: The ADASI score, *Arch. Dermatol.* **127**:1239–1240.

Berth-Jones, J., 1996, Six area, six sign atopic dermatitis (SASSAD) severity score: A simple system for monitoring disease activity in atopic dermatitis, *Br. J. Dermatol.* **135**:25–30.

Bombardier, C., and Tugwell, B., 1982, Methodological framework to develop and select indices for clinical trials: Statistical and judgemental approaches, *J. Rheumatol.* **9**:753–757.

Broberg, A., and Faergemann, J. 1995, Topical antimycotic treatment of atopic dermatitis in the head/neck area, *Acta. Derm. Venerol. (Stockh.)* **75**:46–49.

Broberg, A., Kalimo, K., Lindblad, B., Swanbeck, G., 1990, Parental education in the treatment of childhood atopic eczema, *Acta. Derm. Venereol. (Stockh.)* **70**:495–499.

Clendenning, W. E., Clack, W. E., Ogawa, M., and Ishizaka, K., 1973, Serum IgE studies in atopic dermatitis, *J. Invest. Dermatol.* **61**:233–236.

Costa, C., Rilliet, A., Nicolet, M., and Saurat, J. H., 1989, Scoring atopic dermatitis: the simpler the better? *Acta. Derm. Venereol. (Stockh.)* **69**:41–45.

Endo, K., Sumitsuji, H., Fukuzumi, T., Adachi, J., and Aoki, T., 1997, Evaluation of scratch movements by a new scratch-monitor to analyze nocturnal itching in atopic dermatitis, *Acta Derm. Venereol. (Stockh.)* **77**:432–435.

ETFAD, 1993, Severity scoring of atopic dermatitis: The SCORAD index, Consensus report of the European Task Force on Atopic Dermatitis, *Dermatology* **186**:23–31.

Finlay, A. Y., and Khan, G. K., 1994, Dermatology Life Quality Index (DLQI): A simple practical measure for routine clinical use, *Clin. Exp. Dermatol.* **19**:210–216.

Finlay, A. Y., 1996, Measurement of disease activity and outcome in atopic dermatitis, *Br. J. Dermatol.* **135**:509–515.
Frederiksson, T., and Petterson, 1978, U. Severe psoriasis: Oral therapy with a new retinoid, *Dermatologica* **157**:238–244.
Frezzolini, A., Paradisi, M., Ruffelli, M., Cadoni, S., and de Pita, O., 1997, Soluble CD30 in pediatric patients with atopic dermatitis, *Allergy* **52**:106–109.
Hanifin, J. M., and Rajka, G., 1980, Diagnostic features of atopic dermatitis. *Acta. Derm.* (*Stockh.*) *Suppl.* **92**:44–47.
Hanifin, J., 1989, Standardized grading of subjects for clinical research studies in atopic dermatitis: Workshop report, *Acta. Derm. Venereol.* (*Stockh.*) *Suppl.* **144**:13–14.
Herd, R. M., Tidman, M. J., Ruta, D. A., and Hunter, J. A. A., 1997, Measurement of quality of life in atopic dermatitis: Correlation and validation of two different methods, *Br. J. Dermatol.* **136**:502–507.
Jemec, G. B. E., and Wulf, H. C., 1997*a*, Quality assurance in dermatology: The development of a framework, *Int. J. Dermatol.* **36**:721–726.
Jemec, G. B. E., and Wulf, H. C., 1997b, The applicability of clinical scoring systems: SCORAD and PASI in psoriasis and atopic dermatitis, *Acta. Derm. Venereol.* (*Stockh.*) **77**:392–393.
Kezawa, Z., Ikebe, T., Ogura, H., Odajima, H., Kurosaka, F., Sase, K, Sugiuchi, M., *et al.*, 1991, Clinical effect of hypoallergenic rice HRS-1 in atopic dermatitis, *Jpn. J. Allergol.* **40**:633–642.
Kowakzick, L., Kleinheinz, A., Weichenthal, M. *et al.*., 1995, Low dose versus medium dose UV–A1 treatment in severe atopic eczema, *Acta Derm. Venereol.* (*Stockh.*) **75**:43–45.
Kunz, B., Oranje, A. P., Labrèze, L., Stalder, J. F., Ring, J., and Taïeb, A., 1997, Clinical validation and guidelines for the SCORAD index. Consensus report of the European Task Force on Atopic Dermatitis, *Dermatology* **195**:10–19.
Langeland, T., 1983, A clinical and immunological study of allergy to hen's egg white. I. A clinical study of egg allergy, *Clin. Allergy*. **13**:371–382.
Lever, R., Hadley, K., Downey, D., Mackie, R., 1988, Staphylococcal colonization in atopic dermatitis and the effect of topical mupirocin therapy, *Br. J. Dermatol.* **119**:189–198.
Lewis-Jones, M. S., and Finlay, A. Y., 1995, The Children's Dermatology Life Quality Index (CDLQI): Initial validation and practical use, *Br. J. Dermatol.* **132**:942–949.
Oranje, A. P., Stalder, J. F., Taïeb, A., Tasset, C., and de Longueville, M., on behalf of the ETAC study group, 1997, Scoring of atopic dermatitis by the SCORAD index using a training atlas by investigators from different disciplines, *Pediatr. Allergy Immunol.* **8**:28–34.
Penneys, N. S., 1997, Quality: Its definition, measurement, and application in dermatology, *J. Am. Acad. Dermatol.* **37**:503–507.
Queille-Roussel, C., Raynaud, F., and Saurat, J. H., 1985, A prospective computerized study of 500 cases of atopic dermatitis in childhood, *Acta Derm. Venereol.* (*Stockh.*) *Suppl.* **114**:87–92.
Rajka, G., and Langeland, T., 1989, Grading of the severity of atopic dermatitis, *Acta Derm. Venereol.* (*Stockh.*) *Suppl* **144**:13–14.
Ramsay, B., and Lawrence, C. M., 1991, Measurement of involved surface in patients with psoriasis, *Br. J. Dermatol.* **124**:565–570.
Ring, J., Senter, T., Cornell, R. C., Arroyave, C. M., and Tan, E. M., 1979, Plasma complement and histamine changes in atopic dermatitis, *Br. J. Dermatol.* **100**:521–530.
Ruzicka, T., Bieber, T., Schöpf, E. et al., 1997, A short-term trial of tacrolomus ointment for atopic dermatitis, *N. Engl. J. Med.* **337**:816–821.
Schäfer, T., Dockery, D., Krämer, U., Behrendt, H., and Ring, J., 1997, Experiences with the severity scoring of atopic dermatitis in a population of German pre-school children, *Br. J. Dermatol.* **137**:558–562.
Stalder, J. F., 1995, Atopic dermatitis: From diagnosis to severity scoring. CD-Rom Portfolio, Kodak.
Thurau, A. M., Schylz, U., Wolf, V., Krug, N., and Schauer, U., 1996, Identification of eosinophils by flow cytometry, *Cytometry* **23**:150–158.
Van Neste, D., 1992, Functional relevance of severity scores of atopic dermatitis: Multiparametric *in vivo* evaluation of biophysical parameters of inflamed lesions with noninvasive methods, *Eur. J. Dermatol.* **2**:319–324.
Wahlgren, C. F., Scheynius, A., and Hägermark, O., 1990, Antipruritic effect of oral cyclosporin A in atopic dermatitis, *Acta Derm. Venereol.* (*Stockh.*) **70**:323–329.

Williams, H. C, Burney, P. G. J., Hay, R. J. et al., 1994*a*, The UK working party's diagnostic criteria for atopic dermatitis. I: Derivation of a minimum set of discriminators for atopic dermatitis, *Br. J. Dermatol.* **131**:383–396.

Williams, H. C., Burney, P. G. J., Hay, R. J. et al., 1994*b*, The UK working party's diagnostic criteia for atopic dermatitis. II: Observer variation of clinical variation of clinical diagnosis and signs of atopic dermatitis, *Br. J. Dermatol.* 1994*b*; **131**:397–405.

Williams, H. C., Burney, P. G. J., Pembroke, A. C., Hay, R. J. et al., 1994*c*, The UK working party's diagnostic criteria for atopic dermatitis. III: Independent hospital validation, *Br. J. Dermatol.* **131**:406–416.

Wolkerstorfer, A., Laan, M. P., Savelkoul, H. F., Neijens, H. J., Mulder, P. G. *et al.*, 1998, Soluble E-selectin, other markers of inflammation and disease severity in children with atopic dermatitis, *Br. J. Dermatol.* **138**:431–435.

Zachary, C. B., and MacDonald, D. M., 1983, Quantitative analysis of T lymphocytes subsets in atopic eczema, using monoclonal antibodies and flow cytometry, *Br. J. Dermatol.* **108**:411–422.

Chapter 9

A Review of the Use of Essential Fatty Acid Supplementation in Atopic Dermatitis with Emphasis on the Methodology of Trial Design

JOHN BERTH-JONES AND ROBIN A. C. GRAHAM-BROWN

1. Introduction

The possible beneficial effect of essential fatty acid supplementation in atopic dermatitis has generated considerable controversy among dermatologists worldwide. Some studies apparently demonstrate dramatic benefit, while other have failed to show any convincing response. The fact that this form of dietary supplementation seems completely safe makes it even more important to determine whether it is effective. Even a modest degree of beneficial effect would be clinically valuable since the therapeutic index (the balance of beneficial vs. adverse effects) would be favorable relative to the use of topical corticosteroids.

Much of the controversy has been generated by relatively small trials showing impressive results. Careful examination of these and comparison with larger trials provides some useful insights into the problems of designing and analyzing clinical trials in atopic dermatitis. We conducted a large trial on the use of essential fatty acid supplementation and used this study to compare a number of different methodologies.

When the published data are reviewed critically, the evidence for efficacy of essential fatty acid supplementation in atopic dermatitis is not strong.

John Berth-Jones • Department of Dermatology, Walsgrave Hospital, Coventry CV2 2DX, England. *Robin A. C. Graham-Brown* • Department of Dermatology, Leicester Royal Infirmary, Leicester LE1 5WW, England.

Cutaneous Biometrics, edited by Doris A. Schwindt and Howard I. Maibach, Kluwer Academic/Plenum Publishers, New York, 2000.

2. Rationale for Essential Fatty Acid Supplementation in Atopic Dermatitis

Essential fatty acids (EFA) are long-chain fatty acids, which are required for normal cutaneous function and cannot be synthesized by human metabolism. They can be classified into series according to the position of the first double bond from the *n*-terminal carbon atom. Most important are the *n*6 series, which includes gamma linolenic acid (GLA), and the *n*3 series, which includes eicosapentaenoic acid (EPA). Both of these have been investigated for possible efficacy in atopic dermatitis (AD). The rationale lies not in the correction of any dietary insufficiency, but in the fact that fatty acids are absorbed and incorporated into body lipids unchanged, so that the precise lipid composition of membranes and other structures depends to a considerable degree on the lipids contained in the diet.

Fatty acids serve many functions: they play a structural role in membranes, they are precursors for complex lipids including the ceramides of the stratum corneum, and they are precursors of eicosanoid mediators of inflammation. There are therefore numerous possible mechanisms whereby altering the fatty acid composition of the diet may alter the activity of inflammatory dermatoses.

It has been suggested that patients with AD may have defective conversion of linoleic acid, the major dietary component of *n*6 series EFA, to its metabolites, as a result of a low level of activity of the enzyme—delta-6 desaturase. These patients might therefore be responsive to the administration of fatty acid supplements that would bypass this metabolic step. Supplementation with the correct fatty acids could then provide high levels of the substrates that are preferentially metabolized to relatively noninflammatory prostaglandins and leukotrienes such as PGE1 (Wright, 1991; Burton, 1989). However, prostaglandin E1 levels in serum appear to be no different in patients with AD than in normal controls, arguing against a role for EFA through correction of a deficiency in the synthesis of these noninflammatory prostaglandins (Leonhardt *et al.*, 1997). Other possible mechanisms of action for such supplements include alteration of cell membrane composition with resultant changes in the mobility or conformation of protein components such as enzymes or adhesion molecules (Lu *et al.*, 1995).

Another possible mode of action might be correction of the deficient barrier function of the stratum corneum, which characterizes AD. This barrier depends to a substantial degree on the presence of complex lipid molecules, principally ceramides. There is limited evidence that synthesis or release of stratum corneum lipids is abnormal in AD (Yamamoto *et al.*, 1991; Fartasch *et al.*, 1992) and although it is not known whether this is secondary to the inflammatory process or a primary defect responsible for the disease, there is clearly the possibility that altering the availability of lipid substrates may influence the composition of the stratum corneum lipids, with a consequent alteration in "barrier function."

3. Evidence for Efficacy

Dietary supplementation has been investigated using *n*6 and *n*3 series EFA and, in one study, both. The most thoroughly examined type of supplementation has been with the *n*6 series EFA GLA in the form of evening primrose oil. Other studies have examined supplementation with the *n*3 series EFA's eicosapentaenoic acid and docosahexaenoic acid in the form of fish oil.

3.1. The n6 Series EFA Supplementation

Lovell and Burton (1981) stimulated a great deal of interest in EFA supplementation when they reported a crossover study on 32 adults and children in which evening primrose oil and placebo were each given for 3 weeks. They reported a significant improvement on active treatment but not on placebo. It is important, however, to examine for a statistical difference between the response to placebo and the response to active treatment. The data given allow the reader to perform this analysis and there was no significant difference (*t*-test).

This initial trial was then followed by another larger study from the same center, which yielded more convincing evidence of a response of AD to evening primrose oil. Wright and Burton (1982) performed a crossover study in which 99 adults and children participated. They reported a significant, dose-related response, which reached a 43% overall improvement in patients using 6 g daily. The analysis of this study was more appropriate, although one possible criticism is that no data were provided to establish the comparability of the two treatment groups at baseline.

A response to evening primrose oil was also subsequently reported in a parallel-group study of 12 weeks duration performed by Schalin-Karrila *et al.* (1987) with 25 adult subjects. They reported large and significant reductions (30–50%) in surface area affected and severity of signs and symptoms. The group on active treatment used 60% less topical steroid. In this report, as in the study of Lovell and Burton discussed above, the analysis did not properly address differences in response between treatments. Furthermore, in this study, there were clear baseline differences in severity between the groups so that there was more scope for improvement in the group that received active treatment. This is always a potential problem with randomized trials, although it is less likely to occur in larger studies.

Bordoni *et al.* (1987) performed a parallel-group study on 24 children using 3 g of evening primrose oil daily for 4 weeks. These authors also reported a significant difference between active and placebo groups (confirmed by personal communication with Scotia Pharmaceuticals).

Unfortunately, other investigators have not been able to confirm this response in larger placebo-controlled trials. The largest study yet performed on evening primrose oil in AD was conducted in the United States by Bamford *et al.* (1985) These authors recruited 154 adults and children into a placebo-controlled, crossover trial. No therapeutic effect was confirmed. The results have been called into question (by Scotia, the producers of Epogam® evening primrose oil) as a result of plasma lipid assays raising the possibility that placebo and active treatments were intermixed (Morse *et al.*, 1989), but the investigators have no reason to believe that this occurred (personal communication JTM Bamford).

No response to treatment was seen in a study on 24 subjects performed by Rilliet *et al.* (1988) in which blackcurrant seed oil was used as a source of GLA.

A small placebo-controlled, parallel-group study has subsequently been reported on the use of borage oil in AD (Valsecchi *et al.*, 1996). Borage oil contains a higher concentration of GLA than evening primrose oil. Subjects were treated with five capsules daily (2.5 g of borage oil) containing an estimated 400 mg daily of GLA. Thirty-one subjects were randomized. In this study both treatment groups improved, but there was no significant difference between them.

Three of these published studies and five others were the subject of a meta-analysis performed by Scotia Pharmaceuticals, which showed a dose-related response, increasing progressively with treatment duration (Morse *et al.*, 1989) Unfortunately such an analysis is only as valid as the individual studies of which it is composed, and it is clear that in at least one of these there was a marked difference in baseline severity between treatment groups. In such an analysis it is important to establish baseline comparability of treatment groups in terms of disease severity but this issue was not addressed. We believe that the results of a meta-analysis containing unpublished trials and inadequate baseline data should be regarded with caution.

To resolve this apparent conflict in results from existing reports, we undertook a further broad study (Berth-Jones and Graham-Brown, 1993), which also examined the effect of a combination of *n*6 and *n*3 series EFA, discussed in detail below.

3.2. The n3 Series EFA Supplementation

A rich source of *n*3 series EFA is fish oil. Studies on patients with AD using fish oil supplementation have, as studies on *n*6 supplementation, yielded inconclusive results. Two double-blind, placebo-controlled, parallel-group trials have been reported, as well as an additional study in which corn oil was employed as a control. In the first of these, 31 subjects were treated for 12 weeks with 10 g of fish oil (or placebo) daily. Subjects' grading of symptom severity showed a significant response. The investigators' assessments improved more in the active treatment group than in the placebo group but the difference did not reach statistical significance (Bjorneboe *et al.*, 1987). In the second study, reported only in an abstract, 17 subjects were treated with EPA for 4 months. The size of the placebo group was not given. No significant differences between active and placebo groups were demonstrated (Kunz *et al.*, 1989).

In a larger, more recent study by Soyland *et al.* (1994), 145 subjects were randomized to receive fish oil (6 g daily containing 51% EPA) or corn oil. It is not clear from this report whether corn oil was employed as a placebo or whether it was expected to show activity in AD. Corn oil contains predominantly linoleic acid, a precursor of GLA through the action of delta-6 desaturase. No difference in response was seen between these treatments although the condition improved in both treatment groups.

3.3. The n6 and Combined n6 and n3 Series EFA Supplementation

In order to resolve the question as to whether these treatments were effective, we conducted a further parallel-group study (Berth-Jones and Graham-Brown, 1993). This contained three treatment limbs in which subjects received 6 g daily of evening primrose oil, (Epogam®), 5.2 g of evening primrose oil combined with 1.3 g of fish oil (Efamol Marine®), or matching placebo. Each treatment was given for 16 weeks and subjects were followed for a further 8-week washout period.

This study was used as an opportunity to compare differing methods of monitoring disease activity in AD so numerous parameters were studied to assess the response to treatment. The proportion of skin affected by the disease was monitored using the rule of nines (as used to assess burns injuries) as a guide to the area affected.

A sign score we developed from previous trials was specified, prospectively, to be the primary response criterion. The body was divided into 10 zones: face, neck, abdomen, back, elbows, antecubital fossae, dorsa of hands, palms and wrists, popliteal fossae, feet. Each zone was scored on a scale of 0 (absent) to 3 (severe) for each of the following: erythema, excoriation, dryness, cracking, lichenification. Total scores from each zone were added to give a maximum of 150.

For comparison we also employed a clinical scoring system previously published by Costa *et al.* (1989) In this system, symptoms and signs are assessed at the most severely affected site. Erythema, edema, vesiculation, crusting, excoriation, scaling, lichenification, pigmentation/depigmentation, pruritus, and sleep disturbance are each graded on a seven-point scale and the scores are added together giving a range of 0–70. A further score (range 0–30) is derived from the proportion of skin affected and this is then added to the first score giving an overall possible range of 0–100.

Symptoms were monitored using patient diaries. Each subject was issued with a diary containing a page of 10-cm visual analog scales for baseline and each of the next 24 weeks. Scales were provided for itch, dryness, scaling, redness, and overall impression. All were labeled "none" at the left-hand side end and "worst ever" at the right. Scores were added to give a maximum of 50 cm. Subjects were instructed to record symptoms experienced over the entire week and not just at the time when the diary was completed.

The requirement for topical steroids can also be regarded as a measure of disease activity and was monitored during this study. To facilitate monitoring, topical steroids were prescribed by the investigator. Usage was assessed by weighing returned tubes of medication. Some subjects using potent corticosteroids also used a mild compound for the face, in which case only the potent steroid was monitored.

This study is, to date, the only one published in which the report clearly establishes that a primary response criterion was prospectively defined. This is essential to validate the statistics when multiple signs, symptoms, and "severity scores" are monitored. In addition, this study has the particular strength of being the only published trial on the use of EFA supplementation in AD in which a power calculation is reported to determine the sample size required. This was based on the total sign score (the primary response criterion). The study was designed to have 80% power to detect a treatment response of 20% with a standard deviation (derived from existing data) of 30%, at significance level 0.05. It was estimated that 37 subjects would be required in each limb.

Forty-one adults and children were recruited to each treatment group and the groups were well matched at baseline for demographics and for disease severity. All three groups showed some degree of improvement in most assessments, but no significant differences from placebo were seen in the response of either active treatment limb. Since both treatments consisted predominantly of evening primrose oil in high dosage, it was considered justifiable to perform an analysis with the two "active" groups combined, but this analysis also showed no response.

The dose of fish oil used in the combined treatment group in this study was low relative to other trials. This limb of the trial was not intended to determine whether *n*3 series fatty acids were effective *per se*, but rather to investigate the possibility of a synergistic effect when *n*3 and *n*6 supplementation was combined. No evidence was obtained of any synergism.

4. Observations on the Methodology of Clinical Trials in Atopic Dermatitis

One of the aims of the trial we conducted and described above was to study the methodology of clinical trials in AD. Some useful observations were made regarding the advantages and disadvantages of different methods of monitoring disease activity.

Estimation of the proportion of skin affected is very unsatisfactory in AD since it is much more subjective than it might sound. In contrast to some other dermatoses, such as psoriasis, the lesions of AD are rarely sharply marginated so it is necessary to "judge" how large an area is affected. In many patients large areas of skin are faintly erythematous and dry; some observers would count this as involved skin, while others would not. It is therefore necessary to derive an arbitrary threshold at which these signs become significant. Histologically, even skin that is apparently completely uninvolved in atopic eczema often shows evidence of inflammation (personal observations).

The sign score employed as the primary response criterion proved simple and satisfactory to use although some possible improvements became apparent. It has subsequently proved possible to simplify this score by reducing the number of areas assessed to six instead of ten. Another subsequent modification has been the addition of a sixth sign—exudation, which has been found to be a further helpful sign to monitor in subsequent trials. The simplified version of this score has been named the Six Area, Six Sign Atopic Dermatitis (SASSAD) score and has been described elsewhere together with a discussion of its use (Berth-Jones, 1996).

The clinical score described by Costa *et al.* (1989) as a "simple scoring system" proved far more complicated that the sign score discussed above and offered no advantages over the latter. On the contrary, a number of problems were encountered with it. For example, in the component of the score in which signs and symptoms were assessed, three of the signs proved noncontributory. Vesiculation was observed in fewer than 1% of subjects. Pigmentation was observed to vary inversely with the total severity score, as abnormal pigmentation became more evident when the disease improved. Edema was a difficult sign to define and assess consistently. The addition of scores for symptoms, such as pruritus, to the scores for signs, such as erythema, involved adding an objective measurement by the physician to a subjective measurement by a different observer (the patient). This proved unsatisfactory because two observers do not always agree even about the direction of changes in disease activity. The area component of this compound clinical score was subject to all the limitations discussed above under assessment of area affected.

The use of patient diaries to monitor symptoms was also not without problems. Diaries that could be evaluated were available at the end of the study from 97 out of 123 subjects, a substantial proportion having been lost. This problem could have been reduced by collecting diaries at each visit and issuing new ones, thus minimizing the amount of data lost with each diary. The use of visual analog scales was well understood by most patients, although some tended to vary recordings across a wider range of the scale than others.

Monitoring the requirement for topical steroids by weighing returned tubes of medication proved very difficult. Some patients used more than one tube concurrently. Some tubes were lost. Many subjects used more than one compound concurrently on different body sites. Apparently accurate data were obtained with only 81 of the 123 subjects at the end of the 24 week trial. Even these data are difficult to interpret because of the differing potencies of compounds used by different subjects. It would be highly desirable to standardize the use of topical corticosteroids during trials of this sort, either by forbidding their use altogether or by allowing only one or two compounds to be used. This would be difficult, however, owing to the labile nature of the disease and the tendency for patients to develop strong, and sometimes justified, preferences for one steroid formulation over another. One useful conclusion that could be drawn was that the overall rate of usage of steroids was very low throughout the study so this would be most unlikely to have obscured a response to the EFA supplement.

A further recent development in the design of clinical trials in AD has been the validation of methodologies for assessing the impact of the disease on quality of life (Salek *et al.*, 1993; Finlay and Khan 1994; Lewis-Jones and Finlay, 1995; Chren *et al.*, 1995) The assessment of quality of life can provide an additional parameter to monitor changes in disease activity and while less objective, it is ultimately more important than changes in sign scores.

5. Conclusions

The published evidence for the efficacy of EFA supplementation in AD is far from conclusive. Some of the published trials have involved rather small numbers of subjects, the primary response criterion has not been defined, and no indication has been provided that any power calculation was performed. In some cases the statistical analysis provided was not optimal. Some of the dramatic results from small trials seem incompatible with the lack of response in larger studies. This may be the result of baseline differences between treatment groups in the smaller studies.

We would recommend that in future trials on the use of EFA supplementation in AD, a primary response criterion be defined prospectively and a power calculation be performed to ensure that the sample size is adequate. We recommend the use of a sign score such as the SASSAD score as a primary response criterion. Symptoms such as pruritus and sleep loss should also be monitored, either on the same occasions as other assessments or by using patient diaries replaced at each visit. Quality of life assessments offer an additional and valuable means to assess the benefits of a treatment.

References

Bamford, J. T. M., Gibson, R. W., and Renier, C. M., 1985, Atopic eczema unresponsive to evening primrose oil (linoleic and gamma-linolenic acids), *J. Am. Acad. Dermatol.* **13**:959–65.

Berth-Jones, J., and Graham-Brown, R. A. C., 1993, Essential fatty acid supplementation in atopic dermatitis: A placebo-controlled trial, *Lancet* **341**:1557–60.

Berth-Jones, J., 1996, Six Area Six Sign AD (SASSAD) severity score: A simple system for monitoring disease activity in atopic dermatitis, *Br. J. Dermatol.* **135**(Suppl 48): 25–30.

Bjorneboe, A., Soyland, E., Bjorneboe, G.-E. A., Rajka, G., and Drevon, C. A., 1987, Effect of dietary supplementation with eicosapentaenoic acid in the treatment of atopic dermatitis, *Br. J. Dermatol.* **117**:463–9.

Bordoni, A., Biagi, B. L., Masi, M., Ricci, G., Fanelli, C., Patrizi, A., and Creccolini, E., 1987, Evening primrose oil in the treatment of children with atopic eczema, *Drugs. Exptl. Clin. Res.* **14**:291–297.

Burton, J. L., 1989, Dietary fatty acids and inflammatory skin disease, *Lancet* **1** 27–30.

Chren, M. M., Lasek, R. J., Quinn, L. M., and Zyzanski, S. J., 1995, A measure of the effect of skin disease on quality of life: Conceptual framework and initial tests, *J. Invest. Med.* **43**:348A.

Costa, C., Rilliet, A., Nicolet, M., and Saurat, J. H., 1989, Scoring atopic dermatitis, *Acta. Derm. Venereol.* **69**:41–45.

Fartasch, M., Bassukas, I. D., and Diepgen, T. L., 1992, Disturbed extruding mechanism of lamellar bodies in dry non-eczematous skin of atopics, *Br. J. Dermatol.* **127**:221–227.

Finlay, A. Y., and Khan, G. K., 1994, Dermatology life quality index (DLQI)—a simple practical measure for routine clinical use. *Clin. Exp. Dermatol.* **19**:210–216.

Kunz, B., Ring, J., and Braun-Falco, O., 1989, Eicosapentaenoic acid (EPA) treatment in atopic eczema (AE): A prospective double-blind trial. *J. Allergy Clin. Immunol.* **83**:196.

Leonhardt, A., Krauss, M., Gieler, U., Schweer, H., Happle, R., and Seyberth, H. W., 1997, *In vivo* formation of prostaglandin E1 and prostaglandin E2 in atopic dermatitis, *Br. J. Dermatol.* **136**:337–340.

Lewis-Jones, M. S., and Finlay, A. Y., 1995, The children's dermatology life quality index (CDLQI): Initial validation and practical use. *Br. J. Dermatol.* **132**:942–949.

Lovell, C. R., and Burton, J. L., 1981, Treatment of atopic eczema with evening primrose oil, *Lancet* **1**:278.

Lu, L., Okada, N., Nakatani, S., and Yoshikawa, K., 1995, Eicosapentaenoic acid-induced changes in membrane fluidity and cell adhesion molecules in cultured human keratinocytes, *Br. J. Dermatol.* **133**:217–222.

Morse, P. F., Horrobin, D. F., Manku, M. S., Stewart, J. C. M., Allen, R., Littlewood, S., Wright, S., Burton, J., Gould, D. J., Holt, P. J., Jansen, C. T., Mattila, L., Meigel, W., Dettke, T. H., Wexler, D., Guenther, L., Bordoni, A., and Patrizi, A., 1989, Meta-analysis of placebo-controlled studies on the efficacy of Epogam in the treatment of atopic eczema. Relationship between plasma EFA changes and clinical response, *Br. J. Dermatol.* **121**:75–90.

Rilliet, A., Queille, C., and Saurat, J. H., 1988, Effects of gamma-linolenic acid in atopic dermatitis, *Dermatologica* **177**:257.

Salek, M. S., Finlay, A. Y., Luscombe, D. K., Allen, B. R., Berth-Jones, J.,. Camp, R. D. R., Graham-Brown, R. A. C., Khan, G. K., Marks, R., Motley, R. J., Ross, J, and Sowden, J. M., 1993, Cyclosporin greatly improves the quality of life of adults with severe atopic dermatitis. *Br. J. Dermatol.* **129**:422–30.

Schalin-Karrila, M., Mattila, L., Jansen, C. T., and Uotila, P., 1987, Evening primrose oil in the treatment of atopic eczema: Effect on clinical status, plasma phospholipid fatty acids and circulating blood prostaglandins, *Br. J. Dermatol.* **117**:11–19.

Soyland, E., Funk, J., Rajka, G:, Sandberg, M., Thune, P., Rustad, L., Helland, S., Middelfart, K., Odu, S., Falk, E. S., Solvoll, K., Bjorneboe, G. E. Aa., and Drevon, C. A., 1994, Dietary supplementation with very long-chain fatty acids in patients with atopic dermatitis: A double-blind, multicentre study, *Br. J. Dermatol.* **130**:757–764.

Valsecchi, R., Di Landro, A., Pansera, B., and Reseghetti, A., 1996, Gamma-linolenic acid in the treatment of atopic dermatitis, *J. Eur. Acad. Derm. Venereol.* **7**:77–79.

Wright, S., and Burton, J. L., 1982, Oral evening-primrose-seed oil improves atopic eczema, *Lancet* **2**:1120–23.

Wright, S., 1991, Essential fatty acids and the skin, Br. J. Dermatol. **125**:503–15.

Yamamoto, A., Serizawa, S., Ito, M., and Sato, Y., 1991, Stratum corneum lipid abnormalities in atopic dermatitis. *Arch. Dermatol. Res.* **283**:219–23.

Chapter 10

Severity Scoring of Atopic Dermatitis

ALINE B. SPRIKKELMAN AND WIM M. C. VAN AALDEREN

1. Introduction

Hanifin and Rajka (1980) outlined the diagnostic criteria of atopic dermatitis (AD) in order to apply uniform criteria in diagnosing the disease. At the present time a pathognomonic diagnostic laboratory test is still lacking, and the diagnosis of AD continues to be based on the presence or absence of a combination of signs and symptoms. The significance and validity of the major and minor diagnostic criteria they proposed have constituted a topic of interest in several studies, and adjustments have been made in order to improve the reliability of diagnoses (Kanwar *et al.*, 1991; Kapp *et al.*, 1991; Diepgen and Fartasch, 1992; Williams *et al.*, 1994; Tada *et al.*, 1994; Rudzki *et al.*, 1994; Williams and Pembroke, 1996; Nagaraja *et al.*, 1996). Although, the outline of diagnostic criteria proved useful for practitioners and investigators, a more objective, quantitative, and accurate assessment of the severity of AD was needed, so different methods have been developed over the last years, including laboratory techniques and clinical scoring systems.

It has been suggested that some serological tests might be used as markers of AD severity. It has been shown that eosinophil cationic protein (ECP) levels are increased in patients with moderate to severe AD in comparison with nonatopic controls, and that their ECP levels correlate with AD severity. Furthermore, it has been shown that ECP levels decrease if AD lesions remit (Kapp *et al.*, 1991; Czech *et al.*, 1992). Although these results are promising, further standardization is necessary before clinical use or use in research can be recommended. Less consistent correlations have been found between severity of AD and serum levels of IgE, interleukin-2 receptors, or the number ofeosinophils (Kapp *et al.*, 1991; Czech *et al.*, 1992, Kagi *et al.*, 1992).

Aline B. Sprikkelman and Wim M. C. van Aalderen • Department of Pediatric Pulmonology, Emma Children's Hospital Academic Medical Center/University of Amsterdam, 1100 DD Amsterdam, The Netherlands.

Cutaneous Biometrics, edited by Doris A. Schwindt and Howard I. Maibach, Kluwer Academic/Plenum Publishers, New York, 2000.

AD scoring systems, which are based on clinical evaluation of the skin, are frequently used in research as well as in clinical practice to assess the severity of the disease and its clinical course. The advantages of scoring systems in the assessment of the severity of AD are their noninvasive character, the fact that there is no need for sophisticated equipment, and the relatively quick insight they provide into the disease activity.

At the present time, there are different scoring systems to evaluate the severity of AD. In this chapter we will give an overview of more that are available, focusing on their usefulness and their validity, including intraobserver and interobserver variation. Moreover, we will indicate which system can be used for different purposes.

2. Atopic Dermatitis Severity Scoring Systems

As reviewed by Finlay (1996), several AD severity scoring systems have been developed in the last decade in order to evaluate the activity of the disease. Simple and more elaborating systems are available, each with its own application in clinical practice and research. Severity of AD can be assessed by evaluating the signs, patient symptoms, and the area of involvement. Scoring systems usually consist of a combination of assessment of intensity items and extent of the disease. Subjective patient symptoms such as pruritus and sleep loss are often included.

The choice of a certain scoring system depends on the purpose for which it is to be used. For example, when subtle modifications in the severity of AD need to be detected, such as in drug-effect and intervention studies or for follow-up of the progression of the disease, an elaborate scoring system is needed which takes the different aspects of AD into account. In long-term follow-up studies, a rough estimate of overall severity is often sufficient because AD is a chronically relapsing disease, and a wide variation in the disease state is possible at the time of follow-up (Sprikkelman *et al.*, 1997).

Other aspects that are very important in choosing a specific scoring system are its feasibility, accuracy, sensitivity, and the interobserver and intraobserver reliability of the scoring system. Table 1 gives an overview of the available AD severity scoring systems, divided into systems developed primarily for clinical use and systems for research. The table shows the items (intensity criteria, extent, subjective symptoms) that are taken into account in each system.

2.1. Severity Scoring Systems for Clinical Practice

Queille-Roussel *et al.* (1985) were the first to present a prospective follow-up study of AD in childhood, in which the severity of AD was studied systematically using a severity scoring system (Queille-Roussel *et al.*, 1984). This scoring system is designed to evaluate AD in children in an in- or outpatient department and consists of scoring intensity criteria, topographic sites as a measure of the extent of AD, and subjective criteria, such as pruritus and sleep loss. Each criterion is evaluated at the most severely affected area. The error among observers on the reproducibility of this scoring system is about 5%. As this system was primarily designed to assess the severity of AD in children, there was some question as to whether it could be used with adults.

Costa *et al.* (1989) subsequently developed a new, more elaborate scoring system in which the severity of AD at each affected site was taken into account. They evaluated the severity of AD in adults using the simple scoring system based on that of Queille *et al.* (1984), and their new, more elaborate system, and were the first to compare two AD severity scoring systems for reproducibility and validity (Costa *et al.*, 1989). They found a significant correlation between the two systems in assessing the overall severity of AD. However, the more elaborate scoring system was significantly less reproducible between two observers compared with the simple one, so it was decided to use the simple scoring system.

Although, the latter according to Costa *et al.* (1989) was developed primarily for monitoring the severity of AD in an outpatient clinic, it is being used in clinical trials, because it proved to be easy to use, quick, and reproducible. Nevertheless, it has its disadvantages, among them that it is insensitive to change, because the severity of AD is assessed only at one site, the most severely affected one, and it may not detect changes in disease severity of other parts of the body (Berth-Jones *et al.*, 1995). Furthermore, it appears that the intensity scoring item pigmentation or depigmentation varies inversely with the severity of AD, because changes in this intensity item become more apparent when AD remits (Berth-Jones *et al.*, 1995).

Another disadvantage of the simple system is the fact that it includes symptoms mentioned by the patient as well as signs assessed by the observer. These two may not always coincide. It has been proposed to separate observer-derived scores from patient-derived symptoms in the overall AD severity score in order to clearly interpret changes in severity scores over time and to reduce interobserver variation (Finlay, 1996; Berth-Jones *et al.*, 1995). Recently, we (Sprikkelman *et al.*, 1997) demonstrated good agreement between observers in assessing the overall severity of AD in the lower and higher scoring rates when using the simple system. However, we found significant interobserver variation in assessing isolated intensity items, such as excoriations and scales. Therefore, for research purposes, it seems advisable to have these scores assessed by a single observer.

Another simple grading severity scoring system primarily developed for clinical practice and not for research was proposed by Rajka and Langeland (1989). It involves scoring the extent of AD, the intensity of the disease as assessed by disturbance of sleep by itching and the course based on history. The last parameter gives insight in the long-term activity of AD. This system permits distinction between mild, moderate, and severe AD in order to roughly categorize patients, and can be useful for selecting patients for clinical trials.

2.2. Severity Scoring Systems for Clinical Research

Hanifin (1989) presented a practical framework of parameters for assessing the severity of AD in clinical research at the Third International Symposium on Atopic Dermatitis. He proposed that parameters such as extent of AD at baseline, the course and stability of the disease, and the intensity be taken into account. In the majority of the available scoring systems these parameters are subject to evaluation.

TABLE 1. A Selection of Available Atopic Dermatitis Severity Scoring Systems

	Objective signs										Subjective signs	
	Intensity criteria								Extent			
Reference	erythema	edema	vesicles papules	excoriations	crusts oozing	scales	lichenification	dry-skin	pigmentation	area involved	pruritus	sleeploss
a. Severity Scoring Systems for Clinical Practice												
Queille-Roussel *et al.* (1984)	+	+	+	+	+	+	+	+		+	+	+
Costa *et al.* (1989) (simple scoring system)	+	+	+	+	+	+	+		+	+	+	+
Rajka and Langeland (1989)										+	+	+
Hanifin	+	(induration)			+					+	+	
Harper *et al.* (1991)	+	+	+	+	+	+	+		+	+	+	+
Sowden *et al.* (1991)	+			+	+	+	+	+		+		
Bahmer *et al.* (1991)	+					+	+			+		
ETFAD (1993)	+	+	+	+	+		+	+		+	+	+
Kunz *et al.* (1997)[a]	+	+	+	+	+		+	+		+	+	+
Verwimp *et al.* (1995) and Sprikkelman *et al.* (1997)										+		

Berth-Jones (1996)[b]	+			+			+	+	+		
Grandlund *et al.* (1995)[c]									+	+	+
van Joost *et al.* (1994)[d]	+		+	+	+	+	+	+	+	+	+
Hanifin *et al.* (1993)[e]	+	+	+	+		+	+	+	+	+	
Vernon *et al.* (1991)[f]	+			+			+		+	+	
Jeckler and Larko (1990)[g]	+		+	+		+	+	+	+		
Bjørneboe *et al.* (1987)[h]	+			+		+	+		+	+	
Soyland *et al.* (1994)[i]	+					+	+	+	+	+	

[a] Functional impairment is also taken into account: 10 points could be added onto the initial score for perlèche, fissures on the hands or feet, or genital involvement, when lesions interfere with normal activities.

[b] Exudation and cracking are also taken into account.

[c] Requirement for topical steroids is also taken into account.

[d] Patients assessment of the severity of AD and cracking/fissuring is also taken into account.

[e] Patients assessment of the severity of AD is also taken into account.

[f] Exacerbation and remission rates are also taken into account.

[g] Healing rate is taken into account.

[h] Also taken into account is the effect of AD on normal living.

[i] Also taken into account is induration and the effect of AD on normal living.

Harper *et al.* (1991) used an AD severity scoring system in a clinical trial that was similar to the simple scoring system of Costa *et al.* (1989), differing only in scoring the affected areas. In Harper *et al*'s (1991) system the anterior and posterior area involvement are added together and then averaged. Again, a drawback of this system is that objective intensity and extent scores assessed by the observer and subjective symptoms assessed by the patient are combined in the overall disease activity score. This may lead to an increase in interobserver variation and to difficulties in interpreting changes in severity scores over time.

Sowden *et al.* (1991) presented a scoring system in which intensity items, extent scores, and subjective symptoms are assessed separately and not combined in an overall AD severity score. An advantage of this system is that the assessment of the extent of the disease is simple, being according to the "rule of nine," as is used in the estimation of the extent of burned skin. The extent of AD in this system is measured on each defined site by estimating whether none, a third, two thirds, or all of the skin area is affected. The sum of the scores from each site gives a percentage assessment of the total body area affected by AD. Unfortunately, we do not yet know if this scoring system has been validated.

Bahmer *et al.* (1991) presented the Atopic Dermatitis Area and Severity Index, the ADASI scoring system, which focuses mainly on assessment of the extent of AD in which they introduced a grid reference method in order to reduce the inaccuracies in assessment. However, as this method is complicated, it may lead to a reduced reproducibility.

As the majority of the available scoring systems are not standardized and consensus concerning the assessment of the severity of AD has not been reached, the ETFAD (1993) introduced the Scoring Atopic Dermatitis index, SCORAD, which scores intensity items, the extent, and subjective symptoms such as pruritus and sleep loss. Each item is evaluated at its average intensity. The overall AD severity score is mathematically assessed as the sum of these scores. Although the SCORAD was developed primarily for severity scoring in children, it has also been used for adults. Designed to to be used for research as well as for clinical purposes, the SCORAD is supposed to be comprehensive and reliable. However, several of its features, including its validity, accuracy, sensitivity, and biologic relevance were not fully established when the system was introduced. In contrast to the claim of ETFAD the SCORAD appears to be time-consuming and complicated and, therefore, is at risk for interobserver variation. We (Sprikkelman *et al.*, 1997) demonstrated good agreement in the assessment of overall AD severity between observers in the lower and higher scoring rates using the SCORAD, whereas we found more variation in the middle scoring rates. Moreover, we showed a statistically significant interobserver variation in assessing the isolated intensity items edema/papulation, erythema, and excoriations. To our surprise, no significant interobserver variation was found in assessing the extent of AD. Interobserver variation was suspected, because it is known that assessing the extent of AD is subject to variability.

On behalf of the ETAC Study Group, Oranje *et al.* (1997) presented a training program to standardize the objective SCORAD scoring of investigators and to minimize interobserver variation. However, considerable interobserver variation remained after training with regard to several scoring items. This standardization procedure was carried out by investigators with and without dermatological experience. After training, edema/papulation appeared to be the least under- and overscored item by investigators without dermatological experience compared to investigators with dermatological experience. In contrast, lichenification and oozing were the most overscored and erythema the most underscored intensity item. Between investigators with and without dermatological experience a significant difference was found in scoring erythema, which was better scored by the dermatologically experienced investigators. No significant differences were found in scoring the other intensity items. However, it must be noted, that the standard deviations of the scorings in both groups, nondermatologically experienced and dermatologically experienced investigators, are substantial. Furthermore, a significant difference in scoring the extent of AD was found between the two groups of observers.

On behalf of the European Task Force on Atopic Dermatitis, Kunz *et al.* (1997) recently presented clinical validation and guidelines for the SCORAD index. In contrast with our results (Sprikkelman *et al.*, 1997) they found significant interobserver variation in the assessment of the extent of AD, mostly for patients with lesions of moderate intensity involving 20 to 60% of body surface. The reason for this variation is probably due to ill-defined borders of AD lesions, especially in patients with low-intensity AD, but significant body surface involvement (Kunz *et al.*, 1997). Moreover, they demonstrated significant interobserver variation in the assessment of the intensity items oozing/crusts, and lichenifications. Variation in scoring these items is probably due to differences in the choice of the most representative areas in which the severity is assessed.

Kunz *et al.* (1997) proposed grading the severity of AD according to objective criteria only, because the severity of subjective symptoms has been proven to be a cause of large variations in the overall AD severity score. The proposed severity scale can be used to define patients as having mild, moderate, or severe AD. A scoring item for functional impairment was also included in this proposal, in which the interference of AD lesions with normal daily life activities, such as perlèche, fissures on the hands or feet, or genital involvement, is taken into account.

Kunz *et al.* (1997) now advises using the cumulative SCORAD index in clinical trials, i.e., the overall score, to evaluate patients once the objective criteria have been taken account. Still, it needs to be emphasized that training is necessary to assess the severity of AD using the SCORAD.

Verwimp *et al.* (1995) and our group (Sprikkelman *et al.*, 1997) presented the Basic Clinical Scoring System (BCSS), a very simple, fast, and easy to execute scoring system with excellent agreement between observers in the assessment of the overall severity of AD. The BCSS scores the extent of AD in a simple way by evaluating the number of lesion sites. This system has proved its usefulness in an outpatient clinic, in baby health clinics, and in clinical research. A disadvantage of the BCSS is the lack of scoring intensity items and subjective symptoms. For this reason we recommended that it be used in long-term follow-up studies on the development of allergic disorders, when only a rough estimation of the severity of AD is needed.

Berth-Jones (1996) evolved the Six Area, Six Sign Atopic Dermatitis severity score (SASSAD), which is proposed principally as a severity score to assess response to treatment in clinical trials. It has been widely used in single-center and multicenter trials, and has proved effective in clinical trials of topical and systemic therapies and in an epidemiological study. Moreover, the SASSAD has also proved to be useful in monitoring the progress of AD in a clinical setting. The SASSAD is simple and quick to perform, and can be used in adults and children. An advantage of this system is that symptoms and signs are scored separately in order to reduce interobserver variation. Moreover, in order to reduce interobserver variation in the assessment of the extent of AD, only the lesion sites are scored.

Most of the AD scoring systems described herein only focus on signs, symptoms, and extent of the disease. Hower, assessement of other parameters such as topical steroid usage, response to treatment, exacerbation or remission rates of the disease, and impact of the disease on the patient will also give information about the severity of AD. In some scoring systems attention has been paid to the evaluation of some of these parameters. Grandlund *et al.* (1995) and Munro *et al.* (1994) both included the requirement for topical steroid as parameter apart from the intensity, extent, and subjective scores. In the studies of van Joost *et al.* (1994) and Hanifin *et al.* 1993), observers as well as the patients assessed the severity of AD. Vernon *et al.* (1991) assessed the exacerbation and remission rates in order to evaluate the improvement of AD during a clinical trial. Jekler and Larko (1990) also evaluated the improvement of AD by assessing the healing rate. Bjørneboe *et al.* (1987) and Soyland *et al.* (1994) included a parameter for assessing the effect of AD on normal living.

3. Conclusions and Recommendations

Assessing the severity of AD is difficult. At present no validated (serological) tests are available. A number of scoring systems are being used for clinical and research purposes, but using them to assess the severity of AD is difficult. Even after training, and by dermatologically experienced physicians, there is clearly considerable interobserver variation in assessments. Because owing to the considerable number of available scoring systems, it is not easy to decide which one to use. The choice of a certain scoring system has to depend on its particular purpose, its feasibility, its accuracy, its sensitivity, and whether or not the system has been validated. A scoring system with a minimum intraobserver and interobserver variation is to be preferred. At the present few of the available AD scoring systems have been standardized.

We consider that a warning is necessary, and that some recommendations might be helpful in making a judicious choice. When only a rough estimate of the severity of AD is needed, as in long-term follow-up studies, a more simple, rapid, and reproducible scoring system will suffice. For detection of subtle modifications in the severity of AD a more elaborating system seems more appropriate. However, one should be aware of significant interobserver variation in the assessment of isolated intensity items, the extent, as well as in subjective patient symptoms. We advise using a scoring system in which the interobserver variation is minimized, and this can be accomplished by a system in which the intensity items, the extent, and the subjective symptoms are scored separately. For clinical trials it is necessary to use a scoring system designed primarily for that purpose. Furthermore, for research purposes we recommend that the AD severity scores be assessed by a single observer, in order to avoid interobserver variation.

References

Bahmer, F. A. Schafer, J., and Schubert, H. J. 1991, Quantification of the extent and the severity of atopic dermatitis: The ADASI score [letter], *Arch. Dermatol.* **127:**1239–1240.

Berth-Jones, J. 1996, Six area, six sign atopic dermatitis (SASSAD) severity score: A simple system for monitoring disease activity in atopic dermatitis, *Br. J. Dermatol.* **135:**Suppl 48:25–30.

Berth-Jones, J., Thompson, J., and Graham-Brown, R. A. C., 1995, Evening primrose oil and atopic eczema. *Lancet* **345:**520.

Berth Jones J., Finlay A. Y., Zaki I., Tan B., Goodyear H., Lewis-Jones S., Cork M. J., Bleehen S. S., Salek, M. S., Allern, B. R., Smith S., and Graham-Brown, R. A., 1996, Cyclosporine in severe childhood atopic dermatitis: A multicenter study, *J. Am. Acad. Dermatol.* **34**(6):1016–1021.

Bjørneboe A., Soyland, E., Bjrneboe G. E., Rajka, G., and Drevon, C. A., 1987, Effect of dietary supplementation with eicosapentaenoic acid in the treatment of atopic dermatitis. *Br. J. Dermatol.* **117**(4):463–469.

Costa, C., Rilliet, A., Nicolet, M., and Saurat, J. H., 1989, Scoring atopic dermatitis: The simpler the better? *Acta. Derm. Venereol.* **69:**41–45.

Czech, W., Krutmann, J., Schopf, E., and Kapp, A., 1992, Serum eosinophil cationic protein (ECP) is a sensitive measure for disease activity in atopic dermatitis, *Br. J. Dermatol.* **126:**351–355.

Diepgen, T. L., and Fartasch, M., 1992, Recent epidemiological and genetic studies in atopic dermatitis, *Acta. Derm. Venereol.* **176:**13–18.

ETFAD, 1993, Severity Scoring of atopic dermatitis: The SCORAD index, *Dermatology* **186**:23–31.

Finlay, A. Y., 1996, Measurement of disease activity and outcome in atopic dermatitis, *Br. J. Dermatol.* **135:**509–515.

Granlund, H., Erkko, P., Sinisala, M., and Reitamo, S., 1995, Cyclosporine in atopic dermatitis: Time to relapse and effect of intermittent therapy. *Br. J. Dermatol.* **132**(1):106–112.

Hanifin, J. M., 1989, Standardized grading of subjects for clinical research studies in atopic dermatitis: Workshop Report. *Acta Derm. Venereol.* **144:**28–30.

Hanifin, J. M., and Rajka, G., 1980, Diagnostic features of atopic dermatitis, *Acta Derm. Venereol.* **92:**44–47.

Hanifin, J. M., Schneider, L. C., Leung, D. Y., Ellis, C., Jaffe, H. S., and Izu, A. E., 1993, Recombinant interferon gamma therapy for atopic dermatitis, *J. Am. Acad. Dermatol.* **28:**189–197.

Harper, J. I., Mason, U. A., White, T. R., Staughton, R. C. D., and Hobbs, J. R., 1991, A double blind placebo controlled study of thymostimulin (TP-1) for the treatment of atopic eczema, *Br. J. Dermatol.* **125:**368–372.

Jekler J., and Larko, O., 1990, Combined UVA-UVB versus UVB phototherapy for atopic dermatitis: a paired comparison study, *J. Am. Acad. Dermatol.* **22**(1):49–53.

van Joost, T., Heule, F., Korstanje, M., van den Broek, M. J., Stenveld, H. J., and van Vloten, W. A., 1994, Cyclosporin in atopic dermatitis: A multicentre placebo controlled study. *Br. J. Dermatol.* **130**(5)634–640.

Kagi, M. K., Joller-Jemelka, H., and Wuthrich, B., 1992, Correlation of eosinophils, eosinophil cationic protein and soluble interleukin-2 receptor with the clinical activity of atopic dermatitis, *Dermatology* **185:**88–92.

Kanwar, A. J., Dhar, S., and Kaur, S., 1991, Evaluation of minor clinical features of atopic dermatitis, *Pediatr. Dermatol.* **8:**114–116.

Kapp, A., Czech, W., Krutmann, J., and Schopf, E., 1991, Eosinophil cationic protein in sera of patients with atopic dermatitis, *J. Am. Acad. Dermatol.* **24:**555–558.

Kunz, B., Oranje, A. P., Labreze, L., Stalder, J. F., Ring, J., and Taieb, A., 1997, Clinical validation and guidelines for the SCORAD index: Consensus report of the European Task Force on Atopic Dermatitis, *Dermatology* **195**:10–19.

Munro, C. S., Levell, N. J., Shuster, S., and Friedmann, P. S., 1994, Maintenance treatment with cyclosporin in atopic eczema, *Br. J. Dermatol.* **130**(3):376–380.

Nagaraja, Kanwar, A. J., Dhar, S., and Singh, S., 1996, Frequency and significance of minor clinical features in various age-related subgroups of atopic dermatitis in children. *Pediatr. Dermatol.* **13**:10–13.

Oranje, A. P., Stalder, J., Taieb, A., Tasset, C., and de Longueville, M., 1997, Scoring of atopic dermatitis by SCORAD using a training atlas by investigators from different disciplines, *Pediatr. Allergy Immunol.* **8**:287–34.

Queille, C., Pommarede, R., and Saurat, J. H., 1984, Efficacy versus systemic effects of six topical steroids in the treatment of atopic dermatitis of childhood, *Pediatr Dermatol.* **1**:246–253.

Queille-Roussel, C., Raynaud, F., and Saurat, J. H., 1985, A prospective computerized study of 500 cases of atopic dermatitis in childhood, *Acta Derm. Venereol.* **114**:87–92.

Rajka, G., and Langeland, T., 1989, Grading of the severity of atopic dermatitis, *Acta Derm. Venereol.* **144**:13–14.

Rudzki, E., Samochocki, Z., and Rebandel, P., 1994, Frequency and significance of the major and minor features of Hanifin and Rajka among patients with atopic dermatitis, *Dermatology* **189**:41–46.

Sowden, J. M., Berth-Jones, J., Ross, J. S., Motley, R. J., Marks, R., Finlay, A. Y., Salek, M. S., Graham-Brown, R. A. C., Allen, B. R., and Camp, R. D. R., 1991, Double blind, controlled, crossover study of cyclosporin in adults with severe refractory atopic dermatitis, *Lancet* **338**:137–140.

Soyland, E., Lea, T., Sanstad, B., and Drevon, A., 1994, Dietary supplementation with very long chain *n*3 fatty acids in man decreases expression of the interleukin-2 receptor (CD25) on mitogen stimulated lymphocytes from patients with inflammatory skin disease, *Eur. J. Clin. Invest.* **24**(4):236–242.

Sprikkelman, A. B., Tupker, R. A., Burgerhof, H., Schouten, J. P., Brand, P. L. P., Heymans, H. S. A., and van Aalderen, W. M. C., 1997, Severity scoring of atopic dermatitis: A comparison of three scoring systems, *Allergy* **52**:944–949.

Stalder, J. F., and Taieb, A., 1993, Severity scoring of atopic dermatitis: The SCORAD index, *Dermatology* **186**:23–31.

Tada, J., Toi, Y., Akiyama, H., and Arata, J., 1994, Infra-auricular fissures in atopic dermatitis, *Acta Derm. Venereol.* **74**:129–131.

Vernon, H. J., Lane, A. T., and Weston, W., 1991, Comparison of mometasone furoate 0.1% cream and hydrocortisone 1.0% cream in treatment of childhood atopic dermatitis, *J. Am. Acad. Dermatol.* **24(4)**:603–607.

Verwimp, J. J. M., Bindels, J. G., Barents, M., and Heymans, H. S. A., 1995, Symptomatology and growth in infants with cow's milk protein intolerance using two different whey-protein hydrolysate based formulas in a Primary Health Care setting. *Eur. J. Clin. Nutr.* **49** Suppl 1:S39–48.

Williams, H. C., and Pembroke, A. C., 1996, Infraorbital crease, ethnic group, and atopic dermatitis, *Arch. Dermatol.* **132**:51–54.

Williams, H. C., Burney, P. G. J., Hay, R. J., Archer, C. B., Shipley, M. J., Hunter J. J. A., Bingham, E. A., Finlay, A. Y., Pembroke, A. C., Graham-Brown, R. A. C., Atherton, D. A., Lewis-Jones, M. S., Holden, C. A., Harper, J. I., Champion, R. H., Poyner, T. F., Launer, J., and David, T. J., 1994, The U. K. Working Party's Diagnostic Criteria for Atopic Dermatitis. I: Derivation of a minimum set of discriminators for atopic dermatitis, *Br. J. Dermatol.* **131**:383–396.

Chapter 11

A Metric for Acneiform Eruptions of the Face

Investigations into the Paradox of the Ordinal Scale of Greater Utility and the Interval Scale of Greater Accuracy*

JONATHAN K. WILKIN

1. Introduction

There are multiple contemporary systems for grading and quantifying acneiform eruptions of the face. Clinical pharmacological studies of antiacne drugs incorporate methodology designed to enhance the differences between drugs and placebo. In the clinical setting patients and physicians desire a system for monitoring the progress of the antiacne therapy. However, in both the clinical pharmacological studies and the patient care settings, the contemporary measurement scales of the acneiform eruptions of the face do not directly derive from essential insights from mathematics, statistics, and psychometrics.

Further, the methodology of antiacne clinical pharmacological studies may incorporate significant biases. Inclusion criteria usually emphasize the more severe cases, so that "improvement" can be more easily demonstrated. Exclusion criteria keep mild cases out of studies because improvement is more difficult to measure. Since acne is a process that waxes and wanes unpredictably, and apparently

*Presented in part at the 95th Annual Meeting of the American Society for Clinical Pharmacology and Therapeutics, New Orleans, Louisiana, March 30–April 1, 1994. This study was conducted at the Ohio State University. The opinions stated in this paper reflect the views of the author only and are not official FDA views.

Jonathan K. Wilkin • Division of Dermatologic and Dental Drug Products, Food and Drug Administration, Rockville, Maryland 20851.

Cutaneous Biometrics, edited by Doris A. Schwindt and Howard I. Maibach, Kluwer Academic/Plenum Publishers, New York, 2000.

spontaneously, exclusion of mild cases and inclusion of severe cases often leads to "improvement" with placebo.

The problem of how and what to measure in the assessment of facial acneiform eruptions remains acute and real. Practitioners and patients need the same metric for effective communication and patient–physician relationships. In the clinical setting the patient who has a reduction of 60 lesions from an initial 100 lesions may not be twice as happy about the therapeutic progress as a patient who has a reduction of only 30 lesions from an initial 60. Practitioners are confronted with clinical pharmacological studies of antiacne drugs that demonstrate statistical significance; however, they need studies that accurately describe the clinical significance. Improvement in antiacne drug studies is demonstrated by fewer lesions, either as an absolute number or as a percent reduction from the initial number, and/or a change in an index of global assessment (often an ordinal scale from 0 to 4+). The results from such trials often do not contain information that a physician and a patient might want to know about choosing a treatment (Feinstein, 1977).

The process of measurement is the process of mapping empirical relations into a formal, mathematical model (Stevens, 1958). Weaker forms of measurement scales are abandoned only when stronger forms are discovered. Although discovery of the appropriate measurement procedure is through experimentation (Stevens, 1958), many of the scales used in dermatology are derived by clinicians without the explicit knowledge of psychometric principles or experimentation (Daniels, 1968). Since patients with facial acneiform eruptions and dermatologists act not on reality but on their *perception* of reality, it is important to assess the mathematical, statistical, and psychometric basis for the grading and quantifying of the severity of acneiform eruptions of the face.

The assessment of acneiform eruptions of the face can incorporate each of the four possible measurement scales (in ascending order of mathematically powerful models): nominal, ordinal, interval, and ratio (Stevens, 1958) (Table 1). Distinguishing comedonal acne from cystic acne utilizes a nominal scale. Ranking severity of acne from Grade 4+ down to Grade 0 utilizes an existential-ordinal scale, which

TABLE 1. Scales of Measurement of Acneiform Eruptions[a]

Scale	Empirical operation	Example	Permissible statistics
Nominal	Determination of equality	Assignment of type, e.g., comedones, pustules, etc.	Number of cases
Ordinal	Determination of greater or lesser	Rank ordering of grade of acne, e.g., 0 to 4+	Median, Percentiles, Order correlation
Interval	Determination of the equality of intervals or of differences	Actual number of acneiform lesions or decrease in number (with therapy)	Mean standard deviation, Order correlation, Product moment
Ratio	Determination of the equality of ratios	Numerosity of acneiform lesions	Geometric mean, Harmonic mean, Percent variation

[a]Adapted from Steven (1958).

serves the dual purpose of indicating both existence and severity (Feinstein, 1978). The interval scale involves the actual numbers of lesions in the assessment. Although numbers of lesions confer both unbiasedness and precision, if a patient has 60 pustules and three are resolved, neither the clinician nor the patient are as gratified as when a patient with 3 pustules has complete resolution. The raw interval scale for both is the same, −3, but the perceived improvement is very different.

Accordingly, clinical pharmacological studies of antiacne drugs typically avoid a raw interval scale. Since lesion counts are regarded as more scientific than an ordinal scale of grade 0 to 4+, the change in number of acneiform lesions during treatment is often expressed as a percentage of the initial number of acneiform lesions (Eady *et al.*, 1990). The relationship between the initial number of acneiform lesions and the number that resolve with treatment contains the "initial value bias" when analyzed inappropriately (Gill *et al.*, 1985). The deceptive consequences of regression, comparing numerical improvement with the initial value, are elaborated at length elsewhere (Gill *et al.*, 1985; Blomqvist, 1977; Oldham, 1968). The apparent greater utility of an ordinal scale, despite the enormous precision and unbiasedness of counting lesions for the seemingly less useful interval scale, provides a metrical paradox for acneiform eruptions of the face.

This apparent paradox may be explicable in the context of psychometrics. According to the paradigm of the psychometric model for absolute judgment, as the amount of input information is increased, the observer will make more and more errors. When the amount of input information is increased, the transmitted information will increase at first and will eventually level off at some asymptotic value, the "channel capacity" for the observer (Miller, 1956). In psychometric studies on the discrimination of numbers of items, below seven items the subjects rarely make any errors and above seven they frequently do (Kaufman *et al.*, 1949). Below seven items, subjects are said to *subitize*, and above seven they are said to *estimate*. Some workers postulate about 20 or 30 distinguishable categories of numerousness (Miller, 1956). In estimating acneiform lesions on the face, there may be two dimensions of numerousness, area and density. When the subject can subitize, area and density are not significant variables, but when the subject must estimate, area and density can become significant (Miller, 1956).

The following study was undertaken to characterize the disparity between the actual and the perceived number of acneiform facial lesions. A further goal was to understand the basis of the paradox between the apparent greater utility of the ordinal scale despite the apparent greater precision and unbiasedness of the interval scale. Finally, the information obtained in this study could be the basis for the construction of a new metric for acneiform lesions, based on mathematical, statistical, and psychometric concepts, as a paradigm for future clinical pharmacology studies and clinical assessment in the patient care setting.

2. Materials and Methods

Thirty-three subjects responded to announcements both on bulletin boards at the medical center and in the residence halls on campus and in the local student newspaper. The protocol and the announcements were approved by the local institutional review board. The subjects were escorted to a private room, where they remained for the duration of the experiment. Subjects were told that they would evaluate a standard series of kodachromes for the severity of acneiform eruptions of the face. The subjects were shown a face with no acneiform lesions and told that it was "clear." The subjects were then shown a kodachrome of a face with the maximal number of acneiform lesions in the series tested (either 101 or 200) and told that it was "as bad as can be."

The acneiform lesions were painted by a professional theatrical cosmetic artist on two normal human subjects who did not have any naturally occurring acneiform lesions. Although the intention was to create numbers of acneiform lesions from 0 to 100 in the first model, an error in counting by the artist resulted in one additional lesion over the last half of the sequence terminating at 101. The second model had lesion numbers created from 1 to 200. The 0 to 101 series was always shown before the 1 to 200 series to each test subject. Thus, the sequence was 0, which was described as "clear" (Fig. 1), and 101, which was described as "bad as can be" (Fig. 2). This was followed by 24 kodachromes with the following lesion numbers: 28, 6, 59, 2, 40, 9 (Fig. 3), 81, 20, 49 (Fig. 4), 4, 91, 45, 12, 36, 73, 1, 53, 16, 101, 24 (Fig. 5), 0, 65, 8, 32. The subjects were then told that the second test would be performed, and the kodachromes of acneiform lesions painted on the second model were shown. The first slide was 0, which was described as "clear" (Fig. 6), and the second slide was 200, which was described as "as bad as can be" (Fig. 7). Varying lesion numbers were shown in the following sequence: 50 (Fig. 8), 6, 120, 2, 80, 10, 160, 30, 100 (Fig. 9),

FIGURE 1. The first model without lesions, described to observers as "clear."

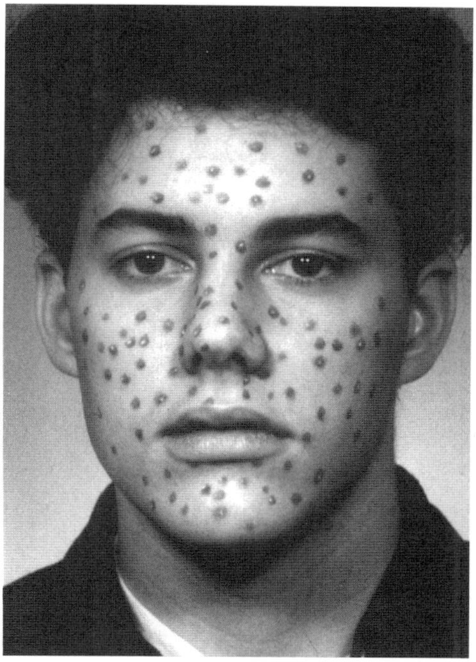

FIGURE 2. The first model with 101 lesions, described to observers as "as bad as can be."

FIGURE 3. The first model with 9 lesions.

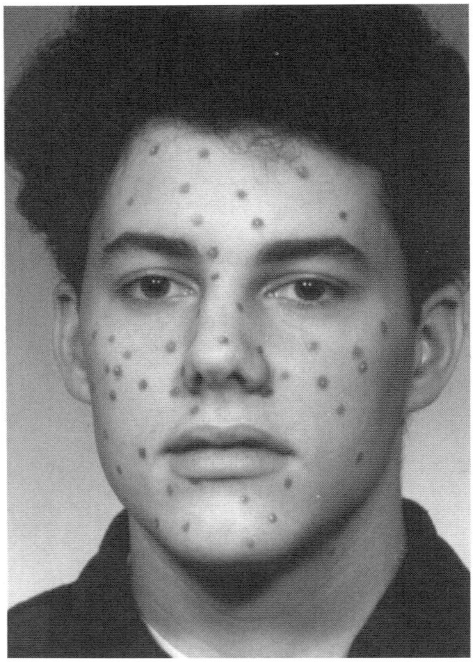

FIGURE 4. The first model with 49 lesions.

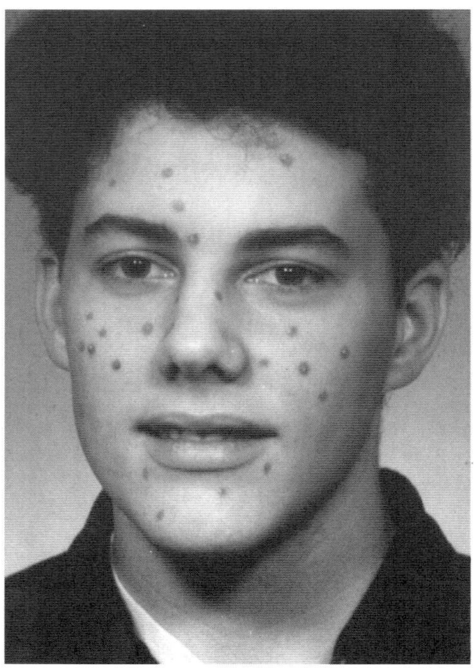

FIGURE 5. The first model with 24 lesions.

FIGURE 6. The second model without lesions, described to observers as "clear."

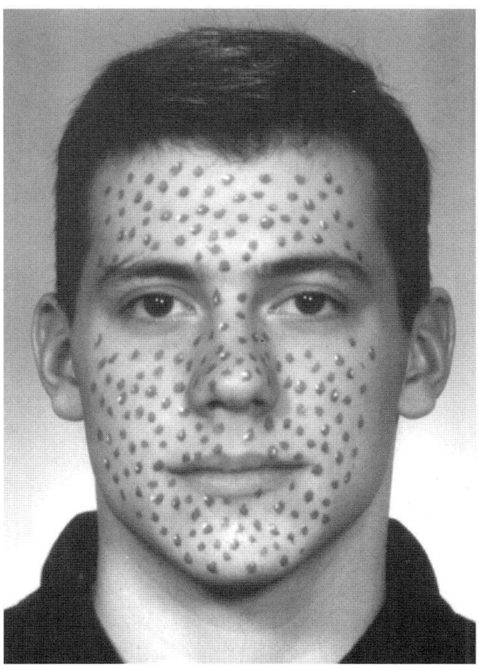

FIGURE 7. The second model with 200 lesions, described to observers as "as bad as can be."

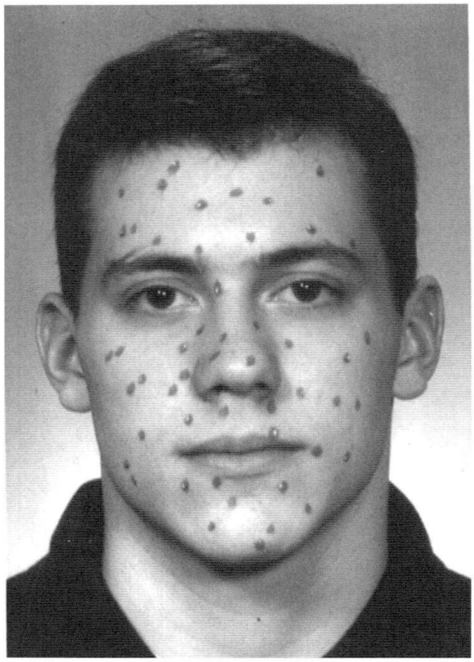

FIGURE 8. The second model with 50 lesions.

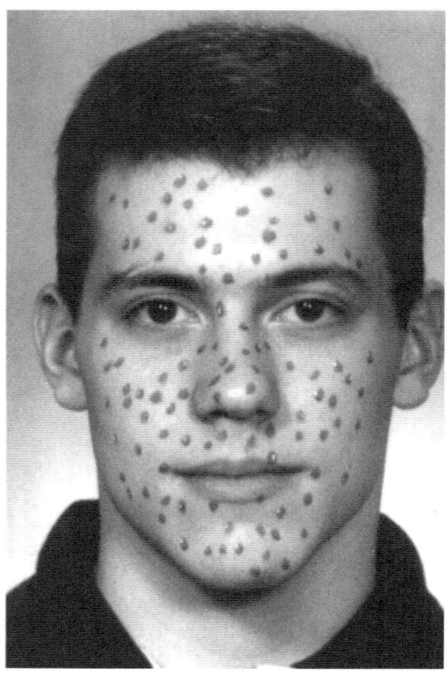

FIGURE 9. The second model with 100 lesions.

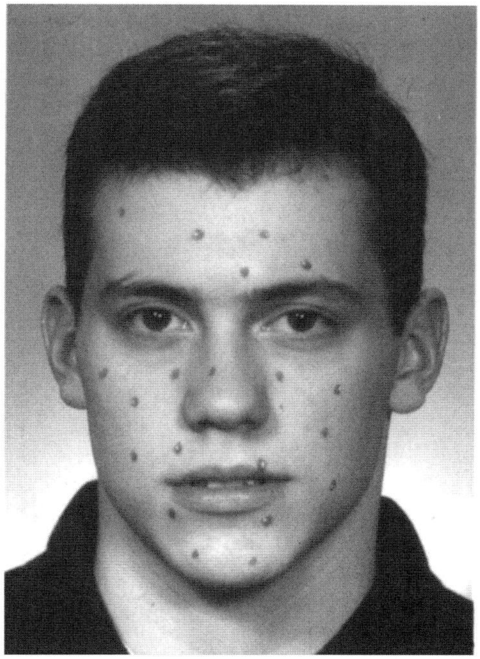

FIGURE 10. The second model with 20 lesions.

4, 180, 90, 15, 70, 145, 1, 110, 20 (Fig. 10), 200, 40, 0, 130, 8, and 60. The subjects had 30 sec to view each slide and estimate the lesion number on a horizontal visual analog scale. The 10-cm linear horizontal visual analog scale had 1-cm gradations (Sriwatanakul *et al.*, 1983), and the subjects were instructed to mark through the horizonal line at the point corresponding to the severity of the acne in the photograph. At the left of the horizontal visual analog scale was the word "CLEAR," and to the right of the scale were the words "AS BAD AS CAN BE." The data from one subject were discarded before analysis, after the subject expressed strong suspicion about the usefulness of such a study and was observed to be marking visual analog scales out of sequence.

The scoring of the visual analog scales was performed electronically by digimatic calipers (Model CD-4″P: 500-350, Mitutoyo Corp.) calibrated to an instrumental error of ± 0.02 mm with a resolution of 0.01 mm. According to the psychometric paradigm, the numbers of acneiform lesions on the kodachrome slides were the Input and the scores on the visual analog scales were the Output. Thus, for each Input there was a corresponding mean Output and standard deviation.

3. Results

The relationship between judged lesion numerosity (mean Output) and actual lesion number (Input) is not linear (Fig. 11). When mean Output is expressed as a

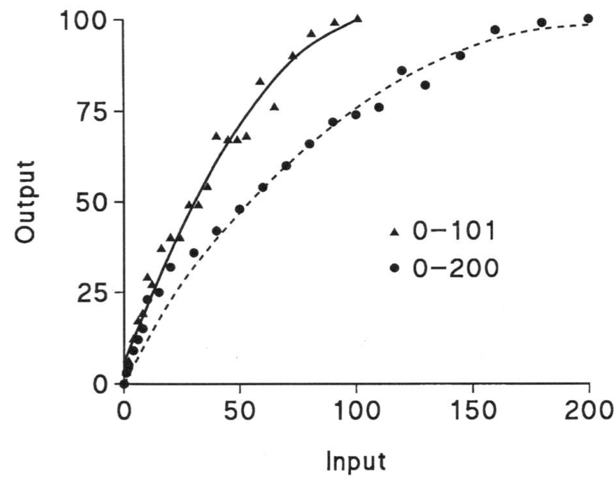

FIGURE 11. Judged lesion numerosity (Output), as millimeters deflection from zero on the visual analog scale, vs. actual number of lesions seen on the kodachromes (Input), showing a nonlinear relationship.

fraction of maximum Output, and Input is expressed as a fraction of maximum Input (I_{max}), the relationships between mean Output and Input are quite similar for both 0–101 and 0–200 series (Fig. 12). The standard deviations of the Output values fall from a plateau at midrange Input values to low standard deviations of Output values corresponding to Input values at either extreme (Fig. 13).

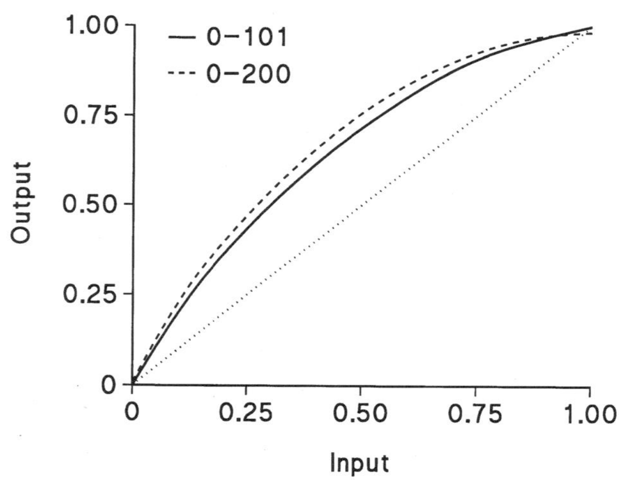

FIGURE 12. Judged lesion numerosity (Output) expressed as a fraction of maximum Output vs. actual number of lesions expressed as a fraction of the maximum number of lesions (Input), showing a curvilinear relationship.

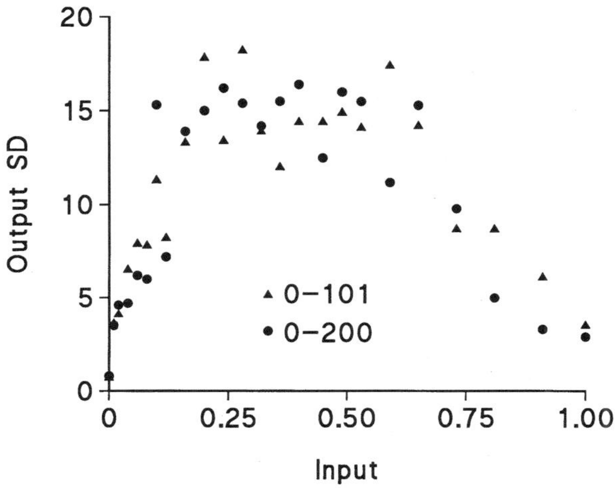

FIGURE 13. Standard deviations of Output values, as millimeters deflection from zero on the visual analog scale, vs. actual number of lesions expressed as a fraction of the maximum number of lesions (Input).

4. Comment

This experimental model is intentionally oversimplified. There is only one lesion type, of only one size and only macular. There are no surrounding halos of erythema, and all lesions are visible *en face*. Despite the absence of true acne lesions, the simplicity of the study permits an analysis of the judged numerosity of facial acneiform lesions.

The relationship between judged lesion numerosity (Output) and actual lesion number (Input) is curvilinear (Figs 11 and 12). Empirical formulas, obtained by observation (Fig. 14), permit useful simplifications of the data to facilitate the postulation of mental constructions in the processing of the Inputs to Outputs. Thus, when the actual lesion number is low (Input $\leqslant 0.25\,I_{\max}$), the judged lesion numerosity (Output) is twice the actual lesion number ($2I$). When the actual lesion number is intermediate ($0.25\,I_{\max} \leqslant$ Input $\leqslant 0.50\,I_{\max}$), the judged lesion numerosity (Output) is linearly related to the actual lesion number, but judged greater by 0.25 of $I_{\max}$ ($I + 0.25I_{\max}$). When the actual lesion number is high (Input $\geqslant 0.5\,I_{\max}$), the judged lesion numerosity (Output) is linearly related to half the actual lesion number, but judged greater by 0.5 of $I_{\max}$ ($I/2 + 0.5\,I_{\max}$).

Consistent with this construction of the cognitive processing of the Inputs to Outputs, the subjects seem to have conceptualized "clear" and "as bad as can be," permitting comparisons of all subsequent Inputs against one of these two endpoints. This process is known as anchoring, i.e., the adjustment from a known value (the anchor) in numerical estimation (Tversky and Kahneman, 1981). Thus, the closer to an endpoint (Inputs of 0 or $I_{\max}$), the lower the variation in the judged lesion number (Output) (Fig. 13). In fact, it is only for actual lesion numbers that are intermediate

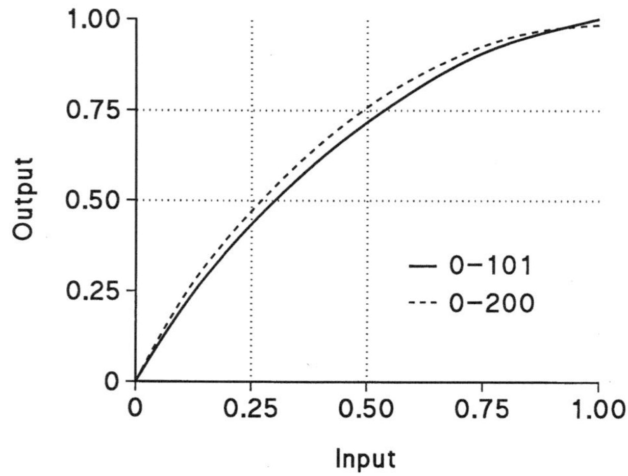

FIGURE 14. The same construction as Fig. 12 with empirical relationships identified for three segments of the curvilinear relationship between Output and Input.

($0.15\,I_{max} \leqslant$ Input $\leqslant 0.60\,I_{max}$) that Output SD rises above 10.0. Similarly, for actual lesion numbers that approach the endpoints, Input of 0 or I_{max} (Input $\leqslant 0.15\,I_{max}$ and Input $\geqslant 0.60\,I_{max}$), the Output SD does not exceed 10.0 (Fig. 13).

Thus, when the actual lesion number is low, the Input may be mentally evaluated as a worsening from the endpoint "clear." Similarly, when the actual lesion number is very high, the Input may be mentally evaluated as an improvement from the endpoint of "as bad as can be." When framed in terms of worsening and improvement from the respective endpoints, the value function proposed by Tversky and Kahneman (1974) predicts that the response to worsening (at low Input values) will be more extreme than the response to improvement (at high Input values).

The domain in which the actual lesion number (Input) is intermediate represents uncertainty as to which of these two cognitive rules to apply. Thus, this is the domain of the highest Output SD and corresponds to the transition between the two domains that approach the endpoints in which there is more "certainty" (lower Output SD), despite consistent biases (Output $= 2I$ for low I; Output $= I/2 + 0.5\,I_{max}$ for high I).

The relationship between judged lesion numerosity (Output) and the logarithm of the actual lesion number (log Input) is not linear. The values for Output rapidly approach a power function of the Input for Input values $\geqslant 0.25\,I_{max}$ (Fig. 15):

$$\text{Output} = 1.1215\,I^{0.583}$$

The value of the exponent, 0.583, suggests that in order for one actual acne lesion count to be considered twice as severe as another, that actual acne lesion count must be about 3.3 times as large. This is consistent with the clinical view that although it is worse to have 100 acne lesions than 50 acne lesions, it is certainly not perceived to be twice as bad.

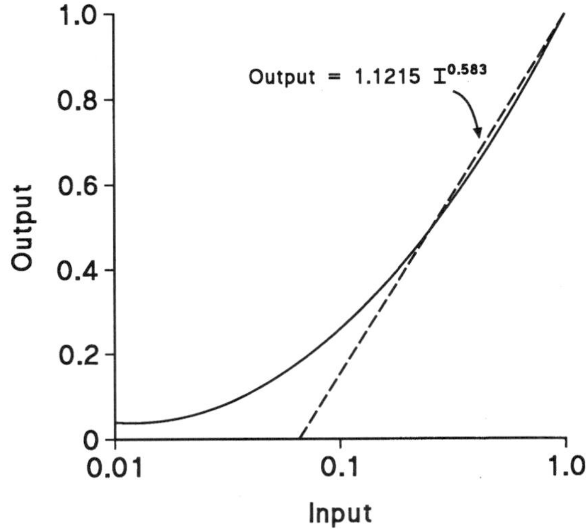

FIGURE 15. Judged lesions numerosity (Output) expressed as a fraction of maximum Output vs. the logarithm of the actual number of lesions expressed as a fraction of the maximum number of lesions (Input) for both model series combined. A power function is approached at higher lesion numbers.

An understanding of the relationship between judged lesion numerosity (Output) and the actual lesion number (Input) is most useful for actual lesion numbers ($\geqslant 40$) encountered in clinical practice and clinical pharmacological trials. Two empirical formulas would be required for actual lesion numbers over this range and would provide for a rather cumbersome metric. The power law operates at moderate to high values, but misses the lower values of actual lesion numbers that would be most useful. A single equation that estimates the judged lesion numerosity (Output) for actual lesion numbers (Input) up to 40 must approximate the two empirical formulas for their respective domains, yet it must be continuous throughout. One such possibility is based on the Michaelis relationship.

The Michaelis relationship identifies the Input that produces the half-maximal Output ($0.5I_{max}$). This value is a cognate of the Michaelis constant, and it may vary with the absolute value of maximal Input (I_{max}). The Michaelis relationship incorporates the notions of "as bad as can be" and "clear" as a self-evident construction. Thus, the Michaelis relationship both provides an equation for the useful range of actual lesion numbers ($\leqslant 40$) and infers the knowledge of both endpoints (0 and I_{max}, or "clear" and "as bad as can be," respectively).

For maximal Inputs of 101 and 200, the $0.5I_{max}$ are 0.27 and 0.25, respectively. Since the actual number of acneiform facial lesions regarded as maximal (I_{max}) determines both $0.5I_{max}$ and the relationship between Input and Output, I_{max} can be projected from the data as approximately 80 acneiform lesions (Fig. 16). At low (less than $0.5\,I_{max}$) lesion counts, the Michaelis relationship is quite good, but at higher lesion counts it is distorted as the maximal Output is approached asymptotically. Since most acne studies will include patients with lesion counts in the domain in

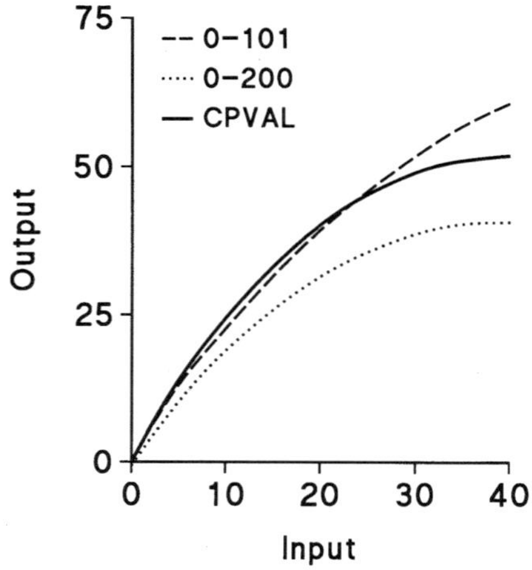

FIGURE 16. Calculated perceived value of acneiform lesions (CPVAL) utilizes the Michaelis relationship to approximate the relationship between judged lesion numerosity (Output) and actual number of lesions (Input) for both model series.

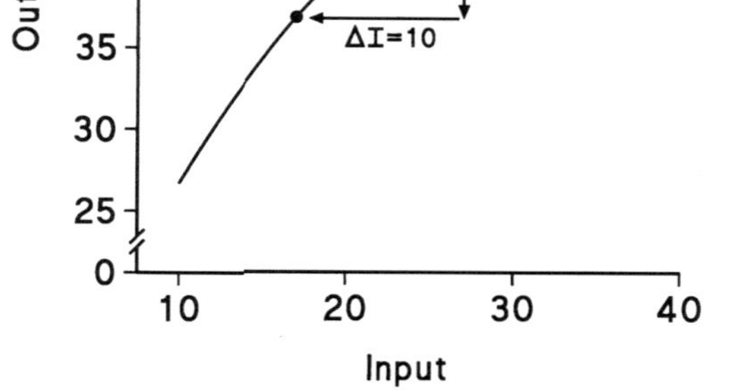

FIGURE 17. Calculated lesion numerosity (Output) by the CPVAL method for actual number of lesions (Input). Changes in actual number of lesions from 37 to 27 and 27 to 17 would give perceived values of decrease in severity of 5.9 and 9.2, respectively.

which the Michaelis relationship is a useful approximation, a new metric for acneiform lesions of the face may be proposed for actual lesion counts $\leqslant 40$:

$$\text{Calculated perceived value of acneiform lesions} = \frac{80 \times \text{Actual number of lesions}}{\text{Actual number of lesions} + 20}$$

Raw data (acne counts) should be transformed to the corresponding calculated perceived values of acneiform lesions (CPVAL) before further data analysis. The new, proposed metric for acneiform eruptions of the face would translate actual lesion count changes of 37 to 27 to a score of −5.9 and changes of 27 to 17 to a score of −9.2, respectively (Fig. 17). Thus, the perceived value of a decrease of 10 lesions from an initial value of 27 is perceived as approximately 50% better than a decrease from 37 to 27.

Although CPVAL does not translate lesion number exactly into an acne grading system, it has the cognate validity of conveying perceived severity. Further, CPVAL has significant reproducibility, since it is calculated directly from lesion counts, which are subject to minimal intraobserver and interobserver variability. In the American Academy of Dermatology Consensus Statement on the Classification of Acne (Pochi *et al.*, 1991), it was pointed out that "quantification of severity is difficult. Although it is possible to count and describe each lesion according to size, such detail is not feasible in day-to-day clinical practice." Thus, the reproducibility and validity (Charlson *et al.*, 1991) of CPVAL provide a useful bridge from the lesion counts in clinical pharmacologic studies of antiacne drugs to the grading systems employed in actual clinical practice.

Although the Michaelis relationship provides CPVAL as a quantitative transformation to a new metric for acneiform eruptions of the face, it is a tool primarily for the clinical pharmacological investigations of antiacne drugs. Most practitioners will agree that lesion counts are not cost-effective in the routine clinical setting. Thus, grading acne by a 0 to 4+ scale is more useful and has recently gained official American Academy of Dermatology support (Pochi *et al.*, 1991). It is important to note that the results of this study provide both a theoretical underpinning and observational support for the psychometric validity of such an ordinal scale.

Although Daniels and van der Leun (1968) believe that an ordinal scale is an attempt to get an interval scale, implying perceived constant intervals that are equally spaced grades throughout the Output domain, Burke and Cunliffe (1984) predicted that the precise relationship between grades is unlikely to be strictly linear. By analyzing the data obtained in this study with the Output domain equally divided into grades up to 4+, the ordinal scale may be seen as an empirical attempt at a ratio scale (100 is twice 50, 50 is twice 25, 25 is approximately twice 10) (Fig. 18). The most powerful model mathematically is the ratio scale (Stevens, 1958). Since the ordinal scale approximates the ratio scale and more accurately portrays the perceived severity of facial acneiform eruptions than the interval scale (actual lesion number count), the 0 to 4+ scale is not only more useful in the clinical setting, it is also more meaningful than raw lesion counts. Unfortunately, lesion counts are still more accurate.

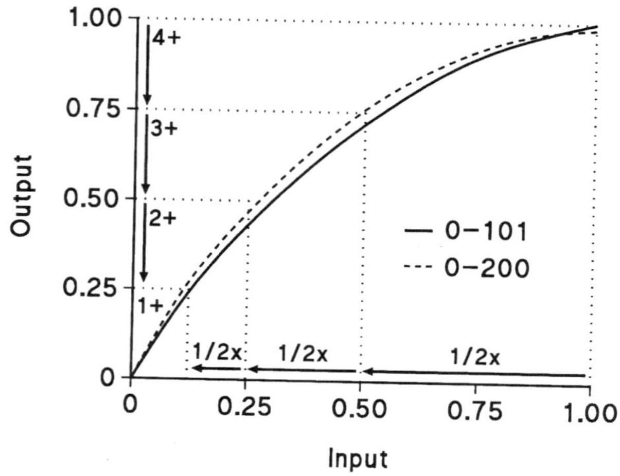

FIGURE 18. The same construction as Fig. 12 with the output domain divided into grades up 4+ and the Input domain subdivided by half values from maximum Input value.

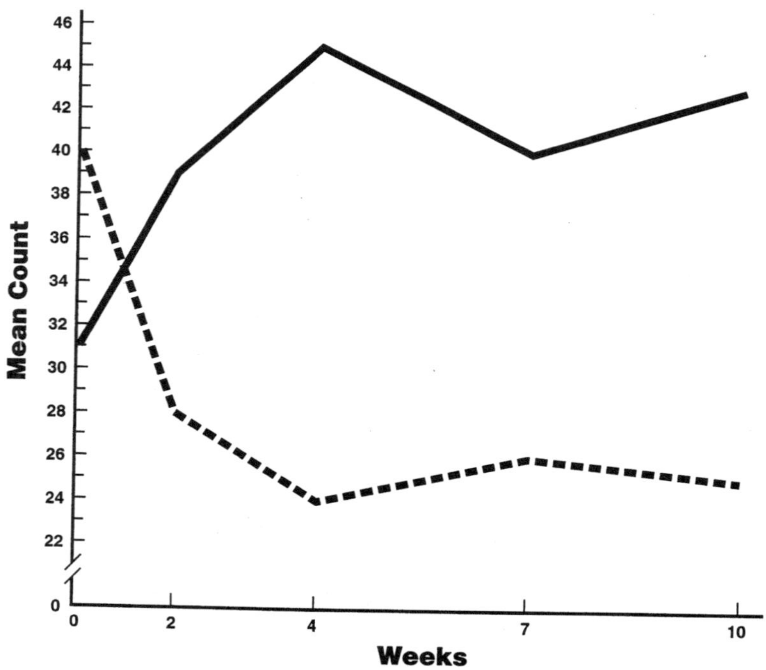

FIGURE 19. Mean comedonal lesion counts during therapy with drug A (dashed line) and drug B (solid line), in the study described in the text, presented graphically with a discontinuous ordinate.

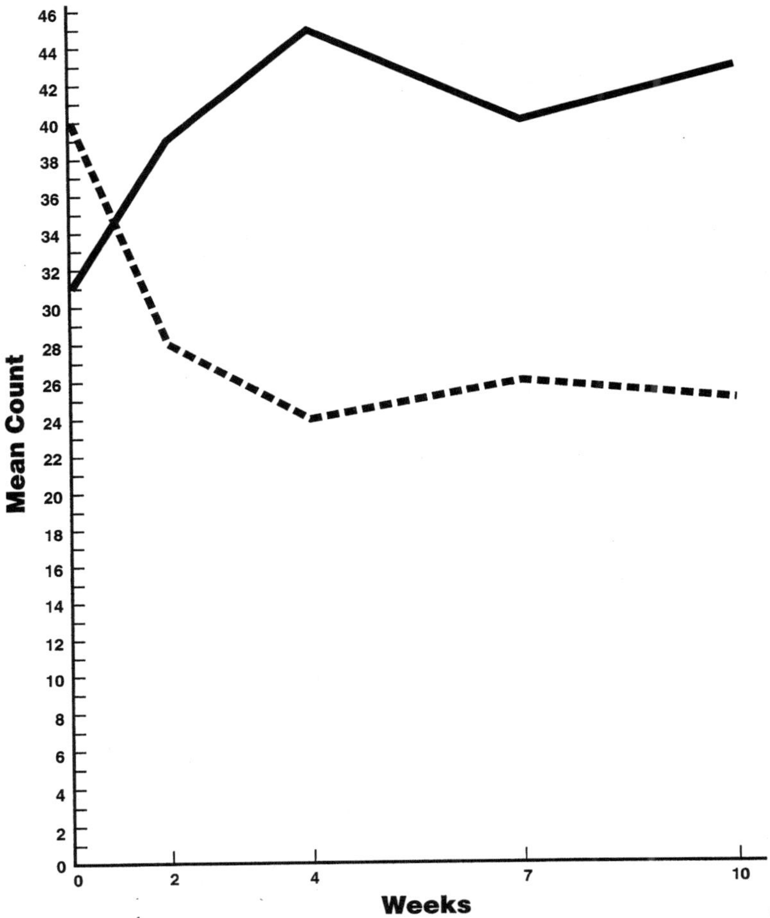

FIGURE 20. Mean comedonal lesion counts during therapy with drug A (dashed line) and drug B (solid line) presented graphically with a continuous ordinate.

The great difficulty is assessing the effectiveness of antiacne drugs and, *a fortiori*, in predicting the satisfaction of patients using these drugs is that a serviceable metric for facial acneiform eruptions that is accurate and conforms to clinical impressions of improvement does not exist. In this report I offer a metric to be employed in parallel with the customary grading methodologies to determine its utility. It is hardly a recondite fact that each of the extant grading methodologies seems to give up an important characteristic, such as accuracy, utility, or meaningfulness. Although lesion counts convey the notion of "hard data" and acne grading scales convey the notion of "soft data," it is far better to have an approximate answer to the right question, viz., perceived severity of acne, than an exact answer to the wrong question (Feinstein, 1977; Tukey, 1962). CPVAL offers an acneiform metric that may translate hard data into the right answer.

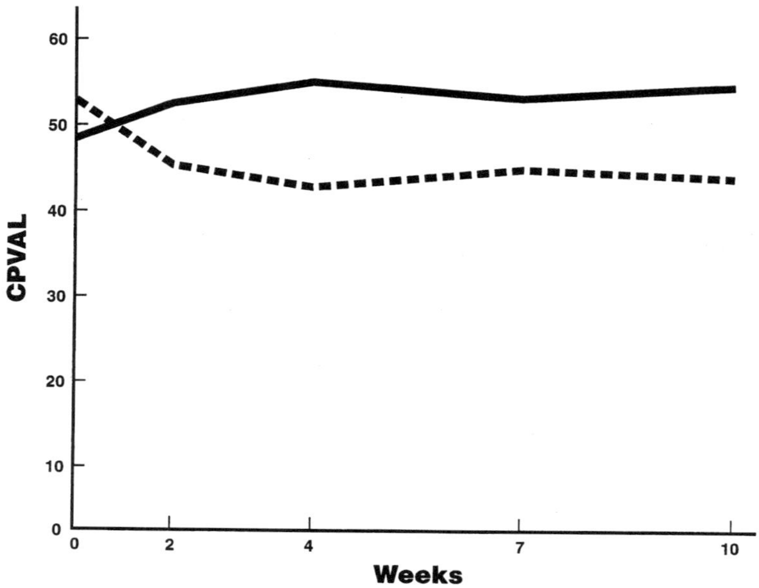

FIGURE 21. Calculated perceived value of acneiform lesions (CPVAL) calculated from mean comedonal lesion counts during therapy with drug A (dashed line) and drug B (solid line).

The CPVAL analysis and the traditional lesion count method can be compared in an example from unpublished data from a 10-week clinical study involving patients aged 23 to 50 years. The mean comedonal counts were determined at baseline (week 0) and weeks 2, 4, 7, and 10 for 60 patients randomly assigned to one of two topical treatment groups. The mean lesion counts for drug A were 39.0, 28.1, 24.0, 25.8, and 24.7 and for drug B 31.8, 39.4, 44.4, 40.0, and 43.4, for weeks 0, 2, 4, 7, and 10, respectively.

Often, lesion count data are presented graphically with a discontinuous ordinate (Fig. 19), which conveys the perception of greater than real differences in effect (Fig. 20). Ethically, data should be encoded by graphical methods so that visual decoding can be performed with greater accuracy (Cleveland *et al.*, 1985). Converting actual lesion counts to CPVAL with the proposed acneiform metric (Fig. 21) indicates that the perceived differences between these two therapies may be even less in the clinical setting than raw lesion counts would imply.

ACKNOWLEDGMENTS: Damita Peace, theatrical makeup artist with the Ohio State University, Department of Theatre crafted the lesions. Christopher and Nathaniel Wilkin, students at Dartmouth College, exhibited filial forbearance by participating as models. Hannah Lim, medical student at the Ohio State University, participated in data acquisition. Clinical study data were generously provided by Dermik Laboratories, Inc.

References

Blomqvist, N., 1977, On the relation between change and initial value, *J. Am. Statist. Assoc.* **72:**746–749.

Burke, B. M., and Cunliffe, W. J., 1984, The assessment of acne vulgaris—the Leeds technique, *Br. J. Dermatol.* **111:**83–92.

Charlson, M. E., Johanson, N. A., and Williams, P. G., 1991, Scaling, scoring, and staging, in: *Principles and Practice in Research: Strategies for Surgical Investigators*, 2nd Ed. (W. Troidl, W. O. Spitzer, B. McPeek, *et al.*, eds.), Springer-Verlag, pp. 192–200

Cleveland, W.S., and McGill, R., 1985, Graphical perception and graphical methods for analyzing scientific data, *Science* **229:**828–833.

Daniels, F., and van der Leun, J. C., 1968, Problems in quantifying skin reactions, *Arch. Dermatol.* **97:**553–565

Eady, E. A., Cove, J. H., Joanes, D. N., and Cunliffe, W. J., 1990, Topical antibiotics for the treatment of acne vulgaris: A critical evaluation of the literature on their clinical benefit and comparative efficacy, *J. Dermatol. Treat.* **1:**215–226.

Feinstein, A. R., 1977, Clinical biostatistics XLI: Hard science, soft data, and the challenges of choosing clinical variables in research, *Clin. Pharmacol. Ther.* **22:**485–498.

Feinstein, A.R., 1978, Clinical biostatistics XLV. The purposes and functions of criteria, *Clin. Pharmacol. Ther.* **24:**479–492.

Gill, J. S., Zezulka, A. V., Beevers, D. G., and Davies, P., 1985, Relation between initial blood pressure and its fall with treatment, *Lancet.* **i:**567–569.

Kaufman, E. L., Lord, M. W., Reese, T. W., and Volkmann, J., 1949, The discrimination of visual number, *Am. J. Psychol.* **62:**498–525.

Miller, G. A., 1956, The magical number seven, plus or minus two: Some limits on our capacity for processing information, *Psychol. Rev.* **63:**81–97.

Oldham, P. D., 1968, Deceptive consequences of regression, in: *Measurement in Medicin: The Interpretation of Numerical Data*, English Universities Press, London, pp. 148–152.

Pochi, P. E., Shalita, A. R., and Strauss, J. S., *et al.*, 1991, Report of the consensus conference on acne classification, *J. Am. Acad. Dermatol.* **24:**495–500.

Sriwatanakul, K., Kelvie, W., Lasagna, L., Calimlim, J. F., Weis, O. F., and Mehta, G., 1983, Studies with different types of visual analog scales for measurement of pain, *Clin. Pharmacol. Ther.* **34:**234–239.

Stevens, S. S., 1958, Measurement and man, *Science* **127:**383–389.

Tukey, J. W., 1962, The future of data analysis, *Ann. Math. Statist.* **33:**1–67.

Tversky, A., Kahneman, D., 1974, Judgment under uncertainty: Heuristics and biases, *Science* **185:**1124–1131.

Tversky, A., and Kahneman, D., 1981, The framing of decisions and the psychology of choice, *Science* **211:**453–458.

Chapter 12

Toward a New Psoriasis Severity Scoring Using Bioengineering Techniques

DORIS A. SCHWINDT AND HOWARD I. MAIBACH

1. Introduction

Psoriasis is a chronic remittent disease of unknown cause that is still not curable. These facts imply the need for new treatments. Unfortunately they also imply that assessing the severity and progress of the disease is problematic. Although advances are being made regarding genetic and pathogenic backgrounds of psoriasis (see Ortonne, 1996, and references therein) we are currently defining the endpoint, thus measuring the outcome of the disease. Reviewing the literature attests to the lack of a reliable measure.

Many clinical trials have evaluated the improvement of the disease resulting from treatment. However, the question what is to be improved, or what are the features of psoriasis that cause severity and which of these signs can be assessed, requires resolution. In their review of the assessment of disease progress in psoriasis Marks and Barton (1989) found great variation in the methods used. Although crucial for the outcome of their study many investigators did not discuss the methods available critically nor comment on those chosen. This complicates the objective judgment of the results, leading the researcher to draw misleading conclusions. A comparison of different clinical trials is therefore impossible.

2. The Psoriasis Area and Severity Index

The most frequently used method in its original or in modified versions is the psoriasis area and severity index (PASI), designed by Fredriksson and Pettersson

Doris A. Schwindt • Department of Dermatology, Rheinisch-Westfälisch Technische Hochschule Aachen, D-52074 Aachen, Germany. Present address: Hofmann-La Roche AG, D-796390 Grenzach, Wyhlen, Germany. *Howard I. Maibach* • Department of Dermatology, School of Medicine, University of California, San Francisco, California 94143-0989.

Cutaneous Biometrics, edited by Doris A. Schwindt and Howard I. Maibach, Kluwer Academic/Plenum Publishers, New York, 2000.

(1978) to evaluate the progress of psoriasis with oral retinoid therapy. PASI is calculated as follows: The areas of involvement of four main sites: the head (A_h), the trunk (A_t), the upper extremities (A_u), and the lower extremities (A_l) are given a numerical value: $0 =$ no involvement; $1 = < 10\%$; $2 = 10 < 30\%$; $3 = 30 < 50\%$; $4 = 50 < 70\%$; $5 = 70 < 90\%$, and $6 = 90–100\%$. For evaluation of the severity of psoriasis lesions erythema (E), infiltration (I), and desquamation (D) are assessed on a scale 0–4, where 0 means a complete lack of cutaneous involvement and 4 represents severest involvement. The sums of the severity ratings for E, I, and D are multiplied by the value of the areas involved and the percentages of the four body sites.

$$\begin{aligned} \text{PASI} = {} & 0.1(E_h + I_h + D_h)A_h + 0.3(E_t + I_t + D_t)A_t \\ & + 0.2(E_u + I_u + D_u)A_u + 0.4(E_l + I_l + D_l)A_l \end{aligned}$$

PASI varies in steps of 0.1 units from 0 to 72, 0 meaning no psoriatic lesions at all, and 72 meaning erythroderma. PASI is the sum of scored psoriatic lesions at four body sites that are judged by their size. As the affected area A is a factor within each body part, it is given highest priority. PASI was a step in the direction of scoring psoriasis more systematically and specifically, has been criticized for its subjectivity.

Assessing the area of involvement is especially subject to a wide interrater variability. Marks and Barton (1989) studied the reproducibility of the area of involvement as assessed by clinicians. Photographs of patients with chronic plaque psoriasis were assessed by clinicians and by computer-assisted planimetry. The clinician's and planimetric assessment differed significantly, and assessment even of experienced clinicians varied considerably. Ramsay and Lawrence (1991) compared the assessment of involved area, based on the rule of nine with computerized image analysis of whole-body photographs and image analysis of plaque tracings. Although none of the methods will measure the involved area accurately, the plaque tracing method is probably the most accurate. The scientists concluded that untrained observers cannot accurately estimate total involved skin surface area, whereas image analysis of photographs shows an accuracy similar to that obtained from traced plaque outlines. Bahmer (1989) suggested that point counting could be a step toward the solution of this PASI problem, but this technique has never been established in practice.

Savolainen *et al.* (1997) reported on a computer image analysis by a color segmentation method. This technique seemed to be reliable and practicable in estimating the area of involvement. However, the disadvantages are that the method is time-consuming and technically demanding. In their study the human eye assessment differed from image analysis assessment, especially when the involved area of psoriasis was under 30%. Overestimates were more common in agreement with the findings of Ramsay and Lawrence (1991). Tiling-Grosse and Rees (1993) investigated individuals' ability to assess disease area using schematic figure outlines. They found that the ability to estimate the area of involvement accurately strongly depends on the total area affected. They observed great inter- and intraobserver variation. Although training might improve the consistency of assessments, the degree of variance makes it unlikely that discrete differences in response to treatment will be detected. They commented that computer-assisted assessment of area might be complicated by artifacts relating to tar, dithranol, or PUVA treatments.

Marks and Barton (1989) stated that the area of involvement might be the single most important clinical parameter, but it is difficult to assess this parameter objectively. Clinical assessment turns out to be too subjective, whereas more objective measures are too troublesome to be used in practice. Nevertheless computer-assisted image analysis is the method of choice for future development of the assessment of area of involvement in psoriasis; digital optical and infrared images may be used to get objective and precise results; and image analysis has the capability of covering large body areas.

Although the area of involvement is undoubtedly important for absolute psoriasis severity, we believe that it is of secondary importance for the assessment of treatment efficacy (changes in severity). If a treatment is successful within randomly chosen areas, differences in the treatment effect between areas may be evaluated in a second step. The other clinical signs that are readily accessible for clinical evaluation and chosen in the PASI are erythema, scaliness, and infiltration. Although less problematic than area of involvement, erythema and scaliness have been criticized for their validity. Exum *et al.* (1996) reviewed the methodological issues of measuring the severity of psoriasis. They analyzed validity and reliability of the PASI, and recommended use of PASI with caution.

3. Assessing Severity with Bioengineering Techniques

In psoriasis, abnormal keratinocyte differentiation, hyperproliferation of keratinocytes, and infiltration of inflammatory components into the skin are discussed as the three pathogenic features of the disease. Different theories exist as to the activation and interaction of the immunocytes and the keratinocytes (Ortonne, 1996; Barker, 1991). Attempting to find an alternative to the PASI many researchers have reported on the disease activity using biological markers of epidermal differentiation and inflammation. PASI has often been chosen as a gold standard to evaluate validity (Griffiths *et al.*, 1996; Schopf *et al.*, 1993; Alkemade *et al.*, 1995). However, in our opinion, though the most common measure of psoriasis severity, it should not be regarded as the standard for evaluation of validity of other methods. In the ideal case, we know "the" pathogenic agent, which in turn could be used to assess the usefulness of the PASI.

As there is no method of quantifying psoriasis severity using biological markers on the molecular level, we believe a promising approach is to assess the components of psoriasis on a cellular level. Generally, the question arises as to which parameters serve our aim of objectivity quantifying severity of psoriasis.

The histological features indicate that desquamation, infiltration, and erythema used in PASI represent significant changes within the psoriatic lesion. Increased desquamation is a measure for increased cell turnover and accelerated differentiation leading to a thickened stratum corneum; infiltration may be correlated with acanthosis and inflammation; erythema is a reflection of an increased blood flow through dilated, tortured vessels.

In order to derive an objective and accurate psoriasis severity index, it is desirable to quantify desquamation, infiltration, and erythema at a given site and individual by a set of physical measurements, which is generally called a measurement array. This measurement array may be subsequently used to calculate a severity index, which, in the ideal case, should not only enable quantification of changes due to treatment at a given site of an individual, but also comparisons between sites and individuals.

An ideal measurement array for the evaluation of psoriasis severity would: (1) cover the major histological components of psoriasis, (2) be objective, (3) be noninvasive, and (4) be feasible, fast, and cover large areas. Thus, the array would have to assess properties of the stratum corneum (scaling and hyperkeratosis), viable epidermis (acanthosis, elongation of rete ridges), and dermis (inflammation). We believe that a combination of several bioengineering measurements would fulfill many of the criteria mentioned above and is currently a promising approach for creating a psoriasis measurement array.

Nevertheless, it should be noted that such a measurement array does not give any information in regard to the pathogenesis of psoriasis. Any of the three components mentioned could be mechanistically correlated with one another; in this case the resulting severity index would contain redundant information, but the only aim of a measurement array is to quantify dermal and epidermal phenomena specific to psoriasis as accurately and feasibly as possible.

Any new method to quantify severity of psoriasis should be validated for its criterion-related validity. As stated by Exum *et al.* (1996) when analyzing PASI validity this is a critical point, as a gold standard is not available. There is a growing understanding of the pathogenesis on molecular level. The ultimative validation of a phenomenological severity index would be its correlation with pathogenic expression on a subcellular or molecular level. However, until this is achieved a new severity index can be validated against invasive histological investigations.

In this section we discuss the measurement devices that are basically capable of quantifying desquamation, infiltration, and erythema and summarize the results, described in the literature, using these techniques, especially in the context of assessing improvement as a result of treatment.

3.1. Infiltration

With the introduction of high-frequency ultrasound into dermatology (Alexander and Miller, 1979) it has become possible to obtain information on skin structures and their sizes noninvasively. Earlier investigations used B-scan sonography up to 25 MHz. A major finding was an echopoor band that was suggested to be a correlate of the acanthotic epidermis (Sondergaard *et al.*, 1985; Overgaard Olsen and Serup, 1993; Vaillant *et al.*, 1994). This was confirmed by El-Gammal *et al.* (1994) and Gupta *et al.* (1996) using higher frequencies that produce pictures of higher resolution. Using 50 MHz El-Gammal *et al.* (1994) were able to study epidermal and dermal changes. Most studies have used histology to validate the sonographic findings.

By comparing histological to ultrasonic findings it was possible to identify different structures in the sonogram: the so-called entry echo corresponds to the irregularly thickened stratum corneum, the echopoor band corresponds to the sum of the acanthotic epidermis and the inflammatory infiltrate in the upper dermis, and the zone of scattered reflexes reflects the dermis underneath the infiltrate (El-Gammal *et al.*, 1994). In addition Gupta *et al.* (1996) compared his ultrasonic results (using 40 MHz ultrasound scanning) to a SET-score, the latter assessing scaliness, erythema, and thickness of a lesion, and he found that the echopoor band could best be correlated to the clinical score. Assuming that scaliness is represented by the A-band, one would expect a lower correlation. Several studies used ultrasound to assess treatment efficacy (Seidenari *et al.*, 1994).

Bangha and Elsner (1996) treated psoriatic areas with betamethasone and calcipotriol, respectively. They measured decreased distances between the skin surface and the interface between the dermis and subcutaneous tissue of treated compared to untreated skin. Hoffman *et al.* (1995) studied changes of psoriatic lesions treated with dithranol (0.05 – 2%), sole bath, and UV-irradiation. They observed that the echolucent area gradually disappears, and that the thickness of the entry echo, as well as the thickness of the dermis and overall skin thickness decreased. Further they found increasing densitometry values correlated with thickness. Vaillant *et al.* (1994) suggested that for ideal evaluation a representative plaque has to be chosen. To overcome inaccuracy of dermal thickness due to hyperkeratotic plaque, El-Gammal (1994) suggested pretreatment of these areas with an ointment. Taken together, ultrasound offers a promising, noninvasive tool for the quantification of hyperkeratosis and acanthosis.

Another promising, upcoming approach to retrieve an image of the epidermis is confocal microscopy with a near-infrared laser. With this new technique it is possible to image the epidermis up to a depth of 400 μm; thus, hyperkeratosis and acanthosis may basically be quantified.

3.2. *Erythema*

The erythema of psoriatic plaques is an indication of altered cutaneous vasculature. Thus, with laser Doppler velocimetry (LDV) it is possible to measure a signal related to the severity of psoriasis. In laser Doppler scanning a plaque is seen as a sharply demarcated hyperemic area. Speight *et al.* (1993) and Speight and Farr (1994) found an increased blood flow is psoriatic plaque that was also present in 2- to 4-mm rims around plaques. Some investigators have attributed the increased LDV signal to an increased number of vessels (Barton *et al.*, 1992). Auer *et al.* (1994), however, showed that the increased perfusion is due to an increased blood flow through vessels being dilated but not increased in number. They also noted that in acanthosis the laser beam is less scattered and has increased sampling depth. On the other hand, as stated by Speight and Farr (1994), thick scaling appears to cause some reduction in the blood flux reading. In other words, the conditions in more supraficial epidermal layers may influence the measurement of blood flow.

Earlier studies had investigated psoriatic plaque by means of single-point measurements. Staberg and Klemp (1993) used this technique to measure treatment efficacy. The correlation of these measures with the clinical psoriasis index (the sum of the three visually assessed parameters: infiltration, erythema, and scaling) yielded a high linear correlation ($r = 0.84$). Analogously to the study by Gupta *et al.* (1996) one could have expected that infiltration and scaling might have yielded a lower correlation. This could be an indication for the three clinical parameters to be highly correlated yielding a high criterion validity. Similarly Berardesca *et al.* (1994) were able to demonstrate the efficacy of calcipotriol versus clobetasol. A visually assessed total sign score (average of the values for erythema, infiltration, and scaling), however, yielded a correlation of only $r = 0.342$.

In contrast to the single-point measurements, laser Doppler scanning should better account for the inhomogeneity in perfusion within a psoriatic plaque. The size of a psoriatic lesion can be determined, and enable quantification of a lesion (Altmeyer *et al.*, 1997). Using mean blood flux within the plaques Speight and Farr (1994) we are able to demonstrate that calcipotriol significantly improves the response of psoriasis to PUVA. Unfortunately one laser Doppler scan is much slower than the natural fluctuations in cutaneous blood flow. Therefore it appears advantageous to measure several spots simultaneously with a multichannel laser Doppler instrument. To further overcome the influence of a thickened stratum corneum and acanthosis on the blood flow readings, we suggest placing several probes on the hyperemic rim.

The usefulness of measuring erythema by means of chromametry has been judged differently. Takiwaki and Serup (1994) correlated their measures of color parameters of psoriatic plaque, as assessed by narrow-band reflectance spectrophotometry and tristimulus colorimetry, with clinically assessed scaliness. Measured skin color was highly negatively correlated with the scaliness of a lesion, indicating that thick scaling masks underlying erythema. Thus, it appears that colorimetry is rather an assessment of scaliness unless, e.g., the psoriatic plaques are initially descaled. Therefore Bangha and Elsner (1996), before starting actual therapy, used a 1-week pretreatment with salicylic acid. Thereafter they were able to assess significant changes in skin color due to treatment. Seidenari *et al.* (1997) reported that the assessment of erythema is not always a valid method for monitoring therapeutic efficacy. Hoffman *et al.* (1995), however, judged colorimetry less useful. Although they could assess slight changes due to treatment, the correlation to clinical evaluation was weak. Seitz and Whitmore (1998), on the other hand, observed the *a* parameter to reflect redness and the *b* parameter to reflect tanning of a lesion. Hoffman *et al.* (1995) stated that colorimetry might not discriminate between redness and tan. It is, however, not clear what influence scaling could have had on their measures. As colorimetry measures increased perfusion phenomenologically as skin color, it is influenced by overlying tissue.

3.3. Desquamation

Measuring desquamation in psoriasis is certainly the most difficult part of a bioengineering severity assessment. A direct approach is to sample the squames from the skin surface and to quantify their thickness and area. Adhesive tape may be used for sampling the squames. Gravimetric analysis, image analysis (Miller, 1995), and colorimetric analysis of dyed proteins (Dreher *et al.*, 1998) quantify the quantity of squames. Although only mildly invasive, sampling with adhesive tape may cause bleeding in psoriatic lesions.

There are two bioengineering measurements that can indirectly assess skin desquamation. The first is skin colorimetry as mentioned above; the second is stratum corneum corneometry, assessing the stratum corneum hydration. The only known study that correlated skin color with desquamation scored the latter visually. For both indirect measures, skin color and stratum corneum hydration, a correlation has to be established with one of the direct desquamation measurements in healthy and psoriatic skin. This has yet to be done, although there are indications for such correlations (Jemec and Serup, 1992). Nevertheless, Serup *et al.* (1987) found that hydration did not correlate with scaling, assessed clinically, indicating that severe scaliness does not necessarily imply dry skin. Some investigators characterized psoriatic stratum corneum measuring transepidermal water loss (TEWL) (Serup and Blichmann, 1987; Tagami and Yoshikuni, 1985). Berardesca *et al.* (1994) showed that TEWL decreased significantly due to two different treatments, and they were able to differentiate between the treatments in their capacity to restore barrier function.

4. Summary

Psoriasis is a decrease of high incidence. Only recently, since more insight about the pathogenesis, more causal therapies have become available (e.g., Chandraratna, 1997). In order to judge the efficacy of any treatment an objective and reproducible assessment of the psoriasis is mandatory. PASI, developed in 1978, was based on clinical assessments of four parameters: area, desquamation, infiltration, and erythema. Although PASI overestimates the importance of the involved area and uses subjectively assessed scores, it is still widely used owing to its practicability.

Many noninvasive biophysical measurements of various skin properties have been developed and put to use during the past decade (e.g., Berardesca *et al.*, 1995). Although these so-called bioengineering techniques are superior in evaluating skin conditions as compared to clinical scoring schemes, they have not yet been used in clinical trials of any consequence. The reasons for this skeptisism are manifold, the most serious one being that bioengineering techniques for physiological skin conditions, e.g., water content and flux and blood flux often reflect a combination of skin properties; therefore it is impossible using only one measurement to assign a change in response to one skin property. An adequate combination of several bioengineering measurements, a measurement array, offers a possible solution to this problem. Like every measurement the results of bioengineering measurements are dependent on the ambient conditions. Therefore it is indeed worthwhile to define standardized procedures that will ensure that different measurements are comparable (e.g., Tupker *et al.*, 1997). However, in our view, this is not the main point in helping these techniques toward a breakthrough in clinical application and trials. The main point is that for any measurement a change in response may be linked to a change in skin condition. This can only be achieved if, firstly, the bioengineering measurement response is carefully validated, and, secondly, a measurement array is developed that is meaningful to the dermatological aspects being investigated. No single measurement can provide a complete picture of skin condition.

In order to be able to compare results from different psoriasis studies and to be able to detect subtle differences, we think it is now possible to use noninvasive biophysical measurements to quantify the severity of psoriasis. The terms desquamation, infiltration, and erythema included in the PASI are valid for the phenomenological description of psoriasis. Thus, these skin conditions are to be quantified by bioengineering methods and combined into a severity index. In this chapter we gave an overview of the status of bioengineering techniques to quantify these terms of psoriasis. This overview is neither intended to be conclusive nor to give a final recipe for quantifying psoriasis severity with bioengineering measurements. We are well on the way to quantifying the relevant skin conditions of psoriasis reliably and thus a combination of bioengineering measurements should now be used and tested in clinical evaluations. The first task is to find the appropriate measurement array. Such measurement arrays should replace the procedure validating any one measurement with PASI.

References

Alexander, H., and Miller, D. L., 1979, Determining skin thickness with pulsed ultra sound, *J. Invest. Dermatol.* **72:**17–19.

Alkemade, H. A. C., de Jongh, G. J., Arnold, W. P., van de Kerkhof, P. C. M., and Schalkwijk, J., 1995, Levels of skin-derived antileukoproteinase (SKALP)/Elafin in serum of severe psoriasis with cyclosporin A, *J. Invest. Dermatol.* **104:**189–193.

Altmeyer, P., Hoffman, K., and Stücker, M. (eds.), 1997, *Kutane Mikrozirkulation*, Springer, Berlin.

Auer, T., Bacharach-Buhles, M., El-Gammal, S., Stücker, M., Panz, B., Popp, C., Hoffman, K., Happe, M., and Altmeter, P., 1994, The hyperperfusion of the psoriatic plaque correlates histologically with dilatation of vessels, *Acta Derm. Venereol.* (*Stockh.*) **186** (Suppl.):30–32.

Bahmer, F., 1989, The size of lesions, or point counting as a step toward the solution of the PASI problem, *Arch. Dermatol. Res.* **125:**1282–1283.

Bangha, E., and Elsner, P., 1996, Evaluation of topical antipsoriatic treatment by chromametry, visiometry and 20-MHz ultrasound in the psoriasis plaque test, *Skin Pharmacol.* **90**:298–306.

Barker, J. N. W. N., 1991, The pathophysiology of psoriasis, *Lancet* **338:**227–230.

Barton, S. P., Abdullah, M. S., and Marks, R, 1992, Quantification of microvascular changes in the skin in patients with psoriasis, *Br. J. Dermatoll.* **126:**569–574.

Berardesca, E., Vignoli, G., Farinelli, N., Vignini, M., Distante, F., and Rabbiosi, G., 1994, Non-invasive evaluation of topical calcipotriol versus clobetasol in the treatment of psoriasis, *Acta Derm. Venereol.* (*Stockh.*) **74:**302–304.

Berardesca, E., Elsner, P., Wilhelm, K.-P., and Maibach, H. I., eds, 1995, *Bioengineering of the Skin: Methods and Instrumentation*, CRC Press, Boca Raton.

Chandraratna, R. A. S., 1997, Tazarotene: The first receptor-selective topical retinoid for the treatment of psoriasis, *J. Am. Acad. Dermatol.* (*Suppl.*) **37:**S12–S17.

Dreher, F., Ahrens, A., Hostynek, J., Jujumba, S., Ademola, J., and Maibach, H. I., 1998, Colorimetric method for quantifying human stratum corneum removed by adhesive-tape-stripping, *Acta Derm. Venereol.* (*Stockh.*) **78:**186–189.

El-Gammal, S., Auer, T., Hoffmann, C. Popp Amd K., Altmeyer, P., Passmann, C., and Ermert, H., 1994, Psoriasis vulgaris in 50 MHz B-scan ultrasound—characteristic features of stratum corneum, epidermis and dermis, *Acta Derm. Venereol.* (*Stockh.*) **186** (Suppl.):173–176.

Exum, M. L., Feldmann, S. R., Fleischer, A. B., Reboussin, D. M., and Clark, A. R., 1996, Measuring severity of psoriasis: methodological issues skin chimeras, *J. Dermatol. Treat.* **7:**119–124.

Fredriksson, T., and Pettersson, U., 1978, Severe psoriasis—oral therapy with a new retinoid, *Dermatologica* **157:**238–244.
Griffiths, C. E. M., Boffa, M. J., Gallatin, W. M., and Martin, S., 1996, Elevated levels of circulating intercellular adhesion molecule-3 (cICAM-3) in psoriasis, *Acta Derm. Venereol.* (*Stockh.*) **76:**2–5.
Gupta, A. K., Turnbull, D. H., Harasiewicz, K. A., Shum, D. T., Watteel, G. N., Foster, S., and Sauder, D. N., 1996, The use of high-frequency ultrasound as a method of assessing the severity of a plaque of psoriasis, *Arch. Dermatol.* **132:**658–662.
Hoffman, K., Dirschka, T., Schwarze, H., el Gammal, S., Matthes, U., Hoffman, A., and Altmeyer, P., 1995, 20 MHz sonography, colorimetry and image analysis in the evaluation of psoriasis vulgaris, *J. Dermatol. Sci.* **9:**103–110.
Jemec, G. B. E., and Serup, J., 1992, Scaling, Dry Skin and Gender, *Acta Derm. Venereol.* (*Stockh.*) **177** (Suppl.):26–28.
Marks, R., and Barton, S. P., 1989, Assessment of disease progress in psoriasis, *Arch. Dermatol.* **125:**235–240.
Miller, D. L., 1995, Sticky slides and tape techniques to harvest stratum corneum material, in: *Handbook of Non-invasive Methods and the Skin* (J. Serup and G. B. E. Jemec, eds.). CRC Press, Boca Raton, pp. 149–151.
Ortonne, J.-P., 1996, Aetiology and pathogenesis in psoriasis, *Br. J. Dermatol.* **135:** (Suppl.):1–5.
Overgaard Olsen, L., and Serup, J., 1993, High-frequency ultrasound scan for non-invasive cross-sectional imaging of psoriasis, *Acta Derm. Venereol.* (*Stockh.*) **73:**185–187.
Ramsay, B., and Lawrence, C. M., 1991, Measurement of involved surface area in patients with psoriasis, *Br. J. Dermatol.* **124:**565–570.
Savolainen, L., Kontinen, J., Röning, J., and Oikarinen, A., 1997, Application of machine vision to assess involved surface in patients with psoriasis, *Br. J. Dermatol.* **137:**395–400.
Schopf, R. E., Naumann, S., Rehder, M., and Morsches, B., 1993, Soluble intercellular adhesion molecule-1 in patients with psoriasis, *Br. J. Dermatol.* **128:**34–37.
Seidenari, S., Magni, R., Coppini, M., and Giannetti, A., 1994, Instrumental evaluation of psoriasis lesions treated with tacalcitol, *Acta Derm. Venereol.* (*Stockh.*) **186** (Suppl):177.
Seidenari, S., Magni, R., and Giannetti, A., 1997, Assessment of the Activity of tacalcitol on psoriatic plaques by means of colorimetry and high-frequency ultrasound: A double-blind intrasubject half-side right-left comparison with betamethasone valerate and placebo, *Skin Pharmacol.* **10:**40–47.
Seitz, J. C., and Whitmore, C. G., 1988, Measurement of erythema and tanning responses in human skin using a tri-stimulus colorimeter, *Dermatology* **177:**70–75.
Serup, J., and Blichmann, C., 1987, Epidermal hydration of psoriasis plaque and the relation to scaling, *Acta Derm. Venereol.* (*Stockh.*) **67:**357–366.
Sondergaard, J., Serup, J., and Tikjob, G., 1985, Ultransonic A- and B-scanning in clinical and experimental dermatology, *Acta Derm. Venereol.* (*Stockh.*) **65** (Suppl.):76–82.
Speight, E. L., and Farr, P. M., 1994, Calcipotriol improves the response of psoriasis to PUVA, *Br. J. Dermatol.* **130:**79–82.
Speight, E. L., Essex, T. J. H., and Farr, P. M., 1993, The study of plaque of psoriasis using a scanning laser-Doppler velocimeter, *Br. J. Dermatol.* **128:**519–524.
Staberg, B., and Klemp, P., 1993, Skin blood flow in psoriasis during Goeckerman or beech tar therapy, *Acta Derm. Venereol.* (*Stockh.*) **73:**185–187.
Tagami, H., and Yoshikuni, K., 1985, Interrelationship between water-barrier and reservoir functions of pathologic stratum corneum, *Arch. Dermatol.* **121:**642–645.
Takiwaki, H., and Serup, J., 1994, Measurement of color parameters of psoriatic plaques by narrow-band reflectance spectrophotometry and tristimulus colorimetry, *Skin Pharmacol.* **7:**145–150.
Tiling-Grosse, S., and Rees, J., 1993, Assessment of area of involvement in skin disease: A study using schematic figure outlines, *Br. J. Dermatol.* **128:**69–74.
Tupker, R. A., Willis, C., Berardesca, E., Lee, C. H., Fartasch, M., Agner, T., and Serup, J., 1997, Guidelines on sodium lauryl sulfate (SLS) exposure tests: A report from the Standardization Group of the European Society of Contact Dermatitis, *Contact Dermatitis* **37:**53–69.
Vaillant, L., Berson, M., Machet, L., Callens, A., Pourcelot, L., and Lorette, G., 1994, Ultrasound imaging of psoriatic skin: A noninvasive technique to evaluate treatment of psoriasis, *Int. J. Dermatol.* **33:**786–790.

Chapter 13

Epidemic of Melanoma: How Real Is It?

TSU-YI CHUANG

1. The Epidemic of Melanoma

The incidence of melanoma between the early 1960s and the mid-1980s increased at a rate of 3 to 7% in countries such as Canada, United States, Denmark, Germany, Hungary, Iceland, Poland, Norway, Slovenia, Sweden, United Kingdom, and New Zealand (Armstrong and Kricker, 1994). In Australia the incidence also increased [but the increase from birth cohorts of ages below 50 seem to have slowed down (Boyle *et al.*, 1995)]. Some investigators in the United States further pointed out that the incidence of melanoma has been up more than 1800% since the 1930s (Rigel *et al.*, 1996). It was thus claimed that the incidence is going to rise in the next 10 to 20 years and that the mortality rate will continue to rise (Rigel *et al.*, 1996). By the year 2000, 1 in every 75 Americans is expected to have melanoma (Rigel *et al.*, 1996).

However, there are still some who believe that a "melanoma epidemic" is not as clear as most people think (Swerlick and Chen, 1996; Rees, 1996). They believe that factors contributing to this "epidemic" include: (1) aggressive surveillance resulting in early harvesting of cases, (2) removal of all pigmented lesions leading to many "thin" or *in-situ* melanomas, and (3) shifting diagnostic criteria leading to more so-called "melanomas" that are biologically benign (Swerlick and Chen, 1996; Rees, 1996; Giles *et al.*, 1996). These factors create a paradox: increasing incidence with a decreasing mortality rate (Giles *et al.*, 1996; Roush *et al.*, 1992). This is in spite of the fact that melanoma treatment has not improved enough to claim the credit for the decrease in mortality.

Tsu-Yi Chuang • Department of Dermatology, Indiana University Medical Center, Indianapolis, Indiana 46202-5267.

Cutaneous Biometrics, edited by Doris A. Schwindt and Howard I. Maibach, Kluwer Academic/Plenum Publishers, New York, 2000.

2. An Analysis of the Epidemic

There are many factors which might inflate the incidence of melanoma without a real change in the incidence, including (1) clinical diagnosis, (2) tumor registry, (3) shifting pathological diagnostic criteria, (4) method of case finding, (5) early detection of melanoma, (6) increased incidence only in certain geographic regions, (7) increased incidence only in certain subtypes of melanoma, (8) birth cohort effect, and (9) paradox of increasing incidence and decreasing mortality.

2.1. Clinical Diagnosis

The accuracy of clinical diagnosis of melanoma at the New York University Medical Center between 1955 and 1982 remained steady at about 60–64 (Grin *et al.*, 1990). In other words, 36–40% of melanoma cases were diagnosed as something else. This situation did not improve much in later years. In 1990, another study done at the same institution found that 36% or 128 cases were not identified as melanoma clinically (Miller and Ackerman, 1992). These patients could go without a biopsy because those lesions would not be recognized even by dermatologists. With better training for clinicians and improved diagnostic criteria, clinically and histologically, we may start witnessing an "increasing" trend of melanoma just because of increasing documentation of an early or a thin melanoma that might have been missed in earlier years.

2.2. Tumor Registry

Underreporting of cutaneous melanoma has been documented from 1 to 24% (Koh *et al.*, 1992), but, in reality, the magnitude of underreporting could be much greater. In Indiana, 800 cases of melanoma were expected based on the nationwide incidence, but fewer than 400 cases were recorded in the State's official report, a 50% difference. Most non-hospital-based laboratories do not have to report cases of melanoma to the registries unless the State law requires it. The underreporting may create several problems. Most importantly, if more cases were reported owing to a better registration system, the "increase" of incidence might only reflect what has been missing in the past.

Further, tumor registries do not register melanoma *in situ*. There has been a sixfold increase in the incidence of melanoma *in situ* in recent years (Elder, 1995). We do not know how many melanomas *in situ* had been registered as invasive melanoma. The tumor registry system thus faces a serious challenge: the registries have no safeguard against the practice of "upgrading" a diagnosis from *in situ* to invasive.

2.3. Shifting Pathological Diagnostic Criteria

Some pathologists may call a lesion melanoma which, in others' view, might represent a lentigo maligna, melanoma *in situ*, severe dysplastic melanocytic hyperplasia, etc. Some new entities such as "borderline melanoma" or "minimum-deviated melanoma," also contribute to a long list of subtypes: in other words, more melanomas being called as such than they were before. In a well-executed study done in 1996 the reliability of diagnostic criteria were put to a test (Farmer *et al.*, 1996). An expert panel of pathologists comprising eight leading authorities in melanoma cannot agree on the diagnosis of melanoma or nevus if they read the "classic" slides without consulting each other, a situation mimicking their own daily practice. Two cases were called melanoma by four experts and benign/atypical/indeterminant nevus by another four. Two cases were called melanoma by five experts and benign/atypical nevus by three. Another three were called melanoma by three experts and benign nevi by five. In fact, among a total of 37 "classic" cases selected by the panelists themselves and examined, there were 24 cases, or 65%, diagnosed differently by the eight experts. The discrepancy in diagnoses in at least 13 cases was so significant that it could cause a serious medical or legal problem. As human nature goes, we tend to call a lesion in a more severe term when in doubt (Swerlick and Chen, 1996). As mentioned above, most of the tumor registries relied upon voluntary reporting with no safeguard against reporting an ever-increasing *in situ* melanoma as a "melanoma."

2.4. The Methods of Case Finding and Documentation

The methods used in investigating the incidence of melanoma can vary in many ways. The surveillance could be prospective or retrospective. The investigation could be an incidence study in which only new patients were eligible or a prevalence study in which patients with a past history of melanoma could be included. It could be an enforced tumor registry with a high turnout or a voluntary reporting with a poor compliance, a medical record review with a lot of missing data or an active interview with physical examination recording every detail that one could think of. It may rely on expertise of dermatologists with special training or just recruit students, residents, or nurses for a short-term summer project. The "incidence" data could just include *persons* with their first melanomas. It could also count a patient with a primary and several subsequent melanomas as *multiple* cases. Thus, e.g., a patient who has one primary and a second subsequent melanoma could be entered as two distinct cases with melanoma. Or, the investigators may decide to count the *number of lesions*, not patients, as the case entries. Therefore a person with three concurrent distinct primary melanomas could be documented as three cases in the incidence reports. If this practice continues, we might see the incidence continue to rise just because of the same person being repeatedly entered as a new case or the incidence of *lesions* increase and yet the number of *patients* remain the same. The tumor registry conducted by the Surveillance, Epidemiology, and End Result Program (SEER) is handled exactly this way. This and shifting diagnostic criteria could inflate the incidence of melanoma. Due to a variety of ways in documenting and reporting the incidence, we have to read the incidence reports very carefully. We had better not compare the incidence reports of today with the reports done years ago. The incidence in one geographic region may not be comparable with the data elsewhere.

2.5. Early Detection and Thin Melanoma

It has been evident in many geographic regions that early detection of melanoma through help from skin cancer screening, public campaign for skin cancer prevention, and physician's awareness of dysplastic nevus syndrome or familial B-K mole syndrome is a reality. This practice would have pushed the diagnosis of melanoma a few months or years earlier. A short-term increase of melanoma incidence would thus be observed but would then level off in the next few years (Burton *et al.*, 1993; Newton and Redburn, 1995). We therefore observe the thickness of melanoma significantly decreased. The survey in Australia in 1989 showed that the thickness of melanoma was thinner than 0.76 mm in more than 50% of the cases (Jelfs *et al.*, 1994).

2.6. Increase of Incidence Only in Certain Areas

The incidence of melanoma may continue to rise in certain areas but might soon plateau in places such as Hawaii or Australia. The incidence in Hawaii from 1978 to 1982 showed a downturn (Armstrong and Kricker, 1994). The incidence in Australia from 1988 through 1990 and the mortality between 1986 and 1990 were quite stable (Burton *et al.*, 1993).

2.7. Increase of Incidence Only in Certain Types of Melanoma

It is believed that the "increase" of melanoma is primarily the result of an increase in thin melanoma (Burton *et al.*, 1993; Burton and Armstrong, 1994). The incidence of superficial spreading melanoma may be increasing due to behaviors such as sunbathing. We have seen the incidence of *in situ* melanoma in the United States increase 600% from 1973 to 1987 (Elder, 1995).

The incidence of thick melanoma did not increase. In fact, as mentioned above, most recent melanoma is thinner than 1 mm (Jelfs *et al.*, 1994). If there is an epidemic of melanoma and because the majority of patients often ignore their "funny moles," we should be witnessing a large number of thick melanoma now. Instead, in the early 1930s, the average thickness of melanoma was more than 3 mm, and now it is less than 1 mm (Van der Esch *et al.*, 1991).

2.8. Birth Cohort Effect

The birth cohorts who constitute those with high incidence and mortality of melanoma are around 60 years of age or older now (Giles *et al.*, 1996). In younger birth cohorts, the incidence of melanoma is already showing a downturn in many places (Giles *et al.*, 1996; Roush *et al.*, 1992).

2.9. The Paradox of Increasing Incidence and Decreasing Mortality

We observed a strange phenomenon: a decreasing mortality of melanoma despite the "increase" of melanoma incidence. We tend to quickly claim that the advance of melanoma treatment has made it so. However, we all know very well that the mainstream treatment for melanoma is the same as it was years ago. Surgery is still the primary treatment for all melanomas. Chemotherapy with DTIC is still the gold standard for advanced melanoma. Therapies with vaccines, interferon, interleukins, and chemoimmunotherapy are still in an experimental stage and not yet showing any exciting results. The "improvement" must be due to the expansion of the melanoma pool, which now includes the thick melanoma and the recently discovered thin and ultrathin melanoma, which often carry no risk or low risk of mortality (Burton *et al.*, 1993; Burton and Armstrong, 1994). In 1989-1993, the case fatality rate of melanoma in Hawaii was 13.6%, while in San Jose–Monterey it was 26.2% (SEER, unpublished data, 1997). If we assume that the treatment for melanoma was similar in these two regions, the difference observed may be due to the detection of early melanoma in Hawaii, where people are more aware of the risk of sun exposure and skin cancer. In the United States, for 20 years, the mortality rate of melanoma has been around 2.0–2.5/100,000 population and in the last 6 years, at 2.5/100,000 (SEER, unpublished data, 1997).

3. The Experience of an Epidemic in Kauai, Hawaii, 1981–1990

We (Chuang *et al.*, 1997) have conducted a continual prospective study of skin cancer incidence among residents of Kauai, Hawaii, since 1983. We present our data to help address these changing perspectives on the melanoma incidence and the "epidemic" issue.

3.1. Kauai and Its Residents

Kauai is the westernmost of the major Hawaiian islands, with a permanent population around 45,000 (38,679 in 1980 census and 51,177 in 1990 census, situated approximately 22°N and 159.5°W). This includes a yearly average of 14,711 Caucasians, 11,540 Filipinos, 9,982 Japanese, 6807 part-Hawaiians, plus residents with other backgrounds in the years 1981–1990. Kauai is isolated from the other islands by approximately 80 miles of open ocean and, accordingly, most medical care is delivered by a small group of local physicians. There was only one dermatologist among them, and the majority of the island's skin cancers were seen in his office. Furthermore, there is only one pathology laboratory that processes tissue. Therefore, any skin biopsies performed will pass through this laboratory, regardless if done in the dermatologist's office or by another doctor.

3.2. Study Design

This project was designed as a retrospective population-based 10-year incidence study using medical records and prospective medical data of 1981 through 1990 housed at the dermatology clinic and pathology lab. Some data were collected in prospective fashion.

Definition of a Case: A case was defined as an individual with new, biopsy-proven melanoma during the years 1981–1990. There could be no history of melanoma prior to 1981 and melanoma could not be a recurrent lesion at a previously treated site. The case had to be a resident of Kauai with a local permanent address and could be counted *only once* in the 10-year study period. All pathological specimens were reviewed by two of the dermatopathologists.

Data Collection: No attempt was made to actively recruit patients. Compliance was assured because all skin biopsies were processed and interpreted in the island's only pathology laboratory. In addition, all dermatologists on the next closest island, Oahu, were contacted and asked to report patients from Kauai with biopsy-proven skin cancers. These logistics facilitated the study of skin cancer incidence in this comparatively isolated setting. With cooperation from the Kauai physicians and the Oahu dermatologists, we have high confidence that we captured every biopsy-proven case of skin cancer on this island.

3.3. Results

Incidence Rates (White Residents Only): Fifty-three patients (29 men and 24 women) are reported with their first melanoma during the years 1981–1990. We calculate the crude incidence rate by dividing the number of melanoma patients by 147,110 person-years (75,159 person-years in men and 71,951 person-years in women), accumulated over 10 years. The overall standardized annual incidence rate for both sexes and all ages combined was 35.9 per 100,000 residents (95% CI: 26.9 to 47.0 per 10^5). The age-standardized annual rate for men, 42.5 per 10^5 residents, was obviously higher than the 31.1 per 10^5 for women. However this difference is not significant statistically owing to the small sample size ($P = 0.3$). More than half of the melanoma was of the *in situ* type. The U.S. standardized incidence rate for 1981-1990 of *invasive* melanoma was 16.3 per 10^5 population (CI: 10.6–23.9 per 10^5) with the world-standardized rate, 11.5 per 10^5 (CI: 7.5–16.8 per 10^5).

Incidence Trend, 1981–1985 vs. 1986–1990 (White Residents Only): The incidence of invasive melanoma showed a small decrease (19.9 per 10^5 in 1981–1985 vs. 15.9 per 10^5 in 1986-1990, $P = 0.7$). By breakdown into types of melanoma, only superficial spreading melanoma showed a 20% increase (9.2 per 10^5 in 1981–1995 vs. 11.0 per 10^5 in 1986–1990). The incidence of *in situ* melanoma increased 1.9 times (12.3 per 10^5 in 1981–1985 vs. 23.2 per 10^5 in 1986–1990), although such an increase is not significant statistically ($P = 0.2$, two-sided).

The statewide incidence of invasive melanoma in Hawaii, based on tumor registry by the SEER, was 24.3 per 10^5 after being standardized to the U.S. population for the period 1990–1994 (SEER, unpublished data, 1997), up from 20.5 per 10^5 for the period 1983–1987. (Hartman and Goldstein, 1993). The data echos a familiar message regarding melanoma: the incidence of melanoma continues to increase. However, the Kauai data paints a different picture: not only was the incidence significantly lower than the statewide rate, but it was *decreasing* during the 10 year period.

We believe the Kauai data is reliable as we can capture every case from the clinic and lab, and do not rely upon voluntary reporting. In addition, each pathological specimen is double- or triple-checked to avoid shifting in diagnostic criteria during the study period.

4. Check the Reality

The mortality from melanoma in Australia has stopped rising and the birth cohorts who were under 60 years of age have shown a downturn in mortality rates (Giles *et al.*, 1996; Burton *et al.*, 1993). This birth cohort effect, i.e., younger populations enjoyed a lower mortality from melanoma, was also observed in the United States (Roush *et al.*, 1992). In fact, the overall mortality rate of melanoma in the United States has been stable at 2.5/100,000 for 6 years now (SEER, unpublished data, 1997; see Fig. 1). In concert with these observations, the mortality of melanoma is predicted to decrease by the second decade of the 21st century (Scotto *et al.*, 1991). The incidence of invasive melanoma in the Kauai population has been stabilized or slightly decreased in recent years. The incidence in the United States has been around 14/100,000 for almost 4 years (SEER, unpublished data, 1997; see Fig. 1). In certain areas of Australia, the incidence of melanoma has not increased since 1988. (Burton *et al.*, 1993). In the United Kingdom, several reports indicated a downturn or stabilized incidence since 1988–1989. (Newton *et al.*, 1995; Ko *et al.*, 1994; MacKie *et al.*, 1992; Melia *et al.*, 1995).

Do we have a melanoma epidemic?

Swerlick and Chen (1996) may be right in their comment on "the melanoma epidemic": "We may be sounding the alarm about the frightening increase in melanoma only to find it is a diagnostic artefact... we have created a problem and then placed ourselves in the position of both offering and providing a solution." Professor Rees (1996) in his article "The Melanoma Epidemic: Reality and Artefact" asked whether the increase was real or a diagnostic artefact. We might not have an epidemic of melanoma after all.

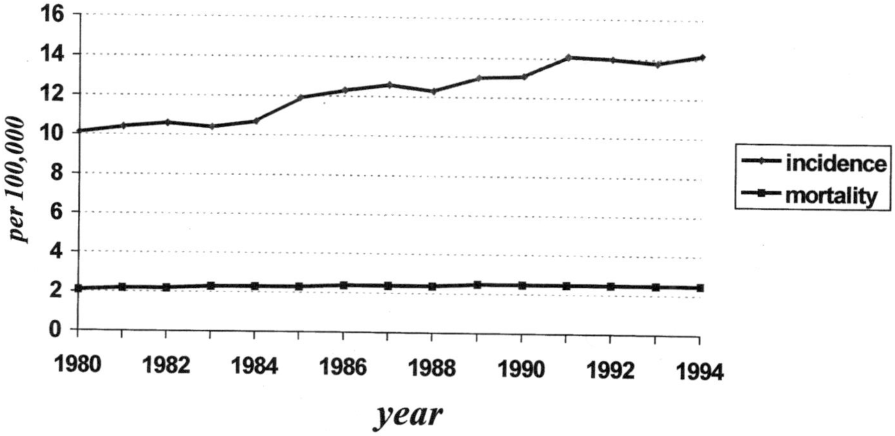

FIGURE 1. Trends in SEER incidence and U.S. mortality, whites.

Refererences

Armstrong, B. K., and Kricker, A., 1994, Cutaneous melanoma, *Cancer Surv.* **19–20:**219–240.

Boyle, P., Maisonneuve, P., and Dore, J.-F., 1995, Epidemiology of malignant melanoma, *Br. Med. Bull.* **51:**523–547.

Burton, R. C., and Armstrong, B. K., 1994, Recent incidence trends imply a non-metastasizing form of invasive melanoma, *Melanoma Res.* **4:**107–113.

Burton, R. C., Coates, M. S., Hersey, P., Roberts, G., Chetty, M. P., Chen, S., Hayea, M. H., Howe, C. G., and Armstrong, B. K., 1993, An analysis of a melanoma epidemic, *Int. J. Cancer* **55:**765–770.

Chuang, T.-Y., Charles, J., Reizner, G. T., Elpern, D. J., and Farmer, E. R., 1999, Melanoma in Kanai, Hawaii, 1981–1990: The significance of in situ melanoma and the incidence trend, *Int. J. Dermatol.* **38**:101–107.

Elder, D. E., 1995, Skin cancer: Melanoma and other specific nonmelanoma skin cancers, *Cancer* **75:**245–256.

Farmer, E. R., Gonin, R., and Hanna, M. P., 1996, Discordance in the histopathologic diagnosis of melanoma and melanocytic nevi between expert pathologists, *Hum. Pathol.* **27:**528–531.

Giles, G. G., Armstrong, B. K., Burton, R. C., Staples, M. P., and Thursfield, V. J., 1996, Has mortality from melanoma stopped rising in Australia? Analysis of trends between 1931 and 1994, *Brit. Med. J.* **312:**1121–1125.

Grin, C. M, Kopf, A. W., Welkovich, B., Bart, R. S., and Levenstein, M. J., 1990, Accuracy in the clinical diagnosis of malignant melanoma, *Arch. Dermatol.* **126:**763–766.

Hartman, A. M., and Goldstein, A. M., 1993, Melanoma of the skin, in: *SEER Cancer Statistics Review: 1973–1990* (B. A. Miller, L. A. G. Ries, B. F. Hankey, C. L. Kosary, A. Harras, S. S. Devesa, B. K. Edwards, eds.), National Cancer Institute, NIH Publication No. 93-2789, pp. XVI. 1–17.

Jelfs, P. L., Giles, G., Shugg, D., Coates, M., Durling, G., Fitzgerald, P, and Ring, I., 1994, Cutaneous malignant melanoma in Australia, 1989, *Med. J. Aust.* **161:**182–189.

Ko, C. B., Walton, S., Keczkes, K., Bury, H. P. R., and Nicholson, C., 1994, The emerging epidemic of skin cancer, *Br. J. Dermatol.* **130:**269–272.

Koh, H. K., Geller, A., Miller, D. R., Clapp, R., W., and Lew, R. A., 1992, Under-reporting of cutaneous melanoma in cancer registries nationwide, *J. Am. Acad. Dermatol.* **27:**1035–1036.

MacKie, R., Hunter, J. A. A., Aitchison, T. C., Hole, D., Mclaren, K., Rankin, R., Blessing, K., Evans, A. T., Hutcheon, A. W., Jones, D. H., Soutar, D. S., Watson, A., C. H., Cornbleet, M. A., and Smyth, J. F., 1992, Cutaneous malignant melanoma, Scotland, 1979–1989, *Lancet* **339:**971–975.

Melia, J., Cooper, E. J., Frost, T., Graham-Brown, R., Hunter, J., Marsden, A., du Vivier, A., White, J., Whitehead, S., Warin, A. P., Wroughton, M., Ellman, R., and Chamberlain, J., 1995, Cancer research campaign health education programme to promote the early detection of cutaneous malignant melanoma. II. Characteristics and incidence of melanoma, *Br. J. Dermatol.* **132:**414–421.

Miller, M., and Ackerman, A. B., 1992, How accurate are dermatologists in the diagnosis of melanoma? Degree of accuracy and implications, *Arch. Dermatol.* **128:**559–560.

Newton, J. N., and Redburn J., 1995, Incidence of melanoma in four English counties, 1989–1992. *Br. Med. J.* **310**:502–503.

Rees, J. L., 1996, The melanoma epidemic: Reality and artefact. Warrants a reappraisal of the relation between histology and clinical behaviour, *Br. Med. J.* **312:**137–138.

Rigel, R. S., Friedman, R. J., and Kopf, A. W., 1996, The incidence of malignant melanoma in the United States: Issue as we approach the 21st century, *J. Am. Acad. Dermatol.* **34:**839–847.

Roush, G. C., McKay, L., and Holford, T. R., 1992, A reversal in the long-term increase in deaths attributable to malignant melanoma, *Cancer* **69:**1714–1720.

Scotto, J., Pitcher, H., and Lee, J. A. H., 1991, Indications of future decreasing trends in skin-melanoma mortality among whites in the United States, *Int. J. Cancer* **49:**490–497.

Swerlick, R. A., and Chen, S., 1996, The melanoma epidemic: Is increased surveillance the solution or the problem? *Arch. Dermatol.* **132:**881–884.

Van der Esch, E. P., Muir, C. S., Nectoux, J., MacFarlane, G., Maisonneuve, P., Bharucha, H., Briggs, J., Cooke, R. A., Dempster, A. G., Essex, W. B., Hofer, P. A., Hood, A. F., Ironside, P., Larsen, T. E., Little, J. H., Phillipps, R., Pfau, M., Prade, M., Pozharisski, K. M., Rilke, F., and Schlafer, K., 1991, Temporal change in diagnostic criteria as a cause of the increase of malignant melanoma over time is unlikely, *Int. J. Cancer* **47:**483–490.

Chapter 14

Prognostic Models in Melanoma

RANJINI NATARAJAN AND MARTIN A. WEINSTOCK

1. Introduction

Multiple regression analyses have found widespread use in melanoma research. Such methods are natural candidates for prognostic factor studies, where it is desired to examine the joint influence of various clinical, pathologic, and demographic variables on indicators of disease. For example, in a recent article, Schuchter *et al.* (1996) consider a four-variable logistic regression model to study association with a binary indicator of survival at 10 years. A more complicated analysis using a Cox proportional hazards model with smooth (nonparametric) functions of prognostic risk factors was considered by Helfenstein *et al.* (1996) to study survival times over a 5-year period. The regression framework is also attractive because predicted disease probabilities for specific risk profiles can be obtained using simple arithmetic calculations (Harrell *et al.*, 1985). Further, computer programs to fit these models are commercially available.

An important assumption of these regression models is that all individuals are identical in their propensity to survive the course of the disease, and that any differences in survival can be explained entirely by certain measurable risk factors. This assumption is a strong one and appears to be at odds with what is known about the biological complexity of melanoma (Vollmer, 1989). Patients are very heterogeneous with respect to several unmeasured variables, such as genetic predisposition, immunity, and nutrition, and these factors also influence their prognosis and ultimate survival. Ignoring such heterogeneity can lead to inaccurate survival estimates. In related work by Natarajan and McCulloch (1997) on survival in wildlife populations, it was found that models that ignore inherent biological variation result in severe underestimation of overall survival.

Ranjini Natarajan • Center for Statistical Sciences, Brown University, Providence, Rhode Island 02912. Present address: Department of Statistics, Division of Biostatistics, University of Florida, Gainesville Florida 32611. *Martin A. Weinstock* • Dermatoepidemiology Unit, VA Medical Center and Department of Dermatology, Rhode Island Hospital and Brown University, Providence, Rhode Island 02912.

Cutaneous Biometrics, edited by Doris A. Schwindt and Howard I. Maibach, Kluwer Academic/Plenum Publishers, New York, 2000.

The purpose of this chapter is to present a flexible parametric model for survival data in melanoma that accounts for unmeasured variability among individuals. Subject heterogeneity that is not explained by covariates is modeled by allowing the survival probability to vary from subject to subject. This is called a mixed regression model as it incorporates both fixed (measured risk factors) and random (unmeasured variability) effects. Overall survival estimates may be calculated for specific risk profiles as in the usual regression models; in this context, however, they are interpreted as estimated survival for an average individual in the population. This interpretation makes them more appropriate for prediction purposes.

2. A Brief Review of Current Methods

There have been many studies of the predictors of the prognosis of melanoma among patients without evidence of metastatic disease at the time of definitive therapy (AJCC stages I and II; clinical stage 1) (Vollmer, 1989). However, few of these have been described in a manner that allows the treating physician to determine the probability of 10-year survival in a practical, i.e., simple and direct manner, a prerequisite for widespread use by physicians in routine clinical care under current circumstances.

The most straightforward of the published, readily usable prognostic models was presented in the form of a table of survival probabilities that depend on the Breslow thickness and anatomic site of the melanoma and on the age and gender of the patient (Schuchter *et al.*, 1996). Other models of prognosis that do not rely on calculations to be performed by the counseling physician include those that are derived from classification and regression trees, so one can simply trace the patient's parameters through the published tree diagrams to determine the survival probability (e.g., Garbe *et al.*, 1995; Aitchison *et al.*, 1995). These can also be reduced to tabular form. Both of these models used Breslow thickness, Clark level, gender, and anatomic site, but one also used age and ulceration, and the other added instead histologic type. A third model using exponential survival trees used only Breslow thickness, anatomic site, and ulceration (Huang, *et al.*, 1997).

More complex but still within the capabilities of a pocket calculator or simple spreadsheet are the regression formulas derived to estimate survival probabilities (Soong, 1992; Garbe, *et al.*, 1995). Both of these formulas included Breslow thickness, Clark level, anatomic site, and gender; one added ulceration and surgical treatment, whereas the other added histologic type and age. Even more complex procedures requiring integration by computer have also been published (Chung, 1994), although this model relies only on tumor thickness. Soong (1992) also described the 10-year survival probability as a simple exponential function of thickness. Although efforts to compare the alternative models have been published (Pritchard *et al.*, 1995), to date this aspect has not been thoroughly explored.

3. A Mixed-Effects Survival Model

In this section we present a flexible parametric model to incorporate biological variability in survival times among individuals. These time endpoints are commonly time to recurrence or time to death. We suppose that there are a total of n patients. For individual i $(i = 1, \ldots, n)$ we observe the number of time-units (maybe day, week, month, or year) since their entry into the study to the time last seen, denoted by T_i; and their survival status δ_i during this period of observation. By convention, $\delta_i = 0$ indicates no failure (no recurrence or death) and $\delta_i = 1$ indicates failure. Associated with each individual is a set of p explanatory variables (risk factors) denoted by $x_{i_1}, x_{i_2}, \ldots, x_{i_p}$. These are called "fixed-effects" and include age, sex, Breslow thickness, Clark level, and so on.

A schematic of the observed data is displayed below:

$$\begin{array}{ccccc} T_1 & \delta_1 & x_{1_1} & \cdots & x_{1_p} \\ T_2 & \delta_2 & x_{2_1} & \cdots & x_{2_p} \\ \vdots & \vdots & \vdots & & \vdots \\ T_n & \delta_n & x_{n_1} & \cdots & x_{n_p} \end{array}$$

3.1. The Model

Our model for these data is a two-stage formulation. The first stage specifies a probability model for the outcome for individual i, i.e., $\{T_i, \delta_i\}$, conditional on their propensity to survive. More mathematically, we let p_i denote individual i's probability of surviving a single time-unit. Then the survival experience of that individual is distributed according to the probability density function f where

$$f(T_i, \delta_i \mid p_i) = \begin{cases} p_i^{T_i} & \text{if } \delta_i = 0 \\ p_i^{T_i - 1}(1 - p_i) & \text{if } \delta_i = 1 \end{cases} \tag{1}$$

which is simply the chance of surviving T_i time-units for individuals that survive or the chance of surviving $T_i - 1$ units and succumbing to a failure in the last time-unit for individuals that fail.

In the second stage, we postulate that there is some underlying survival distribution characterizing the entire population from which each individuals' propensity p_i is drawn. This is accomplished by allowing the p_i to vary randomly according to some distribution, which for convenience, we take to to be the beta distribution (Casella and Berger, 1990):

$$p_i \sim p_i^{\gamma\alpha_i - 1}(1 - p_i)^{\gamma(1 - \alpha_i) - 1} \tag{2}$$

The parameter α_i relates the risk factors x to average survival by the usual logistic formulation:

$$\alpha_i = \frac{1}{1 + \exp(-x_{i_1}\beta_1 - x_{i_2}\beta_2 - \cdots + x_{i_p}\beta_p)} \tag{3}$$

where $\beta = \{\beta_1, \ldots, \beta_p\}^t$ is a (column) vector of regression coefficients representing the effect of the explanatory variables for the average individual in the population.

The parameter γ models the heterogeneity among individuals and larger values indicate less heterogeneity. Equation (2) says that the true survival of all patients that have covariate information $x_{i_1}, \ldots, x_{i_p}$ can be modeled as coming from a beta distribution. Thus, the model acknowledges the extra variability in these data, which the usual logistic regression or proportional hazards models do not. The beta distribution is very flexible and can capture various realistic survival patterns.

Combining Eqs. (1) and (2), the overall likelihood for the mixed model is given by

$$L(\beta, \gamma) = \prod_{i=1}^{N} \underbrace{\left\{ \prod_{j=1}^{T_i} \frac{T_i + \gamma\alpha_i - j}{T_i + \gamma - j} \right\}^{1-\delta_i}}_{\text{survivors}} \underbrace{\left\{ \gamma(1 - \alpha_i) \frac{\prod_{j=2}^{T_i} T_i + \gamma\alpha_i - j}{\prod_{j=1}^{T_i} T_i + \gamma - j} \right\}^{\delta_i}}_{\text{failures}} \tag{4}$$

which is decomposed into a term contributed by survivors and another by failures, just as in the usual survival models. As in logistic or Cox proportional hazards models, we seek values of β and γ such that L is maximized over the entire data set. The resulting equations are nonlinear and hence computer programs must use iterative techniques to solve the problem.

A primary quantity of interest in the application of these models to melanoma is overall survival. This is defined as the probability of surviving a length of J time-units, for example, 10 years. Model-based estimates of overall survival for a particular risk profile is given by:

$$\text{Prob(surviving } J \text{ time-units)} = \left\{ \prod_{j=1}^{J} \frac{J + \hat{\gamma}\hat{\alpha}_i - j}{J + \hat{\gamma} - j} \right\} \tag{5}$$

where $\hat{\gamma}$ and $\hat{\alpha}_i$ denote that estimates of γ and β have been inserted. Thus, predictions for individual patients may be accomplished by simply inserting the appropriate length of time J of interest; and their covariance values x in the expression for α_i using Eq. (3) in Eq. (5). Asymptotic standard errors for these predictions can be calculated using the delta method and should be provided by programs implementing such an analysis.

3.2. Assessing Significance of the Model

A test for the presence of patient heterogeneity is the ratio of the full likelihood function $L(\beta, \gamma)$ to one with $\gamma = \infty$ (i.e., no heterogeneity). This ratio yields a chi-square statistic as follows:

$$\text{chi-square} = -2 \ln \frac{L(\beta, \gamma)}{L(\beta, \infty)} \tag{6}$$

which produces a P-value. If the P-value is low (usually less than 0.05) then the presence of heterogeneity can be concluded to be significant.

TABLE 1. Mixed Model Results for Hypothetical Data Set with Two Prognostic Factors

(a) Estimated coefficients and associated standard errors

Variable	Coefficient	s.e.	P-value
γ	—	—	—
x_1	—	—	—
x_2	—	—	—

(b) Predicted probabilities for risk profiles

Risk profile	Prob(10 year survival)	s.e.
Profile 1	Eq. (5)	—
Profile 2	Eq. (5)	—

(c) Goodness of fit

Likelihood (heterogeneity)	Eq. (6)	*P*-value

A measure of prognostic significance of the model is the ratio of the likelihood function with the regression coefficients, i.e., $L(\beta, \gamma)$ to the one without, i.e., $L(0, \gamma)$. This again yields a chi-square:

$$\text{chi-square} = -2 \ln \frac{L(\beta, \gamma)}{L(0, \gamma)} \tag{7}$$

Similarly, key regression coefficients, say β_1, may be tested using the ratio of likelihoods with and without the prognostic factor corresponding to β_1.

3.3. Reporting Results From a Mixed Model

Analyses of survival data using the mixed models in Section 3.1 should report all the usual statistics as from a logistic regression or proportional hazards model. These include (but are not limited to) the list of factors in the final model, their coefficients and respective standard errors and *P*-values, predicted probabilities of survival for key risk profiles and their standard errors, and the likelihood function. Further, the coefficient of γ along with the chi-square test for heterogeneity [Eq. (6)] should be reported. Table 1 shows the results from the mixed model for a hypothetical data set with two prognostic factors.

4. Future Directions

We believe that future research in prognostic factor studies in melanoma should be directed toward prospective controlled trials of prestated hypotheses, rather than the retrospective analyses that are currently undertaken. While retrospective studies have the advantage of making long-term data available immediately, they suffer from the limitation of missing or incomplete information. For instance, Schuchter *et al.* (1996) found that of 100 randomly reviewed pathology reports, important predictor variables such as histologic subtype were recorded in only 64% of cases, tumor infiltrating lymphocytes and regression in fewer than 20%, and mitotic rate in 5% of the cases. Incomplete covariate information results in a smaller effective sample size and prevents a thorough understanding of an already complicated disease process.

Statistical methods for the design and implementation of clinical trials in other cancers are fairly well established by now. A horizontal shift of these methods can facilitate similar developments in melanoma research. Sample size programs, such as one developed by Natarajan *et al.* (1996) are available for exact power calculations with survival time endpoints. Further, there are available several programs to fit mixed models, such as SAS (Proc Mixed) and Eventreg (Natarajan *et al.*, 1994), which should make the necessary analyses feasible.

References

Aitchison, T. C., Sirel, J. M., Watt, D. C., and MacKie, R. M., 1995, Prognostic trees to aid prognosis in patients with cutaneous malignant melanoma, *Br. Med. J.* **311**:1536–1539.

Casella, G., and Berger, R. L., 1990, *Statistical Inference.* Wadsworth, Belmont, CA, p. 107.

Chung, S. J., 1994, Formula expressing a relationship among lesion thickness and time after diagnosis and survival probability in patients with malignant melanoma, *J. Bio-Med. Comp.* **37**:171–180.

Garbe, C., Büttner, P., Bertz, J., Burg, G., d'Hoedt, B., Dreeper, H., Guggenmoos-Holzmann, I., Lechner, W., Lippold, A., Orfanos, C. E., Peters, A., Rassner, G., Stadler, R., and Stroebel, W., 1995, Primary cutaneous melanoma, *Cancer* **75**:2484–2491.

Harrell, F., Lee, K. L., Matchar, D. B., and Reichert, T. A., 1985, Regression models for prognostic prediction: Advantages, problems, and suggested solutions. *Cancer Treat. Rep.* **69**:1071–1077.

Helfenstein, U., Schüler, G., Morf, S., and Schüler, D., 1996, Action profiles of predictors of death and survival times in stage I malignant melanoma, *Dermatology* **192**:1–7.

Huang, X., Soong, S., McCarthy, W. H., Urist, M. M., and Balch, C. M., 1997, Classification of localized melanoma by the exponential survival trees method, *Cancer* **79**:1122–1128.

Natarajan, R., Turnbull, B. W., Slate, E. H., Wells, M. T., Clark, L. C., and Abu-Libdeh, 1994, A computer program for the statistical analysis of repeated event data using a mixed effects regression model, *Comp. Meth. Prog. Biomed.* **42**:283–294.

Natarajan, R., and McCulloch, C. E., 1999, Modeling heterogeneity in nest survival data. *Biometrics* **55**:553–559.

Natarajan, R., Turnbull, B. W., Slate, E. H., and Clark, L. C., 1996, A computer program for sample size and power calculations in the design of multi-arm and factorial clinical trials with survival time endpoints, *Comp. Meth. Prog. Biomed.* **49**:137–147.

Pritchard, M. L., and Woosley, J. T., 1995, Comparison of two prognostic models predicting survival in patients with malignant melanoma, *Hum. Pathol.* **26**:1028–1031.

Schuchter, L., Schultz, D. J., Synnestvedt, M., Trock, B. J., Guerry, D., Elder, D. E., Elenitsas, R., Clark, W. H., and Halpern, A. C., 1996, A prognostic model for predicting 10-year survival in patients with primary melanoma, *Ann. Int. Med.* **125**:369–375.

Soong, S., 1992, A computerized mathematical model and scoring system for predicting outcome in patients with localized melanoma, *Cutaneous Melanoma* (C. M. Balch, A. N. Houghton, G. W. Milton, A. J. Sober, and S. Soong, eds.), JB Lippincott, Philadelphia.

Vollmer, R. T., 1989, Malignant melanoma: A multivariate analysis of prognostic factors. *Pathol. Ann.* **24**:383–407.

Chapter 15

Photoaging

WILLIAM J. CUNNINGHAM

1. Introduction

In the early part of the past decade it was nearly impossible to convince the conservative, pharmaceutically oriented scientist or physician that photoaging was a worthwhile research or development pursuit. The condition was viewed alternatively as purely cosmetic, not serious enough for pharmaceutical companies, completely unfounded in science or, worse yet, potentially tarnishing to the image of altruistic search for cures for "real diseases." Today it is equally difficult to stem the tide of belief, this time overwhelmingly in the direction of exaggerated claims for a host of treatments all purporting success in the treatment of the now well-recognized process of photoaging. Discovery of the truth in this wide-swinging tide can be achieved only by careful quantitation.

2. Background

Photoaging is herein specifically defined as a dynamic *process* and the term photodamage as a description of pathology or condition at any particular point in time. These definitions reflect the view that change in the skin as a result of exposure to ultraviolet radiation (UVR) is a continuum, initiated very early in life, usually with progression but with potential for at least partial reversal. Interruption, slowing, or reversal of this process appear feasible through avoidance of UVR and by medical or surgical intervention. The basic question addressed in this chapter is how best to evaluate and quantitate the basic condition and its change over time or with therapy. Although there are many observations that have been made relative to photoaging, the primary focus of this chapter will be techniques that have direct clinical

---

William J. Cunningham • CU-TECH, Inc., International Cutaneous Technologies & Development, Mountain Lakes, New Jersey 07046.

Cutaneous Biometrics, edited by Doris A. Schwindt and Howard I. Maibach, Kluwer Academic/Plenum Publishers, New York, 2000.

implications and that have, to at least some extent, proven to be adequate for large-scale clinical trials of pharmaceutical products.

3. Rationale for Biometrics

What can or should be measured is dependent on the goal in mind, and the method should be congruent with both the specific goal and common sense. There is little point in measuring transepidermal water loss if the product is designed to reduce wrinkles. The prospective "customer" of the data is also important. Consumer needs, cosmetic aspects of the product itself, and pharmaceutical requirements each mandate a unique perspective of testing.

The application of biometrics to the study of photoaging should obtain data satisfactory for statistical analysis. Application of standard statistical principles generally suffices for data generated according to most of the techniques to be discussed. The testing must be rigorous to arrive at valid conclusions, and it is important that the conclusions have clinical relevance.

4. Application of Biometrics

4.1. Clinical Aspects

(a) Diagnosis. The perception of age is mostly related to the changes attributable to photoaging (Warren *et al.*, 1991) yet the process is difficult to define specifically because of its lesion complexity. The spectrum of clinical change encompasses fine and coarse wrinkles, irregular and mottled discrete and diffuse pigmentation changes, tactile and visual roughness and dryness, leathery texture, laxity, and sallowness (Drake *et al.*, 1996).

(b) Trial Design. Trial design is critical for accurate, precise, consistent, reproducible, and valid observations in photoaging. The optimum trial is double-blind, placebo-controlled, and multicenter, of at least 6 months duration. Study of a relatively large genetically homogeneous population of a narrow range of similar skin type (usually skin types I-III) is optimum for the same reason that basic research utilizes in-bred animal strains and will better allow small and valid differences to be observed. Seasonal influences, soap, emollient, and cosmetic use, UVR exposure, and sunscreen use should be controlled in a uniform manner.

Examples of adequate trials include several performed for the evaluation of topical retinoids in the treatment of photoaging. Generally, at least 200–300 patients have been enrolled and evaluated at regular intervals for a minimum of 6 months with a focus on the clinical signs of wrinkling, color, and pigmentary irregularity, which signs appear to be the most amenable to effects of topical retinoids. The most convincing trials of topical retinoids have frequently enrolled multiples of this n number in the range of 800 plus patients divided between active and placebo vehicle.

(c) Baseline Evaluation. A number of attempts have succeeded in development of at least partially validated clinical grading systems, which have been widely utilized in clinical trials as entry criteria and baseline evaluation.

One US group used five photographic standards as representative examples of photoaging with assigned grades 0, 4, 6, and 8 for the possibility of a 9-grade scale (Griffiths *et al.*, 1992). Slight modification of the system produced a 10-point scale, with 0 = none, 1 to 3 = mild, 4 to 6 = moderate and 7 to 9 = severe photodamage. The photographic standards were selected to reflect the more easily recognized signs of photoaging of fine and coarse wrinkles, dyspigmentation, elastosis, comedones, and telangiectasia.

A European group developed a 6-point photographic scale in a similar manner with grades of mild, mild/moderate, moderate, moderate/severe, severe, and very severe (Larnier *et al.*, 1994).

Use of a 100-mm visual analog scale rating the overall appearance of the skin as good to poor and specific parameters of fine wrinkles, discrete pigmentation, sallowness, and texture as absent to severe may be used in lieu of either of these systems and has been successfully utilized in trials of topical isotretinoin (Cunningham *et al.*, 1990).

(d) Global Evaluation. Both investigator and patient can make valid observations of the overall change from baseline or global response to therapy. This evaluation is much more valid when a full-face 8 × 10 inch color photograph is meticulously obtained at baseline and used at the various time points of evaluation (e.g., 12, 24, and 36 weeks) for comparison to the present state.

Several rating systems have been utilized, the simplest being a rating of the response as excellent (much improved), good (improved), fair (slightly improved) or poor (no change or worse) (Table 1). This system is simple enough for use without significant prior training and has had widespread applications, although it is somewhat lacking in ability to differentiate small changes. Statistically significant results ($P < 0.001$) were obtained in a trial that demonstrated excellent, good, or fair responses in 5, 21, and 42% of 76 tretinoin-treated and in 0, 11, and 32% of 72 vehicle-treated subjects. Numerical grading as above demonstrated mean baseline of 3.82, mean week 24 of 3.18, mean change from baseline of −0.63, and percent change from baseline of 16.5%, results that were statistically significant and consistent (Table 2) (Olsen *et al.*, 1992).

(e) Visual Analog Scale. Fine discrimination can be obtained by use of a 100-mm balanced visual analog scale with a score of zero designating no change

TABLE 1. Percentage of Subjects with Improvement in Overall Severity of Photodamage after 24 weeks of Therapy[a]

	Tretinoin 0.05% (n = 76)	Vehicle (n = 72)
Improved	51	28
No change	46	69
Worse	3	3

[a]Adapted from Olsen, E. A. *et al.*, 1992, Tretinoin emollient cream: A new therapy for photodamaged skin, *J. Am. Acad. Dermatol.* **26**(2 Pt 1):220.

TABLE 2. Overall Severity of Photodamage after 24 Weeks of Therapy[a]

	Tretinoin 0.05% ($n = 76$)	Vehicle ($n = 72$)
Mean baseline	3.82	4.06
Mean week 24	3.18	3.78
Mean change from baseline	−0.63	−0.28
Percent change from baseline	16.5	6.9
P Value	0.002	

[a]Adapted from Olsen, E. A. *et al.*, 1992, Tretinoin emollient cream: a new therapy for photodamaged skin, *J. Am. Acad. Dermatol.*, **26**(2 Pt 1):220.

from baseline, improvement recorded to the right side of zero (to a maximum of +50 mm), and worsening recorded to the left side of zero (to a minimum of −50 mm) (Sendagorta *et al.*, 1992).

Although investigators or patients may, without training, be capable of using these scales to adequately document photoaging, when comparisons must be made across all patients and all investigative sites, some degree of training in the method is imperative to achieve uniformity of evaluations that can be statistically analyzed without unacceptable standard deviations in the data.

(f) Clinical Panel. Attempts have been made to evaluate the above described baseline condition and change over time using evaluators not directly involved in the subjects' care or observations, potentially allowing a more uniform and non-biased evaluation. The most elaborately developed and validated system was

TABLE 3. Perceived Improvement between Baseline and End of Treatment[a]

	Panelist A	Panelist B	Panelist C	Panelist D
Overall appearance				
Isotretinoin	0.48	0.74	1.21	0.24
Vehicle	0.14	0.21	0.57	0.10
Mean difference	0.34[b]	0.53[b]	0.64[b]	0.14
Fine wrinkles				
Isotretinoin	0.46	0.60	1.10	0.30
Vehicle	−0.02	0.18	−0.10	0.10
Mean difference	0.48[b]	0.42[b]	1.22[b]	0.20[b]
Discrete pigmentation				
Isotretinoin	0.46	0.54	0.92	0.14
Vehicle	0.02	0.12	0.46	0.14
Mean difference	0.42[b]	0.42[b]	0.46[b]	0.00

[a]Adapted from Armstrong, R. B. *et al.*, 1992, Clinical panel assessment of photodamaged skin treated with isotretinoin using photographs, *Arch. Dermatol.* **128**:354.
[b]$P < 0.05$, comparison of the difference in mean change between treatments.

successfully employed in a 17-center study of 776 patients treated with topical isotretinoin (Armstrong *et al.*, 1992).

Very high quality photographs at baseline and evaluation time points are obtained in a rigorous and reproducible manner utilizing specially constructed stereotactic devices and a single emulsion of film for the entire study. Film processing of transparencies, slide duplication, and color correction are carefully controlled.

Pairs of identical slide carousels containing the same patients' slides loaded in identical order are prepared for each center. Matched carousels are placed on two random-access projectors in a rear-screen projection booth, which permits side-by-side comparison of each patient's baseline and end-of-treatment photographic slides. Randomization of sequence of patients is controlled by computer to ensure blinding of evaluations.

Five dermatologists experienced in photodamage trials but uninvolved in the designated trial directly enter responses into the database after comparison of the time-randomized patients' slides and decision as to which side is better. Both global response and specific parameters can be judged including overall appearance, fine wrinkles, and discrete pigmentation (Table 3). Because all photographs are color-corrected in duplication, an assessment of changes in coloring cannot be achieved.

Evaluations use a 13-point balanced categorical scale with zero representing no difference between the two photographs. A score of $+1$ to $+6$ is assigned when the end-of-treatment photograph is perceived to be better and a score of -1 to -6 when the baseline photograph is perceived to be better. Representative mean changes from baseline of global appearance of 0.6, fine wrinkles of 0.4, and discrete pigmentation of 0.5 for isotretinoin-treated subjects were statistically significantly superior to changes of 0.1, 0.1, and 0.1, respectively, with vehicle.

Statistical aspects of this study are informative for the study of photoaging in general. The sample size of 800 patients was selected to detect a 10% difference between the two treatments in mean change from baseline for any parameter with a power of 80% at a significance level of 0.05. These calculations assume a 60% variation in the data for the parameters measured, a variation likely to be encountered in any study of the highly heterogeneous process of photoaging.

(g) Individual Parameters. The many parameters of photoaging variously lend themselves to quantitation by the above-described methodologies. As there are parameters that are best evaluated by visual or tactile methods, not all methods will be adequate for each parameter. All are adequate for evaluation of fine wrinkles. Discrete pigmentation may also be satisfactorily quantitated by the various visually or photographically based systems. Changes in skin texture are more tactile than visual and are best evaluated with the more clinically oriented methodologies or with the instrumental profilometric methods described below than by photographically based methodologies.

Comparison of current state with a baseline photograph or comparison of two or more photographs taken over time will be needed for adequate evaluation of skin color. Chromometry may be satisfactory for accurate and reproducible evaluation of some aspects of this parameter, especially color of erythema or of mottled or diffuse pigmentation.

4.2. Instrumentation

(a) Mechanical Profilometry. A replica of the skin is obtained utilizing silicone rubber. A hard cast is prepared from the replica and a surface profile is generated by means of a stylus passing over the surface of the cast. Analysis of the parameters obtained by the movements of the stylus over the profile of the cast allows deductions as to various aspects of skin contour, including overall roughness and wrinkling.

(b) Optical Profilometry. A high-quality negative impression of the skin surface is obtained utilizing silflo silicon rubber dental impression material (Grove and Grove, 1990). The impression is removed, oriented, coded, and placed under a fiber optic illuminator set at a fixed angle to bring out surface detail. A high-resolution video camera, interfaced with a computer containing image-processing hardware and software, captures the image, which is then digitized. Analysis of the image allows evaluation of parameters of degree of wrinkling, roughness, and other surface markings.

Orientation of light source relative to the "peaks" and "valleys" of the impression allows relative quantitation of wrinkles. By definition, "north" (0 degrees) and "south" (180 degrees) are orientations with the light source perpendicular to major skin lines and "east" (90 degrees) and "west" (270 degrees) are obtained with the light source parallel to the major skin lines. Two parameters found to be particularly useful in evaluation of wrinkling are designated Ra and Rz. Ra is defined as the area of deviation of surface features above and below an average central line of the surface profile. Rz is defined as the difference between the maximum and minimum heights in five equal segments of the surface profile (Fig. 1).

The cheek and "crows feet" areas are frequently chosen for evaluation with this method. In the "crows feet" area, fine wrinkles are principally defined by the

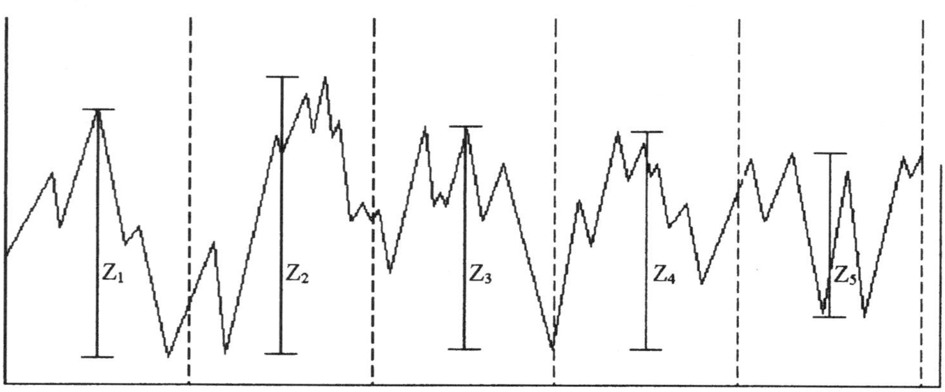

$$R_Z = \frac{Z_1 + Z_2 + Z_3 + Z_4 + Z_5}{5}$$

FIGURE 1. Graphic definition of R_Z (adapted from Grove, G. L. *et al.*, 1990, Effects of tropical retinoids on photoaged skin as measured by optical profilometry, *Meth. Enzymol.* **190**:365).

east–west values. Analysis of a sample is based upon 10 parallel scans and Ra and Rz computations. The overall method has a coefficient of variation of approximately 5%. Reproducible results are obtained in clinical trials of topical tretinoin consistently demonstrating approximately a 10% improvement in fine wrinkles, results which are statistically significantly compared to vehicle (Fig. 2) (Grove *et al.*, 1991).

This technique is highly sophisticated, reproducible, and of great importance, has correlation with clinical ratings. It can be considered to be adequately validated statistically and in clinical trials experience and is objective when performed by an individual blinded to treatment. It is feasible to implement the technique in large clinical trials after a modest amount of training in obtaining the replicas.

(c) High-Resolution Facial Photography. A method has been developed to directly visualize and measure individual wrinkles (Garstein and Shaya, 1986). A full-face photograph of the subject is obtained utilizing a specially designed system of a highly controlled spatially dimensioned apparatus, allowing absolutely precise camera placement and subject positioning. The method has rigid control of parameters of film and processing. Color transparencies are digitized and compensated to control for uniformity of light. Exclusion of features of the eyes, mouth, nose, and eyebrows is achieved by manually masking the image with a graphics tablet. Furrows corresponding to wrinkles can then be defined by their shadows measured by a sophisticated image-analysis technique permitting the detection of all furrows and lines wider than 0.5 mm. Total length is calculated by summing all lines longer than 0.5 cm. Wrinkles are defined as only those skin features with a length-to-width ratio greater than 10:1. Whereas this technique appears to be the most sophisticated available for elaboration of wrinkling, it is essentially a proprietary system and would be exceedingly expensive to develop and utilize in clinical trials.

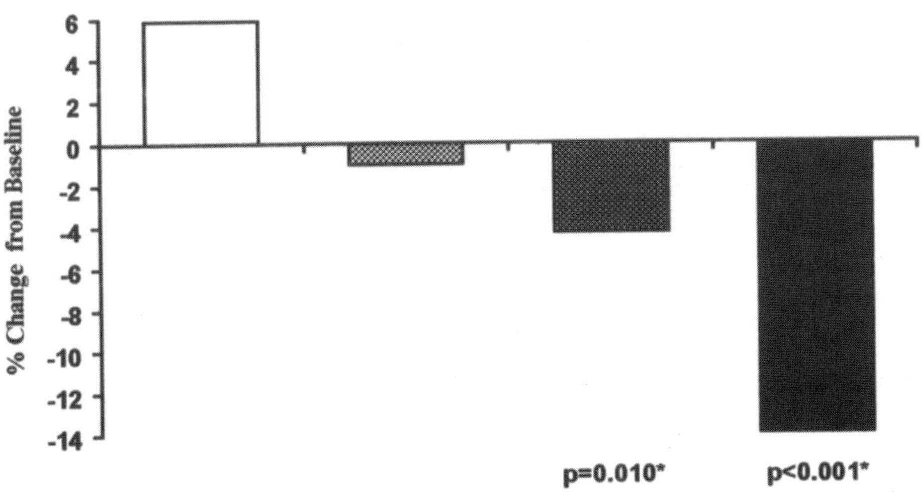

FIGURE 2. Results of optical profilometry performed in the crow's feet area in a 24-week study comparing vehicle and three concentrations of tretinoin emolient cream (TEC) (adapted from Grove, G. H. *et al.*, 1991, Skin replicas analysis of photodamaged skin after therapy with tretinoin emollient cream, *J. Am. Acad. Dermatol.* **25**(2 Pt 1):231–237.

(d) Fluorescence and Polarized Photography. Fluorescence photography is highly useful in evaluation and quantification of the distribution and change of mottled and diffuse hyperpigmentation (Kollias *et al.*, 1997). Photographs are obtained with illumination of UVR of approximately 365 nm. The entering wavelength and subsequent fluorescence emission by collagen are attenuated by epidermal melanin, which is highlighted on black and white photographs as dark areas. Quantitation of discrete macules is performed by visual counting and diffuse areas by evaluation with a 20-point Kodak standard gray scale. Changes from baseline macule count may be analyzed with a two-tailed Student *t* test and consistently are demonstrated after prolonged use of topical tretinoin versus vehicle (31% vs. 11%, e.g.).

Polarized light photography may produce images enhancing selectively, surface or subsurface features of the skin, allowing an alternative method of highlighting wrinkles, pigment, or erythema (Muccini *et al.*, 1995).

(e) Ultrasound. Utilizing high-frequency B-scan sonography, the most consistently observed change in photoaging is an echo-poor band in the upper dermis (Hoffmann *et al.*, 1994). The location corresponds to the area of elastosis observed by histology and is observable in over 90% of highly sun-exposed areas. The density of the band diminishes proportionally by age only up to the age of 40–50 and the width of the band is not related to age (and thus by inference to sun exposure).

(f) Miscellaneous Instrumentation. There exists an impressive array of bioengineering techniques measuring blood flow, transepidermal water loss, skin hydration, etc., which may, when judiciously applied, have significant applicability in future study of photoaging (Serup and Jemec, 1995). While highly sensitive, reproducible, and in some circumstances validated, most of these techniques have not been widely used in clinical trials of therapeutic agents specific for photoaging.

Twistometry, indentometry, levarometry, ballistometry, and other instrumental techniques reflect many of the most clinically obvious functional abnormalities of loss of elasticity and increased extensibility. These techniques have similarly not yet been widely utilized in clinical trials.

4.3. Histologic Assessments

A most specific and characteristic aspect of photoaging is the histologic change of dermal elastosis consisting of a very characteristic band in the upper dermis separated from the epidermis by a "grenz" zone of normal collagen and believed to represent mostly altered and defectively repaired elastic fibers. Irregular epidermal atrophy and keratinocyte atypia may be accompanied by additional dermal changes of alteration of collagen and inflammatory infiltrate.

Many of these qualitative changes such as cellular atypia can be descriptively rated as mild, moderate to severe and graded, e.g., on a scale of 0 to 6. Some changes may be further quantitated by actual measurement of thickness of epidermis. Computer-assisted image analysis may be used to quantitate such parameters as epidermal thickness, melanin content, papillary dermal thickness, and percentage of total dermis occupied by elastic tissue (Bhawan *et al.*, 1991).

Proof of significant dermal change that would reflect substantial remodeling or repair of elastin or collagen has been elusive, although some ultrastructural and immunohistochemical evidence has begun to appear. Statistically significant ultrastructural dermal changes with use of tretinoin have been demonstrated (Yamamoto *et al.*, 1995). For this analysis, the degree of disorganization of collagen fibers was evaluated as: 0, no disorganization; 1, few scattered foci of disorganization; 2, numerous disorganized areas. Degree of microfibril change was rated; 0, none or rare focal degeneration; 1, scattered fragmented or nodularly arranged degenerated microfibrils. Collagen formation measured by antibody to human procollagen I showed 80% increases ($P = 0.006$) in collagen I synthesis after treatment with tretinoin (Griffiths *et al.*, 1993).

5. Summary

Biometrics applied to photoaging follows the same dictums as elsewhere. Measurements must be accurate, precise, reproducible, representative, capable of being validated, and statistically analyzed, and of great importance, should be clinically meaningful. Most of the techniques discussed herein fulfill these requirements only partially. A combination of multiple methods in clinical studies of photoaging is most useful to adequately describe and quantitate the pertinent global aspects and particular parameters of interest. The most useful techniques presently include careful clinical grading, clinical panel assessment, optical profilometry, and histology. Although there is as yet incomplete correlation among the multitude of possible clinical, histological, ultrastructural, immunohistochemical, and instrumentational observations, adequate description and quantitation of the process of photoaging and its reversal may truthfully be achieved.

References

Armstrong, R. B., Lesiewicz, J. Harvey, G., Lee, L. F., Spoehr, K. T., and Zultak, M., 1992, Clinical panel assessment of photodamaged skin treated with isotretinoin using photographs, *Arch. Dermatol.* **128**(3):352–356.

Bhawan, J., Gonzalez-Serva, A., Nehal, K., Labadie, R. Lufrano, L., Thorne, E. G., and Gilchrest, B. A., 1991, Effects of tretinoin on photodamaged skin. *Arch. Dermatol.* **127**(5):666–672.

Cunningham, W. J., Bryce, G. F., Armstrong, R. B., Lesiewicz, J., Kim, H.-J., and Sendagorta, E., 1990, Topical isotretinoin and photodamage, in: *Retinoids: 10 Years On* (J. H. Saurat, ed.), Karger, Basel, pp. 182–190.

Drake, L. A., Dinehart, S. M., Farmer, E. R., Goltz, R. W., Graham, G. F., Hordinsky, M. K., Lewis, C. W., Pariser, D. M., Webster, S. B., Whitaker, D. C., Butler, B., and Lowery, B. J., 1996, Guidelines of care for photoaging/photodamage, *J. Am. Acad. Dermatol.* **35**(3, Part 1):462–464.

Gartstein, V., and Shaya, S. A., 1986, Image analysis of facial skin features, *Proc. Int. Soc. Optic. Eng.* **626**:284–288.

Griffiths, C. E., Wang, T. S., Hamilton, T. A., Voorhees, J. J., and Ellis, C. N., 1992, A photonumeric scale for the assessment of cutaneous photodamage, *Arch. Dermatol.* **128**(3):347–351.

Griffiths, C. E., Russman, A. N., Majmudar, G., Singer, R. S., Hamilton, T. A., and Voorhees, J. J., 1993, Restoration of collagen formation in photodamaged human skin by tretinoin (retinoic acid), *N. Engl. J. Med.* **329**(8):530–553.

Grove, G. L., and Grove, M. J., 1990, Effects of topical retinoids on photoaged skin as measured by optical profilometry, *Meth. Enzymol.* **190**:360–371.

Grove, G. L., Grove, M. J., Leyden, J. J., Lufrano, L., Schwab, B., Perry, B. H., and Thorne, E. G., 1991, Skin replica analysis of photodamaged skin after therapy with tretinoin emollient cream, *J. Am. Acad. Dermatol.* **25**(2, Part 1):231–237.

Hoffmann, K., Dirschka, T. P., Stucker, M., el-Gammal, S., and Altmeyer, P., 1994, Assessment of actinic skin damage by 20-MHz sonography, *Photodermatol. Photoimmunol. Photomed.* **10**(3):97–101.

Kollias, N., Gillies, R., Cohen-Goihman, C., Phillips, S. B., Muccini, J. A., Stiller, M. J., and Drake, L. A., 1997, Fluorescence photography in the evaluation of hyperpigmentation in photodamaged skin, *J. Am. Acad. Dermatol.* **36**(2, Part 1):226–230.

Larnier, C., Ortonne, J. P., Venot, A., Faivre, B., Beani, J. C., Thomas, P., Brown, T. C., and Sendagorta, E., 1994, Evaluation of cutaneous photodamage using a photographic scale, *Br. J. Dermatol.* **130**(2):167–173.

Muccini, J. A., Kollias, N., Phillips, S. B., Anderson, R. R., Sober, A. J., Stiller, M. J., and Drake, L. A., 1995, Polarized light photography in the evaluation of photoaging, *J. Am. Acad. Dermatol.* **33**(5, Part 1):765–769.

Olsen, E. A., Katz, H. I., Levine, N., Shupack, J., Billys, M. M., Prawer, S., Gold, J., Stiller, M., Lufrano, L., and Thorne, E. G., 1992, Tretinoin emollient cream: A new therapy for photodamaged skin, *J. Am. Acad. Dermatol.* **26**(2, Part 1):215–224.

Sendagorta, E., Lesiewicz, J., and Armstrong, R. B., 1992, Topical isotretinoin for photodamaged skin, *J. Am. Acad. Dermatol.* **27**(6, Part 2):S15–S18.

Serup, J., and Jemec, G. B. E. (eds.), 1995, *Handbook of Non-Invasive Methods and the Skin*, CRC Press, Boca Raton.

Warren, R., Gartstein, V., Kligman, A. M., Montagna, W., Allendorf, R. A., and Ridder, G. M., 1991, Age, sunlight, and facial skin: A histologic and quantitative study, *J. Am. Acad. Dermatol.* **25**(5, Part 1):751–760.

Yamamoto, O., Bhawan, J., Solares, G., Tsay, A. W., and Gilchrest, B. A., 1995, Ultrastructural effects of topical tretinoin on dermo-epidermal junction and papillary dermis in photodamaged skin. A controlled study, *Exp. Dermatol.* **4**(3):146–154.

Chapter 16

Measurement in Scleroderma Clinical Trials

JANET E. POPE

1. Introduction

Scleroderma or progressive systemic sclerosis is a chronic fibrosing disease that can involve skin and internal organs. Most interventions intended to modify disease in scleroderma are ineffective. This is different from organ specific drugs that can improve the quality of life in patients with scleroderma, such as the treatment of Raynaud's with calcium channel blockers or proton pump inhibitors used to treat reflux esophagitis. The outcome measurements in scleroderma trials, often include observations of skin involvement, functional status, physical performance (such as grip strength and oral aperture) and assessment of internal organ involvement (including pulmonary, gastrointestinal, renal, and cardiac).

In scleroderma trials, very few outcome measures have detected between- or within-group differences even when an active drug was compared to placebo. Skin measures have successfully demonstrated statistical differences in some published studies. As in other chronic diseases, global assessments have yeilded sensitivity to change in scleroderma trials. Internal organ measures detected differences between groups only rarely, with the pulmonary diffusing capacity being statistically significant in intervention studies infrequently. Physical performance measures (e.g., grip strength and oral aperture) often do not significantly improve in clinical trials published in scleroderma.

In order to demonstrate drug efficacy in future trials in scleroderma, better outcome measures have to be developed and a consensus obtained on which outcomes to use so that potentially effective therapies can be tested in a standardized fashion against a placebo or current therapy. Currently, because of a lack of

---

Janet E. Pope • Department of Rheumatology, University of Western Ontario, London ON N6A 4G5, Canada.

Cutaneous Biometrics, edited by Doris A. Schwindt and Howard I. Maibach, Kluwer Academic/Plenum Publishers, New York, 2000.

clinimetric data on outcome measures, we cannot differentiate therapeutic inefficacy from a lack of sensitivity in the outcome measures used. In future, outcome measures should be chosen on the basis of the adequacy of their reliability, construct, and content validity, and be sensitive to change. Ideally, outcome measures should also have criterion validity. For instance, a strong association between the measure (such as a skin score) and an irrefutable gold standard (such as skin pathology) should be demonstrated. Trials in scleroderma should have good design such as being randomized and double-blinded whether they are placebo-controlled or compared to another agent, and the sample size and trial length should be adequate to detect differences in the outcome measurements of interest.

2. Statistics and Clinical Trial Guidelines

Scleroderma has an estimated mean annual incidence of 2.7 to 12 new cases per million population with a median age of onset between 40 to 50 years (Medsger, 1985). The female-to-male ratio is 3:1. Although the incidence is low, the mortality is not inconsequential and has been reported to be 0.9 to 3.8 per year in men and women, respectively (Silman, 1991). Scleroderma is often accompanied by chronic morbidity with higher disability scores on average as measured by the Health Assessment Questionnaire (HAQ) (Fries *et al.*, 1980) than other inflammatory arthritis such as lupus and rheumatoid arthritis (Poole and Steen 1986, 1991). Scleroderma is difficult to treat. While therapy with D-penicillamine has been suggested, a recently completed randomized trial of high-dose D-penicillamine has been found not to be superior to minidose D-penicillamine (Clements *et al.*, 1997).

Suggested guidelines for the conduct of clinical trials in scleroderma have been proposed by the American College of Rheumatology and include good study design such as randomized, blinded trials, and also suggest some parameters that can be measured in organ-specific trials (White, 1993). The committee suggested that patients with early scleroderma of less than 2 years' duration and diffuse skin involvement be included as subjects. The outcome measurements used should be standardized, reflective of disease activity, sensitive to change, and have little variability. The establishment of these guidelines are important because many scleroderma trials to date have been negative and it may be that the patients selected were not those thought to be most apt to respond even if the experimental treatment was efficacious. It is also difficult to ascertain if patients with scleroderma improve within trials because there is a tendency towards spontaneous improvement; especially after a few years of disease, within certain individuals (Black, 1990). In addition many outcome measurements have not shown between-group differences in any controlled studies in scleroderma (Pope and Bellamy, 1993).

3. Diagnosis of Scleroderma

In order to characterize patients for eligibility in scleroderma clinical trials, it is important to have reliable diagnotic criteria. The American College of Rheumatolgy criteria include one major criterion (sclerodermatous skin changes proximal to the metacarpal-phalangeal joints) or two of three minor criteria: (1) sclerodactyly, (2) digital pitting scars, and (3) bibasilar pulmonary fibrosis on chest radiograph (Special article, 1980). Other criteria for scleroderma have been suggested. One involves defining two subsets of scleroderma, each with a different prognosis. Limited scleroderma has been defined as skin involvement distal to the elbows with or without the involvement of the face. Diffuse disease involves sclerodermatous skin changes proximal to the elbows (LeRoy *et al.*, 1988). Vayssairat *et al.*, 1992 reported the results of a follow-up study of 164 patients with scleroderma and compared three classification schemes. This was a retrospective paper where patients diagnosed with scleroderma between 1984 and 1988 were followed to their last clinic visit at which time their disease severity was estimated. They compared the two above classifications (Special article, 1980 and Vayssairat *et al.*, 1992) to a third that differentiates between early cutaneous involvement using the criteria of Masi (1988), Giordano (1986), and Barnett (1988).

According to the early skin involvement classification, three groups are determined by the extent of cutaneous involvement in the first year of presentation. Digital involvement implied scleroderma solely involving the fingers and or toes. Proximal extremity involvement was defined as involving the proximal arms and/or legs but not the trunk. The third group included those with truncal sclerodermatous skin changes. The purpose of this study was to compare the three classifications of scleroderma at the patients' first clinic visit to their last recorded visit. The severity of the disease was also estimated using a quantitative scale consisting of a point system for various organ abnormalities, including skin involvement (Hughes *et al.*, 1976).

Disease severity scores were higher in those with more skin involvement in all three classification systems. The most discerning classification system for large differences in the severity scale was the diffuse versus limited classification. Strong risk factors for high severity scores included proximal sclerosis, truncal involvement, positive anti-Scl 70 antibody, pulmonary or cardiac involvement, and age. In clinical trials it may be worthwhile to randomize patients with stratification by diffuse versus limited disease, or include only those patients with more severe disease, such as those in the "diffuse sclerosis" category or those with "truncal" involvement. As the prognosis is different in the diffuse type of scleroderma compared to limited scleroderma, the patients may warrant different treatment and should be considered as representing different spectra of systemic sclerosis.

There is no truly effective therapy for halting disease progression in scleroderma. Therefore, there is a need to evaluate further therapeutic modalities. To assess efficacy, it is important to use standardized outcome measures that are valid, reliable, sensitive to change, clinically meaningful, and feasible to use. Several scleroderma studies use ad hoc scales that are not standardized and have not been formally validated.

Most drugs used in scleroderma have demonstrated little impact in modifying the disease. Various agents have been studied to determine if they change disease progression in scleroderma including anti-inflammatories and antimetabolites [5 fluorouracil (Casas *et al.*, 1990), chlorambucil (Furst *et al.*, 1989), colchicine (Alarcon-Segovia *et al.*, 1974, 1979; Steigerwald, 1985; Steigerwald and Lynch, 1977), cyclosporin A (Clements *et al.*, 1993*b*; Zacharie *et al.*, 1990; Vayssairat *et al.*, 1990; Appelboom and Itzkowitch, 1987), lymphoid radiation (O'Dell *et al.*, 1989), plasmapheresis (Weiner *et al.*, 1987), photopheresis (Rook *et al.*, 1992; Freundlich *et al.*, 1990), cyclophosphamide (Silver *et al.*, 1993; Akesson *et al.*, 1994), dexamethasone (Sharada *et al.*, 1994), antithymocyte globulin (Sinclair *et al.*, 1994), methotrexate (Seibold *et al.*, 1994; van den hoogen *et al.*, 1994)], and bone marrow transplant antifibrotic agents [D-penicillamine (Akesson *et al.*, 1992; Clements *et al.*, 1997; Steen *et al.*, 1982, 1985, 1986; de Clerck *et al.*, 1987; Jimenez and Sigal, 1991), *N*-acetylcysteine (Furst *et al.*, 1979), aminobenzoate potassium (Bushnell *et al.*, 1966; Zarafonetis, 1988*a, b*; Clegg, 1994), gamma interferon (Kahan *et al.*, 1989; Hein *et al.*, 1992), 1,25-dihydroxyvitamin D_3 (Humbert *et al.*, 1993)], hormones [cyclofenil (Gibson and Grahame, 1983; Black *et al.*, 1983, Blom-Bulow *et al.*, 1981), D-thyroxine (Winkelmann *et al.*, 1965), and relaxin (Seibold *et al.*, 1997)], cytokines [α-interferon (Stevens *et al.*, 1992)], and various other agents [antiplatelet drugs (Beckett *et al.*, 1984), antihypertensives (Fries *et al.*, 1984), Factor XIII (Guillevin *et al.*, 1985), ketotifen (Gruber and Kaufman, 1991), ketanserin (Ortonne *et al.*, 1989), isotretinoin (Maurice *et al.*, 1989), dimethyl sulfoxide (Williams *et al.*, 1985; Binnick *et al.*, 1977; Tuffanelli, 1966; Scherbel, 1983), and griseofulvin (Giordano *et al.*, 1985)].

There are a number of reasons why earlier studies may have failed, in general, to detect any therapeutic effect. The sample size of many studies is small. A Type II error may occur as a consequence. In addition, there is no consensus on which outcome measures should be used in scleroderma clinical trials, and we have little knowledge of their sensitivity to change. The response potential of participating patients is also in question. For example, many trials enrolled patients with prolonged disease duration. Such patients may not be apt to change if fibrosis has already occurred unless the drug under study is thought to reverse fibrosis, as opposed to preventing inflammation, which occurs earlier in the disease pathway. In contrast, patients with disease of less than a few years' duration may be more responsive to treatment with disease-modifying drugs (Seibold *et al.*, 1992). Finally, most trials have been of relatively short duration; often less than 1 year. This time frame may not be sufficient given the long induction–response interval often seen in scleroderma. Table 1 lists several of the controlled trials in scleroderma, where the intent was to modify disease progression. These trials were often small and had less than optimal study design (often not randomized, with inadequate study duration).

A review of clinical trials in scleroderma summarized the major categories of outcome measures used (Pope and Bellamy, 1993). These included: (1) an estimate of skin involvement including a semiquantitative measurement based on skin scores, weight of skin biopsies, percentage of body surface area that was abnormal, measurements of skin elasticity, and radiographs or ultrasound to quantitate skin thickness, (2) functional outcomes including the disability scale of the HAQ (Fries *et al.*, 1980), and a functional index (Guillevin and Ortonne, 1983) developed for scleroderma, (3) global indices rating the subject's overall status assessed by the individual and/or the physician, (4) physical performance measurements, such as grip strength, oral opening, and palm prints, (5) measurement of joint involvement using goniometric measurements or an articular index, (6) the degree of lung involvement assessed by pulmonary function testing, radiographs, and symptoms, and (7) estimates of other internal organ involvement based on symptoms, signs, laboratory, radiographic and nuclear scan parameters, or composite organ indices.

Many baseline measurement characteristics are significantly different in patients with diffuse compared to limited systemic sclerosis (Pope *et al.*, 1995). Sample sizes can vary widely depending upon the outcome measurement chosen and the desired change of a parameter used within the different scleroderma subsets. Sample size requirements for many outcome measures are extremely large owing to marked variability in the baseline measures. Primary outcome measures in scleroderma trials should be chosen that have adequate power to detect a minimally clinically relevant change in the primary outcome measurements at the sample sizes employed. Skin scores, global assessments, and grip-strength measurements require smaller sample sizes than the several other outcome measurements. Sample sizes in future trials will vary depending upon the proportion of patients with diffuse and limited scleroderma included.

4. Skin Outcome Measurements

Skin involvement is frequently assessed in scleroderma clinical trials (Pope and Bellamy, 1993). Often skin scores are used. One can also estimate the extent of skin involvement by determining the percentage of the surface area with scleroderma. The latter technique involves shading the abnormal skin on a schematic diagram of a mannequin shown both anteriorly and posteriorly. A skin score is a semiquantitative estimate summing the severity of scleroderma in several set areas of skin. The UK Scleroderma Study Group compared the reliability of the two techniques using 12 patients and 6 experienced observers (Brennan *et al.*, 1992). A modified Rodnan skin score with a scale of 0–3 was used to estimate the severity of skin involvement in 17 areas of the body. The percentage of surface area used a standard diagram with 9 sites. Each area was scaled to a maximum of 11% and the extra 1% was for the genitalia. Interpatient variability, interobserver variability, and the remaining "residual" variation were calculated using intraclass correlation coefficients. A large residual variation corresponds to less precision. There was more precision in the skin score method. The mannequin method was especially poor when it measured patients with large surface area involvement.

Other researchers found that physicians with less experience estimated higher average skin scores in 12 patients who were evaluated by three physicians: a consultant, a registrar, and a house officer (Harrison *et al.*, 1992). They also found a lower intraclass correlation coefficient and therefore more variability between observers than what was reported from the Brennan *et al.* (1992) study. Both the modified Rodnan skin score with 26 sites (Steen and Follansbee, 1992) and the British skin score using 17 sites were compared. The agreement was better when rating the latter scale with fewer sites to assess.

Furst *et al.* (1989) published a skin score with limited variability, and high reproducibility having a coefficient of variation for interobserver error of 0.08. Their skin score involves 10 areas with a scale of 0–3 in each area. With their score, higher

TABLE 1. Selected Controlled Studies in the Treatment of Scleroderma Intended to Modify Overall Disease Activity[a]

Study	Random	Blind	Design	Control	Time (months)	n	Effective
Factor X111 (Guillevin *et al.*, 1985	+	DB	Crossover	Placebo	1.5	25	?
Colchicine (Steigerwalk and Lynch, 1977; Steigerwald, 1985)	+	DB	Crossover	Placebo	9	28	No
Colchicine (Alarcon-Segovia *et al.*, 1974)	+	DB	Crossover	Placebo	6	14	+
Photopheresis (Freundlich *et al.*, 1992; Rook *et al.*, 1992)	+	SB	Parallel	D-penicillamine	10	79	No
Apheresis (Weiner *et al.*, 1987)	+	SB	Parallel	No Rx, plasmapheresis, lymphoplasmapheresis	3	16	+
Cyclosporine A (Clements *et al.*, 1993*b*)	No	No	Parallel	Placebo from another study	11	23	+
Radiation (O'Dell *et al.*, 1989	+	SB	Parallel	No Rx	>24	6	No
Chlorambucil (Furst *et al.*, 1989)	+	DB	Parallel	Placebo	36	65	No
5-Fluorouracil (Casas *et al.*, 1990)	+	DB	Crossover	Placebo	6	70	No
DMSO (Williams *et al.*, 1985)	+	DB	Parallel	Placebo and 2% DMSO	4	84	No
DMSO (Binnick *et al.*, 1977)	No	No	Parallel	Other extremity 2% DMSO	5	19	No
DMSO (Tuffanelli, 1966)	No	SB	Parallel	Other extremity	4	24	No
DMSO (Scherbel, 1983)	No	No	Parallel	Other extremity	6	22	No
N-Acetylcysteine (Furst *et al.*, 1979)	+	DB	Parallel	Placebo	12	22	No
D-penicillamine (Clements *et al.*, 1997)	+	DB	Parallel	Low dose	24	143	No

D-penicillamine (Steen *et al.*, 1982)	No	No	Parallel	Historical	36	118	+
D-penicillamine (Steen *et al.*, 1985)	No	No	Parallel	Historical	29	92	+
D-penicillamine (de Clerck *et al.*, 1987)	No	No	Parallel	Historical	55	23	+
Cyclofenil (Gibson and Grahame, 1983)	+	DB	Crossover	Placebo	4	11	+
Cyclofenil (Black *et al.*, 1983)	+	DB	Parallel	Placebo	6	28	?
Cyclofenil (Blom-Bulow, 1981)	+	DB	Crossover	Placebo	6	38	?
Potaba (Bushnell *et al.*, 1966)	+	DB	Parallel	Placebo	6	12	No
Potaba (Zarafonetis *et al.*, 1988*a, b*)	No	No	Parallel	Historical	$\geqslant 33$	20	+
Antihypertensive (Fries *et al.*, 1984)	+	DB	Parallel	Placebo	24	28	No
Antiplatelet (Beckett *et al.*, 1984)	+	DB	Parallel	Placebo	>12	41	No
Ketotifen (Gruber and Kaufman, 1991)	+	DB	Parallel	Placebo	6	24	No
D-thyroxine (Winkelmann *et al.*, 1965)	+	DB	Parallel	Placebo	8	32	No
Ketanserin (Ortonne *et al.*, 1989)	+	DB	Parallel	Placebo	6	24	No
Antithymocyte globulin (Sinclair *et al.*, 1994)	+	DB	Parallel	Placebo	6	16	No
Methotrexate (van de Hoogen *et al.*, 1994)	+	DB	Parallel or crossover	Placebo	11	25	No
Dexamethasone (Sharada *et al.*, 1994)	+	DB	Parallel	Placebo	6	35	Yes
Relaxin (Seibold *et al.*, 1997)	+	DB	Parallel and low dose	Placebo	24	68	Yes*

[a]Symbols and abbreviations: + = yes, DB = double blind, SB = single blind, ? = unknown. *at low dose.

skin scores are also strongly related to mortality (Clements *et al.*, 1990). There is also a rough correlation between the extent of skin involvement and the degree of internal organ involvement (Krieg and Meurer, 1998; Barnett *et al.*, 1988). Therefore a reproducible skin score may be an important outcome parameter in future scleroderma clinical trials. A skin score also has prognostic value. The skin score has been shown to be related to the severity of scleroderma in long-term studies (Clements *et al.*, 1990; Barnett *et al.*, 1988).

There are several other published reports evaluating the inter- and intraobserver variability of skin scores used to assess the severity of skin involvement in scleroderma (Clements, *et al.*, 1993*a*, 1995; Harrison *et al.*, 1992; Pope *et al.*, 1995; Silman *et al.*, 1995). Intraobserver agreement is usually better than that between observers, as would be expected. Perhaps, if the techniques were better standardized and the assessors thoroughly trained, the variability of any skin score would be diminished. Also, in a more homogeneous group of patients, such as those with early diffuse scleroderma or only those with limited disease, there may be better agreement between observers. One cannot definitively recommend which skin scoring system to use in a clinical trial, but training the outcome assessors is important to decrease the variability in any clinical study. In general, the variability is less when using a skin scoring method as compared to a drawing of skin involvement or skin map (Brennan *et al.*, 1992).

TABLE 2. Skin Outcome Measurements That Have Been Used in Scleroderma Trials

Type	Grade	Body areas	Maximum score
Rodnan (Rodnan, 1981)	0 to 4	26	104
Modified Rodnan (Giordano *et al.*, 1985)	0 to 3	26	78
Percent body surface area (Rook *et al.*, 1992)	—	—	100%
Variations of the modified Rodnan			
Kahaleh *et al.*, 1986)	0 to 3	22	66
Photopheresis trial (Rook *et al.*, 1992)	0 to 3	15	45
Ketotifen trial (Gruber and Kaufman, 1991)	0 to 3	19	57
Chlorambucil trial (Furst *et al.*, 1989)	0 to 3	11	33
Reliability study (Brennan *et al.*, 1992)	0 to 3	17	51
Other skin measurements			
Biopsy—wet or dry weight, pathology			
Excretion of cytokines important in skin turnover, inflammation or fibrosis			
Ultrasound and radiographic thickness measurements			
Thermography			
Measurements of elasticity and distensibility			

In one review of published trials in scleroderma, several different skin measurements were used (Pope and Bellamy, 1993). Table 2 illustrates the types of skin assessment made in different studies. Most commonly skin scoring is performed. Other skin measures have been used, including a visual analog scale (VAS) for tightness, skin temperature, a skin map, the amount of proline in skin biopsies and use of a durometer to assess skin distensibility. Some studies have used skin biopsies (wet and dry weight) as an outcome measure. All three studies using the Rodnan skin score detected statistical differences in this outcome between the active treatment and control groups (Weiner *et al.*, 1987; Steen *et al.*, 1982, 1985). While this outcome measure appears capable of detecting significant change, in only one trial were the participants randomized. The skin score developed by Furst *et al.* (1989) detected significant between-group differences in one of two studies (Clements *et al.*, 1993*b*; Furst *et al.*, 1989). A measure of percent of body surface area involved with scleroderma was used in the 10-month trial comparing photopheresis and D-penicillamine (Rook *et al.*, 1992; Freundlich *et al.*, 1990). In that trial, a significant improvement was detected by the surface area measure in both treatment groups (within-group comparisons) although no significant between-group difference was observed. A trial of ketotifen versus placebo used another skin score totalling 57 (Gruber and Kaufman, 1991). Since the drug was pronounced ineffective, it is impossible to evaluate the sensitivity to change of this assessment technique.

Of the other skin measures used, few detected significant between- or within-group differences. However, skin measurements, especially with a clinical skin scoring method, are sensitive to change in the disease process and are correlated to other morbidity and mortality, so they are important in the study of interventions that may modify disease activity in scleroderma patients.

Other techniques are being developed for specific subsets of the scleroderma spectrum. For instance, children with linear scleroderma were evaluated by thermography to try to predict active and inactive skin lesions (Birdi *et al.*, 1992). Thermography is a noninvasive technique that uses an infrared camera to detect skin temperature changes. A positive study was considered a lesion that was at least 0.5°C warmer than the surrounding skin or opposite limb. Six of eighteen skin lesions were positive. The patients with positive and negative thermography results were not different with respect to lab parameters, age, and disease duration. All new and increasing linear scleroderma lesions were positive with thermography and all resolving lesions were negative. Further testing is needed to estimate the sensitivity and specificity of this technique before it is recommended for clinical trials or widespread clinical use.

4.1. Functional Assessment

Physical function has been assessed in some trials using either the HAQ disability scale (Fries *et al.*, 1980), or the functional index (Guillevin and Ortonne, 1983). The functional index includes 11 questions concerning upper extremity activities. A statistically significant treatment effect has been detected only rarely in four studies that used functional assessment techniques (Guillevin *et al.*, 1985; Fries *et al.*, 1984; Gruber and Kaufman, 1991; Ortonne *et al.*, 1989). However, in the last three studies, where such assessments were used, the therapy was considered ineffective, and in the Guillevin *et al.* (1985) study, the efficacy of therapy was uncertain. Therefore, the sensitivity of these two questionnaires in scleroderma cannot be ascertained from the current trials literature.

4.2. Global Assessments

Global assessments by patient or physician have frequently been used in scleroderma trials and sometimes detected significant treatment effects. For instance, in a previous publication reporting the seventeen occasions where either patient or physician global assessments were performed, a P-value $\leqslant 0.05$ was detected seven times (Pope and Bellamy, 1993). Patient global assessments, or patient preference measures in some crossover trials detected significant differences favoring active treatment in three trials; but not in six others. Physician global assessment yielded significant differences in four of eight studies. One trial employed global measures, but provided no corresponding data (Scherbel, 1983).

4.3. Physical Performance Assessments

The oral opening or distance between upper and lower teeth with the mouth fully open has been used in many scleroderma trials (Pope and Bellamy, 1993). The relevance of this assessment is that as the skin softens around the lips, the oral opening potentially could improve. No significant treatment effect was detected in any study except a within-group comparison in the photopheresis versus D-penicillamine trial (Rook *et al.*, 1992; Freundlich *et al.*, 1990).

The distance between the third finger to distal palmar crease in full flexion is the finger–palm distance or flexion index. Only rarely has a significant treatment effect been reported in this measure (Pope and Bellamy, 1993). Other hand measures such as palm prints were seldom used and did not detect significant treatment effects. Change in hand closure was significant in a within-group comparison in one trial comparing photophoresis to D-penicillamine (Freundlich *et al.*, 1990). Grip strength was measured in six studies and in only one did the change from baseline reach statistical significance (Steigerwald and Lynch, 1977; Scherbel, 1983; Furst *et al.*, 1979; Blom-Bulow *et al.*, 1981; Bushnell *et al.*, 1966; Beckett *et al.*, 1984; Ortonne *et al.*, 1989).

Often joint measurements were used including the following: range of motion as assessed by a goniometer, tender joint count, change in joint radiographs, and rating joint pain as being "better," "same," or "worse." These measures detected significant treatment effects in one (Clements *et al.*, 1993*b*) of the fourteen studies in which they were employed (Pope and Bellamy, 1993). It is difficult to know whether the measurements are not very sensitive to change or the treatment is ineffective or both. Probably to date, these measurements are not very good and the treatment has also been largely ineffective.

4.4. Visceral Assessments

Respiratory outcomes frequently include pulmonary function tests (PFTs). The diffusing capacity (DLCO) is often decreased early in patients with pulmonary involvement in scleroderma. This measurement certainly can worsen over time in someone with active interstitial lung disease and may possibly improve in reversible disease. A 15% change in DLCO in each subject may be the minimum clinically important treatment effect, as the intrapersonal changes may be $\pm 10\%$. Other pulmonary measurements include parameters on PFTs that demonstrate interstitial lung disease (flows and volumes), dyspnea scales and chest radiographs noting the presence or absence of fibrosis. Of the fifteen studies measuring pulmonary outcomes, only two reported a significant between-group difference (Pope and Bellamy, 1993). Both studies were nonrandomized, comparing D-penicillamine to other treatments, with the difference in the percent predicted diffusing capacity being statistically different (Steen *et al.*, 1985; de Clerck *et al.*, 1987).

The sensitivity of high-resolution pulmonary computed tomography (CT) to detect early pulmonary involvement in 18 patients with scleroderma was compared to other investigations including chest radiographs, pulmonary function testing, and ventilation perfusion scans (Pignone *et al.*, 1992). A grading system was used for the CT scans from 0 to 3. High-resolution CT was significantly correlated to chest radiographs. However, the CT was more sensitive than radiographs and lung scans in the early detection of interstitial lung involvement. Some 72% of the patients had an abnormal CT whereas the chest radiograph was abnormal in only 55%. High-resolution CT could be used as an outcome measurement in studies aimed at detecting early lung involvement.

Klein *et al.* compared esophageal dysfunction in scleroderma using manometry, cine-esophagraphy, and esophageal transit scintigraphy in 17 patients with scleroderma (Klein *et al.*, 1992). There were no significant differences in esophageal findings between those with diffuse and limited disease. None of the esophageal function parameters correlated with symptoms. The parameters were strongly correlated with each other. The authors concluded that the three diagnostic modalities were not significantly different in either detecting esophageal dysmotility or measuring its severity.

Studies have used gastrointestinal outcome measures including upper and lower GI radiographs, esophageal manometry, amount of fecal fat, upper GI endoscopy, and symptoms of dysphagia (Pope and Bellamy, 1993). These outcome measurements have not demonstrated clinically relevant differences when used in scleroderma trials.

Renal outcome measures for efficacy of treatment included creatinine clearance, 24-hr protein excretion, and the presence of renal crisis. The onset of renal crisis is an important event in patients with diffuse scleroderma and is a relevant outcome measurement in trials studying subjects with early diffuse scleroderma. This outcome measurement was used in a recently reported large 2-year trial of D-penicillamine (Clements *et al.*, 1997). Treatment was no different in the high-dose and low-dose groups but renal crisis did occur in some subjects during the trial.

The measurement of cardiac involvement has included symptoms of congestive heart failure, signs of right or left heart failure on physical examination, cardiomegaly on chest radiograph, ECGs, echocardiograms, and Holter monitoring. No significant treatment effects were observed in six studies monitoring cardiac parameters (Pope and Bellamy, 1993). In one study comparing cyclofenil to placebo, the work capacity was lower after active treatment compared to placebo, at $P < 0.05$ using a one-sided test (Blom-Bulow, 1981). Therefore, in general, measures of internal organ involvement did not detect significant treatment effects, with the possible exception of the diffusing capacity and rate of new renal involvement.

Several trials have reported the frequency and size of skin ulcers, frequency and severity of Raynaud's phenomenon, and biochemical and immunological parameters such as the ANA, ESR, interleukin-2 receptors, and urinary excretion of hydroxyproline as outcome measures. None of these measures detected significant between-group differences (Pope and Bellamy, 1993).

Survival was employed as an outcome measure in three studies. Two analyses were retrospective and compared other treatments to either D-penicillamine or potaba. Both drugs were associated with improved survival when compared to other therapies (Steen *et al.*, 1985, Zarafonetis *et al.*, 1988*a*, *b*). However since these studies were nonrandomized, the differences in survival could have been due to differences in preintervention health status rather than differences in drug efficacy. The large 2-year low-dose compared to high-dose D-penicillamine trial did not demonstrate differences in survival (Clements *et al.*, 1997).

5. Raynaud's Phenomenon in Scleroderma

A 6-year follow up study of 28 patients with Raynaud's phenomenon who initially had a scleroderma capillary pattern on nailfold microscopy but did not fulfill the ARA criteria for scleroderma were evaluated to assess the prognostic significance of this abnormal capillary pattern (Zufferey *et al.*, 1992). Of the patients, 8 (28%) developed a scleroderma spectrum disorder. Avascularity and two or more megacapillaries were highly predictive of developing a connective tissue disease, and 15 patients with only a few enlarged capillaries at entry did not develop scleroderma. The capillary abnormalities resolved in 14 of the 15 patients. Capillary microscopy was found to have good prognostic significance and has been included in the above criteria for differentiating between primary and secondary Raynaud's phenomenon (LeRoy and Medsger, 1992). However, although the presence of abnormal periungal capillaries can differentiate between primary and secondary Raynaud's, this is not a likely outcome measurement in clinical trials as it is not apt to change over time.

Several agents have been studied in the treatment of Raynaud's in scleroderma, including calcium channel blockers and prostacyclin analogs. Calcium channel blockers are effective in Raynaud's (Ferri *et al.*, 1992; Rademaker *et al.*, 1989). However, subjects with secondary Raynaud's often have more severe symptoms, including digital ulcers. Thus the treatment effect is less beneficial in those with more severe Raynaud's. Iloprost was effective in a randomized placebo-controlled trial for the treatment of ischemic ulcers from scleroderma (Wigley *et al.*, 1992). Most of the ulcers in the iloprost group healed completely at 10 weeks whereas none in the placebo group had healed in that time. Another randomized crossover trial of iloprost for the treatment of Raynaud's phenomenon included four patients with scleroderma (Kyle *et al.*, 1992). Iloprost significantly reduced the frequency of attacks but was not significantly different from placebo in the other parameters measured.

A randomized study of cicaprost, which is an oral prostacylin analog was performed in scleroderma patients with Raynaud's phenomenon (Lau *et al.*, 1993). The severity of attacks was slightly less than placebo for higher doses of cicaprost. This result was not significant at the end of the trial (10 days) but was significant at the 2-week posttreatment follow-up. The other outcome measurements were not significantly different between the active and placebo groups. It may be that a larger trial of longer duration using higher doses of cicaprost will be necessary to determine its efficacy in the treatment of Raynaud's phenomenon. Outcome measurements including Raynaud's frequency and severity and possibly assessment of digital ulcers may all be valid and sensitive to change in scleroderma-associated Raynaud's. Many temperature measurements have been used but they are not standardized.

One must also use good trial design (randomization and blinding) because there appears to be a large placebo effects in Raynaud's trials (which may be explained partially by regression to the mean and psychological improvement in a trial), which may improve Raynaud's in individuals to a greater degree than a usual expected placebo effect of 20%. Often patients are entered when the Raynaud's is worsening and the natural trend can be for improvement over time, illustrating regression to the mean.

6. Discussion

Many of the outcome measures used in scleroderma trials did not detect statistically significant differences in the treatment effect of an active drug, even when compared to placebo. This is a dilemma as we cannot tell if most measures used to date are responsive to change. There are five possibilities why the majority of outcome measures did not detect differences: (1) The treatments of scleroderma may have been inefficacious. (2) Patients may have been chosen who are largely unresponsive to therapy. (3) The duration of the trials may have been too short. (4) There may have been inadequate power to detect clinically meaningful differences (Type II error). (5) Finally, many of the outcome measures used in these studies may not have been sufficiently sensitive to change.

Several investigators have reported that those with disease duration of less than 5 years are more likely to respond to therapy compared to those with longer disease duration, especially if drugs are studied to reverse the early features of the disease (White 1993; Seibold *et al.*, 1992). Those with higher skin scores are also likely to be more responsive to treatment compared to those with limited disease and lower skin scores (Seibold *et al.*, 1992). Thus it would seem reasonable to include only patients with diffuse, early disease in future scleroderma clinical trials, if the goal of therapy

TABLE 3. Selected Outcome Measures for Scleroderma Trials Where the Intent Is to Modify the Disease

Primary
Skin score—seems more reliable than the percent body surface area
Global asssessment (blinded)—performed by assessor and/or patient
Mortality (if the trial is large enough and for long enough)

Secondary
Functional assessment—ex. modified HAQ, functional index
Physical parameters—ex. grip strength, oral aperture, flexion index, presence of friction rubs
Pulmonary involvement—ex. change in percent predicted DLCO, other flows, and volume measurements on PFTs set *a priori*, dyspnea scales
Renal involvement—ex. presence of new renal involvement, new hypertension, renal crisis
Cardiac involvement—ex. thallium scan or echocardiogram, presence of congestive heart failure

Future outcome measurements which may be of value
A "scleroderma specific" disease activity index
Specific cytokines or fibroblast proliferation markers which are strongly correlated to disease activity

[a]These would be outcomes of interest but the sample size may not be large enough to detect statistically significant differences.

is to halt or improve the progressive component of scleroderma. If, however, an antifibrotic drug were being tested, perhaps patients with stable late disease would be studied to determine if scarring or fibrosis could be reversed.

Which outcome measures should be used in future scleroderma trials is controversial. There is no current consensus, and the clinimetric properties of many outcome measures are unknown. Ideally outcome measures should reflect several aspects of scleroderma including skin involvement, a global assessment, physical function, internal organ involvement, and mortality. Table 3 suggests potential outcome measurements that can be chosen in future scleroderma trials. We advocate choosing responsive measures that are clinically relevant, reliable, and valid. For example, a clinical skin score would be most useful if it corresponded to some aspect of skin histology and if its value changed as a patient's skin involvement changed.

A recent editorial discussed some of the issues in selecting outcome measures in scleroderma and noted that skin involvement seemed to correlate with visceral involvement (Seibold *et al.*, 1992), and Steen *et al.* (1982) found that high skin scores roughly correlated with more severe internal organ involvement. The skin score may be a proxy for severity of disease in trials of relatively short duration (Kreig and Meurer, 1988). A higher skin score is related to mortality (Clements *et al.*, 1990), and at least one skin score has less observer bias than using the percent body surface area (Brennan *et al.*, 1992). Most skin measures, aside from skin scores or estimations of the amount of body surface area involved, did not detect change in any of the trials. However, Weisman *et al.* (1991) stated that using a skin measurement as the primary efficacy variable can result in bias given the natural history of scleroderma, where a gradual softening of the skin may occur in some patients after an initial thickening over the first couple of years of the disease. This bias should be equal in treatment groups provided a randomized controlled trial is performed, so any between-group differences detected would be beyond the natural history of skin softening in some patients.

Both patient and physician global assessments seem useful in detecting between-group differences. It is noteworthy that these measures have been useful in several other chronic diseases including other rheumatic diseases such as rheumatoid arthritis (van der Heide *et al.*, 1992; Anderson *et al.*, 1989). These measures could also be used as primary outcome measures.

Many patients have functional impairments from their scleroderma. Of the five trials employing functional outcome measures only one detected a difference between an active drug and placebo. Given limited experience, one cannot tell whether the HAQ (Fries *et al.*, 1980), Modified HAQ (Poole and Steen, 1991), or Functional Index (Guillevin and Ortonne, 1983) are sensitive measures in scleroderma trials. The HAQ has been validated in several diseases, and in scleroderma higher disability scores are correlated to higher skin scores and lower grip strengths (Poole and Steen, 1991), whereas the functional index has been used in scleroderma but we have not found published reliability or validity characteristics for this measure.

Secondary outcomes could also include physical performance measures such as grip strength, estimates of internal organ involvement, and differences in mortality if the sample size is small or of short duration. The secondary outcome measures would be used only to detect trends and would be thought to have inadequate power to demonstrate clinically important changes in most sample sizes conventionally used in scleroderma trials.

Organ involvement in scleroderma causes morbidity and is related to the mortality in scleroderma. When assessing organ involvement, it may be difficult to detect change if a trial is of relatively short duration (from 1 to 3 years) or if the number of patients entered is small. Proxies for organ-specific involvement can be developed and used as secondary outcome measurements if they are thought not to have adequate power to show statistically significant differences within the constraints of a clinical trial. For instance, diffusing capacity can be used as a proxy for lung involvement, and the development of renal crisis or onset of renal insufficiency defined by an arbitrary rise in serum creatinine can be as proxies for renal involvement. Outcome assessments for cardiac involvement could include echocardiogram and ECG information or changes on cardiac thallium perfusion scans, both of which have been shown to correlate to mortality (Clements *et al.*, 1991; Steen and Follansbee, 1992).

Mortality is an important endpoint in scleroderma, but the likelihood of mortality being significantly different between groups in a clinical trial of only 1 to 2 years duration is unlikely, unless an extremely efficacious therapy was found. Therefore, mortality too would be a secondary outcome measurement unless a trial was quite large and of long duration ($\geqslant 5$ years). Table 3 provides suggestions for outcome measures that could be included in future clinical trials.

Future measurements may include a validated scleroderma disease activity index (i.e., a composite of several measures). Perhaps biological disease markers will be discovered that are strongly related to disease activity. Once outcome measures are chosen, there will still not be a rational basis for sample-size calculation as it is based on the minimum clinically important amount of change as decided by the investigators, and the variance in an outcome measure at baseline (Colton, 1974). It is necessary to establish the reliability, validity, responsiveness, and sample size requirements of those core measures which by consensus adequately captures the dimensionality of the disorder and its potential response to pharmacotherapy.

Treatments in scleroderma in the future should be tested in a double-blinded fashion with the use of recommended guidelines. One can see that several drugs tested in uncontrolled intervention studies are seemingly positive and when a randomized blinded trial is actually performed, the drug is no better than the control intervention, which is usually a placebo. This could be due to reporting bias (only reporting subjects who do well) or biases due to unblinded investigators and subjects who are hopeful that the intervention will be helpful. The uncontrolled studies do not allow us to determine what the natural improvement effect in the study would be and therefore, randomization, blinding, and a mandatory control group are all imperative in study design before any drug use for scleroderma can be interpreted as having positive effects. The duration of any trial should be adequate to demonstrate the hypothesized biological effect on the disease process. This may help us to show appropriate disease-modifying treatment in scleroderma when drugs with efficacy are studied.

The treatment of system-specific complaints in scleroderma is improving over time with the addition of new drugs and techniques that are better tolerated and more efficacious than in the past. As the pathophysiology of scleroderma becomes better understood more specific therapies may be developed for the treatment of scleroderma and proven of value with rigorous study design.

References

Akesson, A., Forsberg, L., Hederstrom, E., and Wollheim, F. A., 1986, Ultrasound examination of skin thickness in patients with progressive systemic sclerosis (scleroderma), *Acta Radiol Diagn.* **27**:472–477.

Akesson, A., Blom-Bulow, B., Scheja, A., Wollmer, P., Valind, S., and Wollheim, F. A., 1992, Long-term evaluation of penicillamine or cyclofenil in systemic sclerosis: Results from a two-year randomized study, *Scand. J. Rheumatol.* **21**:238–244.

Akesson, A., Scheja, A., Lundin, A., and Wollheim, F. A., 1994, Improved pulmonary function in systemic sclerosis after treatment with cyclophosphamide, *Arthritis Rheum.* **37**:729–735.

Alarcon-Segovia, D., Ibanez, G., Hershenobich, D., and Rojkind, M., 1974, Treatment of scleroderma by modification of collagen metabolism: A double blind trial with colchicine-placebo (abstr). *J. Rheum.* **1**:S97.

Alarcon-Segovia, D., Ramos-Niembro, F., Ibanez de Kasep, G., Allcocer, J., Perez Tamayo, R., 1979, Long-term evaluation of colchicine in the treatment of scleroderma, *J. Rheumatol.* **6**:705–712.

Anderson, J. J., Felson, D. T., Meenan, R. F., and Williams, H. J., 1989, Which traditional measures should be used in rheumatoid arthritis clinical trials? *Arthritis Rheum.* **32**:1093–1099.

Appelboom, T., and Itzkowitch, D., 1987, Cyclosporine in successful control of rapidly progressive scleroderma, *Am. J. Med.* **82**:866–867.

Barnett, A. J., Miller, M. H., Littlejohn, G. O., 1988, A survival study of patients with scleroderma diagnosed over 30 years (1953–1983): The value of a simple cutaneous classification in the early stages of the disease, *J. Rheumatol.* **15**:276–283.

Beckett, V. L., Conn, D. L., Ruster, V., Osmundson, P. J., Strong, C. G., Chao, E. Y. S., Chesebro, J. H., O'Fallon, W. M., 1984, Trial of platelet-inhibiting drug in scleroderma: Double-blind study with dipyridamole and aspirin, *Arthritis Rheum.* **27**:1137–1143.

Binnick, S. A., Shore, S. S, Corman, A., and Fleischmajer, R., 1977, Failure of dimethyl sulfoxide in the treatment of scleroderma, *Arch Dermatol.* **113**:1398–1402.

Birdi, N., Short A., Rush P., Laxer R. M., Silverman, E. D., and Krafchik, B., 1992, Childhood linear scleroderma: a possible role of theramography for evaluation, *J. Rheumatol.* **19**:968–973.

Black, C., 1990, Systemic sclerosis: Is there a treatment yet?, *Ann. Rheum. Dis.* **49**:735–737.

Black, C. M., Jayson, M. I. V., White, A. G., and Million, R., 1983, A double-blind comparison of cyclofenil and placebo in systemic sclerosis (SS) (abstr), *Ann. Rheum. Dis.* **42**:229.

Blom-Bulow, B., Oberg, K., Wollheim, R. A., Persson, B., Jonson, B., Malmberg, P., Bostrom, H., and Herbai, G., 1981, Cyclofenil versus placebo in progressive systemic sclerosis: A one-year double-blind crossover study of 27 patients, *Acta. Med. Scand.* **210**:419–428.

Brennan, P., Silman, A., Black, C., Bernstein, R., Coppock, J. Maddison, P., Sheeran, T., Stevens, C., and Wollheim, F., 1992, Reliability of skin involvement measures in scleroderma, *Br. J. Rheumatol.* **31**:457–460.

Bushnell, W. J., Galens, G. J., Bartholomew, L. E., Thompson, G. *et al.*, 1966, The treatment of progressive systemic sclerosis: A comparison of para-amino-benzoate and placebo in a double blind study (abstr), *Arthritis Rheum.* **9**:495.

Casa, J. A., Saway, P. A., Villarreal, I., Nolte, C. Menajovsky, B. L., Escudero, E. E., Blackburn, W. D., Alarcon, G. S., Subauste, C. P., 1990, 5-fluorouracil in the treatment of scleroderma: A randomised, double blind, placebo controlled international collaborative study, *Ann. Rheum. Dis.* **49**:926–928.

Clegg, D. O., Reading, J. C., Mayes, M. D., Seibold, J. R., Harris, C., Wigley, F. M., Ward, J. R., Pisko, E. D., Weisman, M. H., Lee, P., Moreland, L. W., Neuner, R., Gall, E. P., Clements, P., Spindler, J., and Lally, E. V., 1994, Comparison of aminobenzoate potassium and placebo in the treatment of scleroderma, *J. Rheumatol.* **21**:105–110.

Clements, P. J., Lachenbruch, P. A., Cheung, S., Ng, S. C., Simmons, M., Sterz, M., Furst, D., 1990, Skin score: A semiquantitative measure of cutaneous involvement that improves prediction of prognosis in systemic sclerosis, *Arthritis Rheum.* **33**:1256–1263.

Clements, P. J., Lachenbruch, P. A., Furst, D. E., Paulus, H. E., Sterz, M. G., 1991, Cardiac score: A semiquantitative measure of cardiac involvement that improves prediction of prognosis in systemic sclerosis, *Arthritis Rheum.* **34**:1371–1380.

Clements, P. J., Lachenbruch, P. A., Seibold, J. R., Zee, B., Steen, V. D., Brennan, P., Silman, A. J., Allegar, N., Varga, J., Massa, M., Wigley, F. M., Ingenito, F., Weisman, M., White, B., Martin, R. W. McCloskey, D., Moreland, L. W., Mayes, M., Lally, E. V., Unanue, M., Collier, D. H. Weiner, S., Weinstein, A., Medsger, T. A. Jr., Andrews, B., Dixon, M., and Furst, D. 1993*a*. Skin thickness score in systemic sclerosis: An assessment of interobserver variability in 3 independent studies, *J. Rheumatol.* **20**:1892–1896.

Clements, P. J., Lachenbruch, P. A., Sterz, M., Danovitch, G., Hawkins, R., Ippoliti, A., Paulus, H. E., 1993*b*. Cyclosporine in systemic sclerosis: Results of a forty-eight-week open safety study in ten patients, *Arthritis Rheum.* **36**:75–83.

Clements, P., Lachenbruch, P., Seibold, J., White, B., Weiner, S., Martin, R., Weinstein, A., Weisman, M., Mayes, M., Collier, D., Wigley, F., Medsger, T. Jr., Steen, V., Moreland, L., Dixon, M., Massa, M., Lally, E., McCloskey, D., Varga, J., Ingenito, F., and Furst, D., 1995, Inter- and intra-observer variability of total skin thickness score (modified-Rodnan) in systemic sclerosis (SSc), *J. Rheumatol.* **22**:1281–1285.

Clements, P. J., Wong, W. K., Seibold, J. R., Furst, D. E., Mayes, M., White, B., Wigley, F., Weisman, M., Barr, W., Weinstein, A., Medsger, T. Jr., Steen, V., Martin, R., Moreland, L., Collier, D., Lally, E., Varga, J., Andrews, B., Weiner, S., and Abeles, M., 1997, High-dose (HI-DPA) vs. low-dose (LO-DPA) penicillamine in erly diffuse systemic sclerosis (SSc) trial: Analysis of trial, *Arthritis Rheum.* **40**(suppl):S173.

Colton, T., 1974, *Statistics in Medicine*. Little, Brown, Boston, pp. 142–146.

de Clerck, L. S., Dequeker, J., Francx, L., and Demedts, M., 1987, D-penicillamine therapy and interstitial lung disease in scleroderma: A long-term followup study, *Arthritis Rheum.* **30**:643–650.

Ferri, C., Cecchetti, R., Cini, G., Gambini, I., La Civita, L., Bernnini, L., Bombardieri, S., Pasero, G., 1992, Slow-releasing nicardipine in the treatment of Raynaud's phenomena without underlying diseases, *Clin. Rheumatol.* **11**:76–80.

Freundlich, B., Rook, A. H., Edelson, R., Wintroub, B., Perez, M., Barr, W., Massa, M., Jimenez, S. A., Varga, J., Steen, V. D., Rietschel, R., Kahaleh, M. D., Perniciaro, C., Istfan, M., Medsger, T. A., and Jegasothy, B., 1990, Extracorporeal photochemotherapy in the treatment of systemic sclerosis (abstr), *Arthritis Rheum.* **33**:S35.

Freundlich, B., Jiminez, S. A., Steen, V. D., Medsger, T. A. Jr, Szkolnicki, M., and Jaffe, H. S., 1992, Treatment of systemic sclerosis with recombinant interferon-γ, *Arthritis Rheum.* **35**:1134–1141.

Fries, J. F., Spitz, P., Kraines, G., Holman, H. R., 1980, Measurement of patient outcome in arthritis, *Arthritis Rheum.* **23**:137–145.

Fries, J. F., Wasner, C., Brown, J., and Reigenbaum, P., 1984, A controlled trial of antihypertensive therapy in systemic sclerosis (scleroderma), *Ann. Rheum. Dis.* **43**:407–410.

Furst, D. E., Clements, P. J., Harris, R., Ross, M., Levy, J., and Paulus, H. E., 1979, Measurement of clinical change in progressive systemic sclerosis: A 1-year double-blind placebo-controlled trial of N-acetylcysteine, *Ann. Rheum. Dis.* **38**:356–361.

Furst, D. E., Clements, P. J., Hillis, S., Lachenbruch, P. A., Miller, B. L., Sterz, M. G., Paulus, H. E., 1989, Immunosuppression with chlorambucil, versus placebo, for scleroderma: Results of a three-year parallel, randomized, double-blind study, *Arthritis Rheum.* **32**:584–593.

Gibson, T., and Grahame, R., 1983, Cyclofenil treatment of scleroderma: A controlled study, *Br. J. Rheum.* **22**:218–223.

Giordano, M., Ara, M., Capelli, L., Tirri, G., Vatti, M., and Valentini, G., 1985, Griseofulvin in scleroderma, in: *Systemic Sclerosis (Scleroderma)* (C. M. Black and A. R. Myers, eds.), Gower Medical, New York, pp. 446–448.

Giordano, M., Valentini, G., Migliaresi, S., Picillo, V., and Vatti, M., 1986, Different antibody patterns and different prognoses in patients with scleroderma with various extent of skin sclerosis, *J. Rheumatol.* **13**:911–916.

Gruber, B. L., and Kaufman, L. D., 1991, A double-blind randomized controlled trial of ketotifen versus placebo in early diffuse scleroderma, *Arthritis Rheum.* **34**:362–366.

Guillevin, L., and Ortonne, J. P., 1983, Traitement de la sclerodermie, *Ann. Med. Interne (Paris)* **134**:754–765.

Guillevin, L., Chouvet, B., Mery, C., De Gery, A., Thivolet, J., Godeau, P., and Delbarre, F., 1985, Treatment of progressive systemic sclerosis using Factor XIII, *Pharmatherapeutica* **4**:76–80.

Harrison, A., Lusk, J., and Corkill, M., 1992, Reliability of skin score in scleroderma, *Br. J. Rheumatol.* **32**:170.

Hein, R., Behr, J., Hundgen, M., Hunzelmann, N., Meurer, M., Braun-Falco, O., Urbanski, A., and Krieg, T., 1992, Treatment of systemic sclerosis with γ-interferon, *Br. J. Dermatol.* **126**:496–501.

Hughes, P., Holt, S., Rowell, N. R., and Dodd, J., 1976, Thymus-dependent (T) lymphocyte deficiency in progressive systemic sclerosis, *Br. J. Dermatol.*, **95**:469–473.

Humbert, P., Dupond, J. L., Agache, P., Laurent, R., Rochefort, A., Drobacheff, C., de Wazieres, B., and Aubin, F., 1993, Treatment of sceroderma with oral 1,25-dihydroxyvitamin D3: Evaluation of skin involvement using non-invasive techniques: results of an open prospective trial, *Acta Derm. Venereol. (Stockh.)* **73**:449–451.

Jimenez, S., Sigal, S., 1991, A 15-year prospective study of treatment of rapidly progressive systemic sclerosis with D-penicillamine, *J. Rheumatol.* **18**:1496–1503.

Kahaleh, M. B., Sultany, G. L., Smith, E. A., Huffstutter, J. E., Loadholt, C. B., *et al.*, 1986, A modified scleroderma skin scoring method, *Clin. Exp. Rheumatol.* **4**:367–369.

Kahan, A., Amor, B., Menkes, C. J., and Strauch, G., 1989, Recombinant interferon-gamma in the treatment of systemic sclerosis, *Am. J. Med.* **87**:273–277.

Klein, H. A., Wald, A., Graham, T. O., Campbell, W. L., Steen, V. D., 1992, Comparative studies of esophageal function in systemic sclerosis, *Gastroenterology* **102**:1551–1556.

Krieg, T., and Meurer, M., 1988, Systemic scleroderma: Clinical and pathophysiologic aspects, *J. Am. Acad. Dermatol.* **18**:457–481.

Kyle, M. V., Belcher, G., and Hazleman, B. L., 1992, Placebo controlled study showing therapeutic benefit of iloprost in the treatment of Raynaud's phenomenon, *J. Rheumatol.* **19**:1403–1406.

Lau, C. S., Belch, J. J. F., Madhok, R., Cappell, H., Herrick, A., Jayson, M., and Thompson, J. M., 1993, A randomised, double-blind study of cicaprost, an oral prostacyclin analogue, in the treatment of Raynaud's phenomenon secondary to systemic sclerosis, *Clin. Exp. Rheumatol.* **11**:35–40.

LeRoy, E. C., and Medsger, T. A., 1992, Raynaud's phenomenon: A proposal for classification, *Clin. Exp. Rheumatol.* **10**:485–488.

LeRoy, E. C., Black, C., Fleischmajer, R., Jablonska, S., Krieg, T., Medsger, T. A. Jr, Rowell, N., and Wollheim, F., 1988, Scleroderma (systemic sclerosis): Classification, subsets and pathogenesis, *J. Rheumatol.* **15**:202–204.

Masi, A. T., 1988, Classification of systemic sclerosis (scleroderma: Relationship of cutaneous subgroups in early disease to outcome and serologic reactivity, *J Rehumatol.* **15**:894–898.

Maurice, P. D. L., Bunker, C. B., and Dowd, P. M., 1989, Isotretinoin in the treatment of systemic sclerosis, *Br. J. Dermatol.* **121**:367–374.

Medsger, T. A. Jr., 1985, Systemic sclerosis (scleroderma), eosinophilic fasciitis, and calcinosis, in: *Arthritis and Allied Conditions* (10th Ed.) (D. J. McCarty, ed.), Lea and Febiger, Philadelphia, pp. 994–1036.

O'Dell, J. R., Steigerwald, J. C., Kennaugh, R. C., Hawkins, R., Holers, V. M., and Kotzin, B. L., 1989, Lack of clinical benefit after treatment of systemic sclerosis with total lymphoid irradiation, *J. Rheumatol.* **16**:1050–1054.

Ortonne, J. P., Torzuoli, C., Dujardin, P., and Fraitag, B., 1989, Ketanserin in the treatment of systemic sclerosis: A double-blind controlled trial, *Br. J. Dermatol.* **120**:261–266.

Pignone, A., Matucci-Cerinic, M., Lombardi, A., Fedi, R., Fargnoli, R., De Dominicis, R., and Cagnoni, M., 1992, High resolution computed tomography in systemic sclerosis: Real diagnostic utilities in the assessment of pulmonary involvement and comparison with other modalities of lung investigation, *Clin. Rheumatol.* **11**:465–472.

Poole, J., and Steen, V., 1986, The use of the health assessment questionnaire (HAQ) to determine physical disability in systemic sclerosis (abstr), *Arthritis Rheum.* **29**:S152.

Poole, J. L., and Steen, V. D., 1991, The use of the health assessment questionnaire (HAQ) to determine physical disability in systemic sclerosis, *Arthritis Care Res.* **4**:27–31.

Pope, J. E., and Bellamy, N., 1993, Outcome measurement in scleroderma clinical trials, *Sem. Arthritis Rheum.* **23**:22–33.

Pope, J. E., Baron, M., Bellamy, N., Campbell, J., Carette, S., Chalmers, I., Dales, P., Hanly, J., Kaminska, E. A., Lee, P., Sibley, J., and Stevens, A., 1995, Variability of skin scores and clinical measurements in scleroderma, *J. Rheumatol.* **22**:1271–1276.

Rademaker, M., Cooke, E. D., Almond, N. E., Beachman, J. A., Smith, R. E., Mant, T. G. K., and Kirby, J. D., 1989, Comparison of intravenous infusions of iloprost and oral nifedipine in treatment of Raynaud's phenomenon in patients with systemic sclerosis: A double blind randomised study, *Br. Med. J.* **298**:561–564.

Rodnan, G. P., 1981, Progressive systemic sclerosis and penicillamine, *J. Rheumatol.* **8**(suppl 7):116–120.

Rook, A. H., Freundlich, B., Jegasothy, B. V., Perez, M. I., Bart, W. G., Jiminez, S. A., Ritschel, R. L., Wintroub, B., Kahaleh, M. D., Varga, J., Heald, P. W., Steen, V., Massa, M. C., Murphy, G. F., Perniciaro, C., Istfan, M., Bellas, S. K., and Edelson, R. L., 1992, Treatment of systemic sclerosis with extracorporeal photochemotherapy: Results of a multicenter trial, *Arch. Dermatol.* **128**:337–346.

Scherbel, A. L., 1983, The effect of percutaneous dimethyl sulfoxide on cutaneous manifestations of systemic sclerosis, *Ann. NY Acad. Sci.* **411**:120–130.

Seibold, J. R., Furst, D. E., Clements, P. J., 1992, Why everything (or nothing) seems to work in the treatment of scleroderma, *J. Rheumatol.* **19**:673–676.

Seibold, J. R., McCloskey, D. A., Furst, D. E., 1994, Pilot trial of methotrexate (MTX) in treatment of early diffuse scleroderma (abstract), *Arthritis Rheum.* **37**(suppl 16):R35.

Seibold, J. R., Korn, J., Simms, R., Clements, P. J., Moreland, L., Mayes, M. D., Furst, D. E., Rothfield, N., Steen, V., Weisman, M., DeMarco, P., Collier, D., Wigley, F., Merkel, P. A., Csuka, M. E., Rocco, S., Erikson, M., Hannigan, J., Harkonen, W. S., Sanders, M. E., 1997, Controlled trial of recombinant human relaxin (rhRlxn) in diffuse scleroderma (DS), *Arthritis Rheum.* **40**(suppl):S123.

Sharada, B., Kumar, A., Kakker, R., Adya, C. M., Pande, I., Uppal, S. S., Pande, J. N., Sunderam, K. R.,

and Malaviya, A. N., 1994, Intravenous dexamethasone pulse therapy in diffuse systemic sclerosis: A randomized placebo-controlled study, *Rheumatol. Int.* **14**:91–94.
Silman, A. J., 1991, Mortality from scleroderma in England and Wales 1968–1985, *Ann. Rheum. Dis.* **50**:95–96.
Silman, A. J., Harrison, M., and Brennan, P., 1995, Is it possible to reduce the observer variability in skin score assessment of scleroderma? *J. Rheumatol.* **22**:1277–1280.
Silver, R. M., Warrick, J. H., Kinsella, M. B., Standt, L. S., Baumann, M. H., Strange, C., 1993, Cyclophosphamide and low-dose prednisone therapy in patients with systemic sclerosis (scleroderma) with interstitial lung disease, *J. Rheumatol.* **20**:838–844.
Sinclair, H. D., Williams, J. D., Rahman, M. A., Denton, C., and Black, C. M., 1994, Clinical efficacy of anti-thymocyte globulin in systemic sclerosis: Results of a placebo-controlled trial (abstr), *Arthritis Rheum.* **36**(suppl):S217.
Special Article, 1980, Preliminary criteria for the classification of systemic sclerosis (scleroderma). Subcommittee for scleroderma criteria of the American Rheumatism Association Diagnostic and Therapeutic Criteria Committee, *Arthritis Rheum.* **23**:581–590.
Steen, V., and Follansbee, W. P., 1992, Thallium perfusion abnormalities predict survival and cardiac dysfunction in patients with systemic sclerosis (abstr), *Arthritis Rheum.* **35**:S37.
Steen, V. D., Medsger, T. A., and Rodnan, G. P., 1982, D-penicillamine therapy in progressive systemic sclerosis (scleroderma): A retrospective analysis, *Ann. Int. Med.* **97**:652–659.
Steen, V. D., Owens, G. R., Redmond, C., Rodnan, G. P., and Medsger, T. A. Jr., 1985, The effect of D-penicillamine on pulmonary findings in systemic sclerosis, *Arthritis Rheum.* **28**:882–888.
Steen, V. D., Blair, S., and Medsger, T. A. Jr., 1986, The toxicity of D-penicillamine in systemic sclerosis, *Ann. Int. Med.* **104**:699–705.
Steigerwald, J. C., 1985, Colchicine vs. placebo in the treatment of progressive systemic sclerosis, in: *Systemic Sclerosis* (*Scleroderma*) (C. M. Black and A. R. Myers, eds.), Gower Medical, New York, pp. 415–417.
Steigerwald, J. C., and Lynch, D., 1977, Colchicine therapy versus placebo: A double-blind study in progressive systemic sclerosis, *Abstracts of the XIV Int. Cong. on Rheumatology*, San Francisco:163.
Stevens, W., Vancheeswaran, R., Black, C. M., and the UK Systemic Sclerosis Study Group, 1992, Alpha Interferon-2a (Roferon-A) in the treatment of diffuse cutaneous sytemic sclerosis: A pilot study, *Br. J. Rheumatol.* **31**:683–689.
Tuffanelli, D. L., 1966, A clinical trial with dimethyl sulfoxide in scleroderma, *Arch, Derm*, **93**:724–725.
van den Hoogen, F. H. J., Boerbooms, A. M. R., Van Lier, H. J. J., and van de Putte, L. B. A., 1994, Methotrexate in systemic sclerosis: Preliminary 45 week results of a placebo controlled double blind trial (abstr), *Arthritis Rheum.* **36**(suppl):S217.
van der Heide, A., Jacobs, J. W. G., Kinant, H. J., and Bijlsma, J. W. J., 1992, The impact of endpoint measures in rheumatoid arthritis trials, *Sem. Arthritis Rheum.* **21**:287–294.
Vayssairat, M., Baudot, N., Boitard, C., Fiessinger, J. N., and Bach, J. F., 1990, Cyclosporine therapy or severe systemic sclerosis associated with the anti-Scl-70 autoantibody, *J. Am. Acad. Dermatol.* **22**:695–696.
Vayssairat, M., Baudot, N., Abuaf, N., and Johanet, C., 1992, Long-term follow-up study of 164 patients with definite systemic sclerosis: Classification considerations, *Clin. Rheumatol.*, **11**:356–363.
Weiner, S. R., Kono, K. H., Osterman, H. A., Levy, J., Paulus, H. E., Pitts, W. H., 1987, Preliminary report on a controlled trial of apheresis in the treatment of scleroderma (abstr), *Arthritis Rheum.* **30**:S27.
White, B., 1993, For the ACR Committee on study design and response parameters in SSc., Guidelines for clinical trials with disease-modifying interventions in systemic sclerosis (SSc), *Arthritis Rheum.* **36**(suppl):S131.
Wigley, F. M., Seibold, J. R., Wise, R. A., McCloskey, D. A., and Dole, W. P., 1992, Intravenous iloprost treatment of Raynaud's phenomenon and ischemic ulcers secondary to systemic sclerosis, *J. Rheumatol.* **19**:1407–14.
Williams, H. J., Furst, D. E., Dahl, S. L., Steen, V. D., Marks, C., Alpert, E. J., Henderson, A. M., Samuelson, C. O. Jr., Dreyfus, J. N., Weinstein, A., MacLaughlin, E. J., Alarcon, G. S., Kaplan, S. B., Guttadauria, M., Luggen, M. E., Reading, J. C., Egger, M. J., Ward, J. R., 1985, Double-blind, multicenter controlled trial comparing topical dimethyl sulfoxide and normal saline for treatment of hand ulcers in patients with systemic sclerosis, *Arthritis Rheum.* **28**:308–314.

Winkelman, R. K., Kierland, R. R., Perry, H. O., Muller, S. A., 1965, Treatment of scleroderma with sodium dextrothyroxine: A double-blind experimental trial. *Arch. Derm.* **91**:66–69.

Zachariae, H., Helkier-Sorensen, I., Heickendorff, L., Zachariae, E., and Hansen, H. E., 1990, Cyclosporin A treatment of systemic sclerosis, *Br. J. Dermatol.*, **122**:677–681.

Zarafonetis, C. J. D., Dabich, L., Negri, D., Skovronski, J. J., De Vol, E. B., and Wolfe, R., 1988*a*, Retrospective studies in scleroderma: Effect of potassium para-aminobenzoate on survival, *J. Clin. Epidemiol.* **41**:193–205.

Zarafonetis, C. J. D., Dabich, L., Skovronski, J. J., De Vol., E. B., Negri, D., Yuan, W., and Wolfe, R., 1988*b*, Retrospective studies in scleroderma: Skin response to potassium para-aminobenzoate therapy, *Clin. Exp. Rheumatol.* **6**:261–268.

Zufferey, P., Depairon, M., Chamot, A.-M., and Monti, M., 1992, Prognostic significance of nailfold capillary microscopy in patients with Raynaud's phenomenon and scleroderma-pattern abnormalities: A six-year follow-up study, *Clin. Rheumatol.* **11**:536–541.

Chapter 17

Quantification of Wound Healing

LAURA BOLTON, ANN-JEANETTE VASKO, AND KARYN MONTE

1. Why Measure Healing?

Scientific study has improved cutaneous wound healing outcomes more since 1948 than at any time in recorded history. To accomplish this, scientists and clinicians measured the outcomes of healing and its intermediary processes and carefully documented what treatments yielded the best results and what measured changes were correlated with clinical healing (e.g., Bates-Jensen, 1997). In order to measure effects of any treatment on healing, one must quantify the aspect of healing affected. This chapter summarizes reliable, valid techniques for quantifying cutaneous repair and its mediating events so that clinicians and scientists can use these same principles to further elucidate and improve wound healing outcomes based on the best available evidence.

Cutaneous healing consists of interacting stages of hemostasis, inflammation, granulation, epithelization, and maturation, which have been described elsewhere (e.g., Bolton and van Rijswik, 1991). Many events comprise each stage. For example, volumes have been written about the complex clotting cascade and about the pathways of inflammation. This chapter presents examples of measurement techniques for key events in each healing stage, that have documented reliability and have been validated against overall healing results. This is not an exhaustive literature review, but a resource to guide readers toward appropriate literature for the measures of major healing events. If a given stage is completed, it can be assumed that the key events necessary for that stage were also completed, though the quality may vary, resulting in deviations in morphology or function of the healed skin. Normal healing depends on successful completion of each stage appropriately and in the right time span.

Laura Bolton, Ann-Jeanette Vasko, and Karyn Monte • ConvaTec, A Bristol-Myers Squibb Company, Skillman, New Jersey 08558.

Cutaneous Biometrics, edited by Doris A. Schwindt and Howard I. Maibach, Kluwer Academic/Plenum Publishers, New York, 2000.

2. Overall Healing Measures

Clinical wound healing is usually measured as either rate or total time of closure and wound resurfacing with a functional epithelium. Many clinical wounds fail to heal during standard 12-week clinical studies, particularly if the underlying etiology is not corrected. This may be due to continual breakdown of tissue rather than impaired healing. Longer studies become impractical as patients drop out or are lost to follow-up. In such cases, rates of wound closure measured as percent area reduction per week and percents of wounds healed during the study or life-table statistics such as the Kaplan-Meier, Mantel-Haenzel, or Cox proportional hazards statistics are useful in determining the effects of different treatments or associated variables on healing time or probability. In general, percent reduction in area during the first two weeks of wound observation is strongly correlated with the overall healing outcome in chronic leg ulcers (van Rijswik, 1993), venous ulcers (Margolis, 1993), and pressure ulcers (van Rijswik and Polansky, 1994). The rates of contraction reported in human chronic wounds are on the same order of magnitude as those reported in swine (Pirone *et al.*, 1990) or guinea pigs (Bolton *et al.*, 1985). We could find no evidence that the healing process described here is different from one mammalian species to another, with the exception of certain regenerating models, such as deer antlers and the rabbit full-thickness ear punch (Goss, 1972).

In general, site-to-site variability on the body accounts for wider variance in healing rates than species differences. When using an animal model for healing, it is wise to match skin tightness, thickness, and hair growth patterns to those of the human wound modeled, as well as other key variables such as pathology and nutrition. Both full- and partial-thickness wounds heal more rapidly in moist environments (e.g. Hinman and Maibach, 1963; Pirone *et al.*, 1990), so environmental factors should be carefully controlled in studies measuring any of the stages of healing or maturation. Wound contraction or percent reduction in area offers a simple and reliable measure of gross repair (Grillo *et al.*, 1958; Van Winkle, 1967), which is similar in humans and other mammals (Catty, 1965).

3. Quantifying the Stages of Healing

3.1. Hemostasis

The end result of hemostasis is blood clotting. Sample measures listed in Table 1 commonly involve time for blood to clot sufficiently so that fresh blood flow ceases. These are usually measured in a standardized bleeding model and analyzed using parametric statistics for continuous data.

Bleeding time (BT) varies with skin and room temperature as well as blot size and type and size of the vessels pierced. Efforts to standardize have improved its usefulness, however, the methods are still controversial. A review of the literature suggests that there is little evidence to support the predictive value of the bleeding time test for healing outcomes. Although tests performed on individuals with normal BT yield reproducible results, there is a wide range of variability among individuals. There is also little evidence to suggest that the BT is a predictor of bleeding

TABLE 1. Reliable, Clinically Validated Measures of Hemostasis[a]

BT test device measurement technique	Clinical interpretation of normal values	Reference
Lancet ear lobe puncture blotted q 30 sec. Decreased blot size = decrease in hemorrhage	Clinical test to diagnose functional platelet disorders. Normal BT (1–3 min); severe anemics (3–5 min)	Duke, 1910
Lancet forearm puncture. Application of cuff pressure; puncture site blotted q 10 sec (number blood drops × 10 sec)	Eliminates capillary venous tonus. BT 240 sec/site; 125 sec/10 sites. Increased sensitivity over Duke method	Ivy *et al.*, 1941
Automatic stylet wound of terminal phalanx. No renewal of blood flow into physiologic saline indicates hemostasis. Clot resistant test	Controlled test differentiates arteriole vs. veniule bleeding. Normal BT range 30–180 sec.	Copley and Lalich, 1942
Duke/Ivy method with Bard-Parker No. 11 scapel. Hemoglobin concentration via saline immersion	Measures blood loss (>0.22 ml) and duration (7 min). Controls skin and room temperature variables	Adelson *et al.*, 1957
SBT: Gillette surgical blade. Renewed bleeding of incision 20–24 hr postinjury by removal of wound crust	SBT (22 min) vs. BT (7.5 min). Prolonged bleeding suggests usefulness as a diagnostic test for hemophilia	Borchgrevink and Waaler, 1958
Ivy method with Spring lancet: Eel photoelectric colorimeter	Measures BT (3.6 min) and blood loss via hemoglobin concentration (6.8 cmm/min)	Willoughby and and Arlington 1961
Duke method with collection of blood drops in capillary tubes	Measures BT and intensity in normal, hemorrhagic diseases; evaluates drug interaction	De Nicola, 1962
Ivy method with template device: wound blotted w/filter paper q 30 sec. Measured hemostatic plug formation	Standardizes depth and length of the Ivy method. Normal BT (mean 5 min) and its prolongation to aspirin (mean 9 min)	Mielke *et al.*, 1969
Ivy using Mayo automatic lancet: hemorrhagometry via flow cube	Automated technique measures BT (196 sec), intensity (12–15 μl/sec), patterns (3 identified).	Sutor and Bowie, 1971
Ivy using Mayo lancet, measuring electrical conductance	Electronic method for measuring BT (6 min) and blood loss (5.3 μl).	Bowie *et al.*, 1972
Mielke method: seconds to cessation of blood flow to filter paper	Provides duplicate standardized incisions. Normal BT (4.1 min) vs. aspirin (6.2 min).	Babson and Babson, 1978
Incision–canine/rat prostate using surgical blade: measure of spontaneous bleeding: visual assessment q 2.5 min	Materials evaluation microcrystalline collagen (2–5 min); purified gelatin (3–10 min).	Hanisch *et al.*, 1975
Ivy/Mielke using microlance/ template: unopettes/ haemoglobinometer	Ivy microlance BT (160 sec), blood loss (12.1 μl) vs. Mielke template BT (255 sec), Blood loss (42 μl)	Bernal-Hoyos and Caen, 1975
Incision—canine hind limb using Modifier template: time until blood no longer stains filter paper	BT method established in a canine model subjected to hyperthermia/ cardiac surgery (1.31– < 30.00 min).	Thomas *et al.*, 1979
Mielke method measuring platelets on filter paper placed on incision q 20 sec by electronic particle counter	Standardizes BT in neonates (3.6 min), newborns (3.4 min), children (3.4 min) vs. adults (2.8 min)	Feusner, 1980

continued

TABLE 1. Continued

BT test device measurement technique	Clinical interpretation of normal values	Reference
Excision—canine chest/spleen using grafting knife: gravimetric determination of BT and blood loss	Materials evaluation: partial compressed collagen fleece (78.1 sec); loose collagen (131.9 sec).	Silverstein and Chvapil, 1981
FBT venipuncture: timed drop rate (citrated blood) through filter, under pressure until aggregates occlude filter	FBT 2.8 min (1–3 hr postvenipuncture). BT for infants, patients w/edema or dermatologic disorders	Uchiyama *et al.*, 1984
Incision—thumb using autolet: Cessation of blood flow to filter paper measured w/stop watch	Standardizes BT in newborns. 97% below 3.5 min	Rennie and Gibson,
Ivy method using simplate I device: timed cessation of blood flow to filter paper >15 min	IBT range 1–5 min; SBT range 6–>15 min. Simplate technic may detect disorders of secondary hemostasis	Smith *et al.*, 1985
Mielke using hemalet device vs. simplate II device: timed cessation of blood flow to filter paper, to nearest 30 sec	Hemalet (5.8 min); simplate (5.4 min) $r = 0.81$. Devices disposable, clinically reliable, and cost-effective	Toomey *et al.*, 1986
Incision—pig ear using Bard–Parker No. 11 scapel: saline immersion: timed, hemostatic plug formation	@ 10 sec—small platelet plugs; 30 sec—many plugs; 60 sec—fibrins strands appear; 120 sec—arrest of bleeding	Sawada *et al.*, 1986
Cannula/cubital vein: timed cessation of blood flow from cannula onto filter paper and retraction of blood by 1 mm	HT (46s–6:38s) compared to Ivy/Duke. Method addresses variables associated with incisions/punctures	Janzarik *et al.*, 1986
Canine skin flaps abrasions: gravimetric determination of BT and blood loss	Hemostasis of LMW heparin to conventional heparin in a canine model	Ljungberg and Johnsson, 1988
Mielke using simplate II double lancet: Modified *in vivo* platelet retention test	Control IVPR @ 1 min 17.2%, 2 min 19.2%, 3 min 33.1%	de la Cuadra *et al.*, 1988
Burns: cessation of hemorrhage following debridement, comparing thrombin, pressure with plastic film dressing	Small clots result in postremoval of plastic film dressing; clot disruption/hemorrhage occurs with thrombin-soaked gauze	Sawada and Yotsuyanagi, 1992
Canine oral buccal mucosa using a template: number of residual bleeding points posttreatment	Alginates 1–13 bleeding points; Gauze 6–16 bleeding points	Matthew *et al.*, 1994
Mielke method using surgicutt: gravimetric determination of peak bleeding rate, time to peak bleeding, and total bleeding	Coronary bypass patient: baseine BT 5.7 min; total bleeding 13.7 mg; peak BR 3.2 mg/30 sec; time to peak bleeding 1 min	De Caterina *et al.*, 1994
Donor sites using Padgett dermatome: direct observation of coagulation; cessation of free-flowing blood	K–Y jelly/epinephrine mixture-1.64 min; mineral oil-8 min	Netscher *et al.*, 1996

[a]Abbreviations: BT = bleeding time, SBT = secondary bleeding time, FBT = filter bleeding test, IBT = ivy bleeding time, HT = hemostasis time, LMW = low molecular weight, IVPR = *in vivo* platelet retention, BR = bleeding rate.

disorders. Application of parametric statistics may strengthen the usefulness of BT tests in research and distinguish differences among patient populations.

3.2. Inflammation

Inflammation is measured grossly as pain, erythema, and edema (Table 2) or bioassayed at the cellular level (Table 3). The gross measures include rating scales or visual analog scales. A pain thermometer is simpler and provides more reliable results than the standard horizontal 10-cm visual analog scale when gauging pain in more easily confused patients. (Choiniere, 1994). Data from discrete rating scales of pain, erythema, and edema require nonparametric analysis. Measurements from visual analog scales for pain, colorimetric measures of erythema, or ultrasonic

TABLE 2. Reliable, Clinically Validated Measures of Pain, Erythema and Edema[a]

Model	Measurement technique	Clinical interpretation/ Normal values	Reference
Chronic wounds	Subjective evaluation of color, temperature, induration and edema	Redness, warmth and swelling are early signs of inflammation and in later stages of the wound classic signs of infection	van Rijswijk, 1996
Pain, acute—from trauma	100-mm visual analog scale at intervals of 20 min for 2 hr	Patient's reporting less than a 13-mm change in pain severity may not be clinically significant	Todd *et al.*, 1996
Pain—peripheral nerve injury	Computer controlled infusion of plasma lidocaine concentrations to study the effect of i.v. lidocaine on (1) pain scores, (2) current perception thresholds, (3) side effects, and (4) pain distribution	Pain perception was plasma-concentration-dependent for the lidocaine group starting at 1.5 g/ml. Saline infusion had no effect on pain perception. No serious adverse events were recorded	Wallace *et al.*, 1996
Pressure ulcers	PSST—erythema and edema as well as 11 other wound assessments are scored on a modified Lickert scale from 1 to 5	The PSST scales clinical data from the pressure ulcer 1–5, healthiest to most severe, in order to determine the severity of the ulcer	Bates-Jensen, 1997
Pressure ulcer, porcine model	IR imaging equipment to measure hyperemia through skin color and temperature	The IR diagnostic color data showed that there is a clear differentiation of wounds based on wound category	Hansen *et al.*, 1996

[a]Abbreviations: PPST = Pressure Sore Status Tool^{R}, IR = infrared.

TABLE 3. Reliable, Clinically Validated Measures of the Cellular Inflammatory Response[a]

Model	Measurement technique	Clinical interpretation/ normal values	Reference
Burn patients exhibiting SIRS	Harvest neutrophils to determine whether apoptosis is delayed. Plasma from patients inhibits apoptosis in normal neutrophils	Measuring the expression of apoptosis shows that delays in apoptosis may lead to prolonged inflammation	Marshall and Watson, 1997
Catheter, split-thickness wound, healthy elderly volunteers	A subcutaneously implanted catheter was used to measure hydroxyproline, total protein, and DNA content. The first two were measured using HPLC while DNA was measured using a buffered assay system	An implanted catheter was used for collection of fluids, protein, and DNA. Novel way of assessing healing modalities at the cellular level	Kirk *et al.*, 1993
Surgical injury, thermal injury, mechanical trauma	Immunoquantitation and bioassay for detecting IL-6	The IL-6 response can be used as a measure of severity of tissue damage, as a marker of postinjury complications, and delayed host immunosuppression. Magnitude of elevation is directly correlated to extent of tissue damage	Biffl *et al.* 1996
Trauma-related and ischemic tissue damage (review)	Enzyme immunoassay to quantify complement activation products C3 and TCC, requiring precise sample preparation	Detection of C3 and TCC are sufficient to determine degree of activation. Monitoring levels of C3 and TCC differentiate therapeutic treatment effects, as well as harmful complement system activation	Mollnes and Fosse, 1994

[a]Abbreviations: SIRS = systemic inflammatory response syndrome, HPLC = high pressure liquid chromatography, C3 = complement fraction 3, TCC = total complement complex.

measures of edema are continuous and usually are sufficiently robust for parametric statistics, such as *t*-tests or analyses of variance. It should be noted that a patient's perception of pain is influenced by psychological variables (Orbach *et al.*, 1996) and age (Van der Does, 1989) and differs from that reported by nurses.

With the development of immunoquantitative and bioassay techniques, the inflammatory response can now be measured at the cellular level. Table 3 lists cells and molecules that can be assayed to determine the level of expression of inflammatory mediators. These may be useful in staging response and treatment, especially in the case of chronic wounds.

3.3. Granulation

Deep partial-thickness and full-thickness wounds heal by granulation, which can be measured as reduction in wound depth or area or tensile strength. Examples of clinically validated measures are presented in Table 4. Granulation tissue is created as fibroblasts form a collagen matrix and endothelial cells engage in angiogenesis, the formation of new blood vessels. Then new granulation tissue contracts, drawing full-thickness wounds closed, so the wound cavity is both filled with new granulation tissue and closed by contraction. This process is subject to the laws of geometry (du Nuoy, 1919), as well as artifacts of shape, cutaneous architecture, and morphology (Kennedy and Cliff, 1979). Percent wound contraction is a simple, validated measure of full-thickness wound healing progress in mammals, which is acceptable for analysis as parametric data (Pirone *et al.*, 1990).

Tensile studies quantify wound strength, which approaches but never reaches that of original skin at the wound site. Tensile or breaking strength can be artificially increased by excessive inflammation and scarring and so should be interpreted with caution, ideally with corollary measures of inflammation (Bolton *et al.*, 1988).

3.4. Epithelization

Epithelial resurfacing has been quantified using evaporimetry and validated against clinical appearance (Yosipovitch *et al.*, 1998). Examples of measures of epithelization are presented in Table 5. Swine partial-thickness wounds epithelialize at approximately the same rate as those of similar depth on humans (Pirone *et al.*, 1990; Rohrich and Pittman, 1991). Epithelization has been quantified as percent return of normal barrier function (Bolton and van Rijswik, 1991), moisture vapor transmission rate (Pirone *et al.*, 1990; Yosipovitch *et al.*, 1998), percent of wounds reepithelialized after a given number of days (Eaglstein, 1993), or donor site reepithelialization measured by optical planimetry (Bettinger *et al*, 1995).

3.5. Maturation and Complications

Wounds may either fail to heal or heal at an abnormally slow pace, fail to stop healing when the level of the normal skin has been reached, or produce abnormal scars. Failure to heal may result when tissue breaks down faster than it heals. This may result from infection, prolonged ischemia, repeated trauma, or when nutritional or metabolic deficiencies or toxins prevent some stage of healing from progressing. Failure to stop healing may result in hypergranulation, which requires clinical intervention only if pedunculated (Zitelli, 1984); or in a hypertrophic scar, which protrudes above the original wound margins after epithelization then gradually subsides; or in a keloid, which progressively expands beyond the original wound margins after epithelization. Measures of these complications in the healing and/or maturation process are presented in Table 6.

TABLE 4. Reliable, Clinically Validated Measures of Granulation Phase Healing[a]

Wound or model	Measurement technique	Clinical interpretation: normal values	Reference
Amputation sites	$TcPO_2$ at site	$TcPO_2 \geqslant 15$ predicted healing outcome	Yabon *et al.*, 1995
Amputation sites	LDSPP	LDSPP > 30 mv predicted healing outcome 90% sensitivity	Adera *et al.*, 1995
Burns	Laser doppler-derived measures of burn perfusion	Laser doppler-derived function predicted healing with 94% accuracy	Green *et al.*, 1992; Young *et al.*, 1996
Implants: hamster	Fluorescence microscopy after dye injection	Differences in microvessel density (angiogenesis) day 10	Menger *et al.*, 1992
Incisions: rat (quick frozen)	TS measures, collagen cross-linking, matrix MP activity	TS increased over time. MP-9 peaked at 48 hr paralleling inflammation. MP-2 peaked at 72 hr	Paul *et al.*, 1997
Incisions: rat	Tensiometry; hydroxyproline in polyvinyl alcohol sponges	Higher tensile strength, but not more hydroxyproline in growth hormone-releasing factor treated rats	Garrel *et al.*, 1991
Incisions partial thickness: on Yucatan swine	Tensiometry, oximetry, hydroxyproline, rating scales for granulation tissue and vascularization, DNA measurement of cell density	DNA density changed across studies, with advent of occlusive dressings. Other measures were correlated with healing	Byl *et al.*, 1994
Ischemic wounds	$TcPO_2$ or LDHP	$TcPO_2 \geqslant 11$ or LDHP 50 mv predicted healing outcome	Padberg *et al.*, 1992
Mouse ear	CDP	Neovascularization complete in 15 days	Kjolseth *et al.*, 1994
Pressure ulcers	Area image analysis, 4-point scale for deepest depth, color of surrounding skin, necrotic tissue, granulation tissue	Area reduction had high precision (coefficient of variation = 0.06) and preceded wound healing. Kaplan–Meier wound survival times and Cox proportional hazards for healing probability	Riet *et al.*, 1996
Pressure ulcers	Successive wound tracings or photography	Wound contraction measurements are reliable and occur at about the same rate in healing full-thickness pressure ulcers and leg ulcers as observed in swine	Thomas and Wysocki, 1990; van Rijswijk and Polansky, 1994; Pirone *et al.*, 1990
Pressure ulcers	Volume measurements using dental impression material	Wound impressions take about 15 min to make, but can measure volume not reflected by wound area measurements	Covington *et al.*, 1989; Cliff *et al.*, 1989

TABLE 4. Continued

Wound or model	Measurement technique	Clinical interpretation: normal values	Reference
Teat wounds: cow	LDF, tensiometry, hydroxyproline content and histopathology	LDF was more rapid, less invasive and less painful than the other three measures and correlated with histology	Ghamsari *et al.*, 1996
Venous ulcers	Minolta CR 200 photographic colorimetry	Increasing red scale is a validated measure of clinical granulation	Romanelli, 1997

[a]Abbreviations: $TcPO_2$ = transcutaneous oximetry, LDSPP = laser Doppler site perfusion pressure, TS = tensile strength, MP = metalloprotease, LDHP = laser Doppler heated probe, CDP = computerized digitized planimetry, LDF = laser Doppler flowmetry.

TABLE 5. Reliable, Clinically Validated Measures of Epithelialization[a]

Model	Measurement technique	Clinical interpretation/ normal values	References
Biopsy, healthy volunteers	Histology used to measure the level of expression of i_1, i_4, and i_v integrins during wound healing	During reepithelialization modifications of integrin expression pattern were observed on migrating keratinocytes	Cavani *et al.*, 1993
Normal skin and partial thickness wounds	Assay barrier function of the skin. TEWL, stratum corneum hydration, skin surface pH, and skin temperature	Skin permeability is higher in the evening and night than in the morning, suggesting that barrier function is linked to circadian rhythms. Lower numbers in TEWL indicate greater barrier function	Pirone *et al.*, 1990; Rohrich and Pittman, 1991; Yosipovitch *et al.*, 1998
Surgical incision	Biopsy taken for histology to evaluate keratinocyte expression at the wound edge	Keratins 6 and 16 in postmitotic keratinocytes located at the wound edge are induced to accumulate 6 hr after injury in human epidermis	Paladini *et al.*, 1996
Venous stasis and decubitus ulcers	Photographs, acetate wound tracings, and Kundin measuring device	All three methods are valid for measuring wound area; however, acetate tracings were determined to be the most accurate.	Thomas and Wysocki, 1990

[a]Abbreviations: TEWL = transepidermal water loss.

TABLE 6. Reliable, Clinically Validated Measures of Maturation and Complications[a]

Wound or model	Measurement technique	Clinical interpretation/ normal values	Reference
Clinical scars	Alginate molds and clinical ratings	Both molds and clinical ratings were reliable	Phillips, 1994
Clinical scars	Shear wave velocity of propagation	Higher shear wave velocity in scars, verifying increased stiffness	McHugh *et al.*, 1997
Clinical scars	Scar elasticity measurements and ultrasonography	Serial monthly evaluations were correlated with clinical scar grading. Ultrasonography localized the scar tissue and elastometry reliably measured scar viscoelasticity	Fong *et al.*, 1997
Clinical scars	LDF blood flow measurements of scar vascularity	Hypertrophic scars had threefold greater LDF vascularity values than normal skin and fourfold greater values than normal scars 16 to 18 weeks after healing	Ehrlich and Kelly, 1992
Clinical scars	Skin surface high-frequency conduction, reflecting skin surface water content	Noninvasive, reliable detection of moisture that was correlated with functional abnormalities of the stratum corneum	Suetake *et al.*, 1996
Rabbit ear full-thickness wounds	Hypertrophic index = ratio of granulation tissue dimensions to epithelium	Sufficiently reliable to identify significant effects of TGFB3 vs. vehicle control. Clinical validity remains to be tested	Wu *et al.*, 1997; Morris *et al.*, 1997
Venous ulcers	Minolta CR-200 photographic colorimetry of fibrin slough or necrotic tissue on wound surface	More yellow on the yellow-blue scale was validated against clinical observations of wound surface slough and sufficiently reliable to differentiate topical gel effects	Romanelli, 1997
Clinical burn infections	Surface quantitative swabs vs. biopsies to identify wounds at risk of infection vs. classic clinical signs of infection	Clinical infection is associated with >10e5 CFU/ml in both swab and biopsy specimens, but many patients with >10e5 CFU/ml do not develop clinical infection	Thomson and Smith, 1994
Clinical burn scars	A pneumatonometer measured scar compliance	The pneumatonometer differentiated between scar and normal tissue at different body sites, but there was wide variability from site to site	Spann *et al.*, 1996

TABLE 6. Continued

Wound or model	Measurement technique	Clinical interpretation/ normal values	Reference
Rat muscle	ELISA immunoassay to quantify collagenase activity	More collagenase activity indicates more remodeling	Chakraborty and Eghbali, 1989
Clinical scars	Magnetic resonance imaging spin lattice relaxation times (T_1) in scar tissue	T_1 values returned to steady state faster in hypertrophic scars than in keloid tissue	Babu *et al.*, 1993

[a]Abbreviations: LDF = laser Doppler flowmetry; $TGFB_3$ = transforming growth factor β_3, CFU = colony forming units.

4. Conclusion

Armed with these techniques for monitoring all aspects of healing, researchers can determine effects of agents on healing and scarring and identify the stage of healing affected by each agent. Clinicians and scientists can build from this vocabulary of healing measures conclusive evidence on which to base quality clinical practice to improve outcomes for their patients.

ACKNOWLEDGMENT: The authors gratefully acknowledge the assistance of Robbin Kelly in preparing this manuscript for publication.

References

Ådelson, E., and Crosby, W. H., 1957, A new method to measure bleeding time: The "immersion" method, *Acta Haematol* (*Basel*) **18**:281–289.

Adera, H. M., James, K., Castronuovo, J. J., Byrne, M., Deshmukh, R., Lohr, J. M., 1995, Prediction of amputation wound healing with skin perfusion pressure, *J. Vasc. Surg.* **21**(5):823–828.

Babson, S. R., and Babson, A. L., 1978, Development and evaluation of a disposable device for performing simultaneous duplicated bleeding time determinations, *Am. J. Clin. Pathol.* **70**:406–408.

Bates-Jensen, B. M., 1997, The pressure sore status tool a few thousand assessments later, *Adv. Wound Care* **10**(5):65–73.

Babu, M., Bai, R. P., Suguna, L., Ramachandran, K., and Ramakrishnan, K. M., 1993, Differentiation of keloid and hypertrophic scar, *Phyiol. Chem. Phys. Med. NMR* **25**(2):113–120.

Bernal-Hoyos, E. J., and Caen, J. P., 1975, A new and simple method of quantifying the bleeding time, *BioMed Express* (*Paris*) **23**(7):258–262.

Bettinger, D., Gore, D., and Humphries, Y., 1995, Evaluation of calium alginate for skin graft donor sites, *J. Burn Care Rehabil.* **16**:59–61.

Biffl, W. L., Moore, E. E., Moore, F. A., and Peterson, V. M., 1996, Interleukin-6 in the injured patient: Marker of injury of mediator of inflammation? *Ann. Surg.* **224**(5):647–664.

Bolton, L. L., and van Rijswijk, L., 1991, Wound healing: Meeting clinical needs, *Dermatol. Nurs.* **3**:146–160.

Bolton, L., Oleniacz, W., Constantine, B., Kelliher, B. O., Jensen, D., Means, B, and Rovee, D., 1985, Repair and antibacterial effects of topical antiseptic agents *in vivo*, in: *Models in Dermatology* (H. Maibach and N. Lowe, eds.) Karger, Basel, pp. 145–158.

Bolton, L. L., Pines, E., Rovee, D. T., 1988, Wound healing and the intergumentary system, in: *Experimental Surgery and Physiology: Induced Animal Models of Human Disease* (M. Swindel and R. J. Adams, eds.) Williams & Wilkins, Baltimore, pp. 1–9.

Borchgrevink, C. F., and Waaler, B. A., 1958, Secondary bleeding time: A new method for the differentiation of hemorrhagic diseases, *Acta Med. Scand.* **162**:361.

Bowie, E. J. W., Owen, C. A. Jr., Hansen, R. J., and Isaacson, J., 1972, Electronic method for quantification of bleed time, *Am. J. Clin. Path.* **58**:255–259.

Byl, N. N., McKenzie, A. L., West, J. M., Whitney, J. D., Hunt, T. K., Hopf, H. W., Scheuenstuhl, H., Robinson, A. J., and Snyder-Mackler, L., 1994, Pulsed microamperage stimulation: a controlled study of healing of surgically induced wounds in Yucatan pigs, *Phys. Ther.* **74**(3):201–218.

Catty, R. H. C., 1965, Healing and contraction of experimental full-thickness wounds in the human, *Br. J. Surg.* **52**:542–548.

Cavani, A., Zambruno, G., Marconi, A., Manca, V., Marchetti, M., Giannetti, A., 1993, Distinctive integrin expression in the newly forming epidermis during wound healing in humans, *J. Invest. Dermatol.* **101**(4):600–604.

Chakraborty, A., and Eghbali, M., 1989, An immunoassay for qualitative estimation of collagenase activity in mammalian tissues, *Clin. Biochem.* **24**(6):455–461.

Choiniere, M., Auger, F. A., and Latarjet, J., 1994, Visual analogue thermometer: A valid and useful instrument for measuring pain in burned patients, *Burns* **20**(3):229–235.

Clifft, J. K., 1989, Measurement of pressure ulcer volume using dental impression materials, *Phy. Ther.* **69**(8):690–695.

Copley, A. L., and Lalich, J. J., 1942, Bleeding time, lymph time, and clot resistance in men, *J. Clin. Invest.* **21**:145.

Covington, J. S., Griffin, J. W., Mendius, R. K., Tooms, R. E., and Clifft, J. K., 1989, Measurement of pressure ulcer volume using dental impression materials: Suggestions from the field, *Phys. Ther.* **69**(8):690–695.

De Caterina, R., Lanza, M., Manca, G., Strata, G. B., Maffei, S., and Salvatore, L., 1994, Bleeding time and bleeding: An analysis of the relationship of the bleeding time test with parameters of surgical bleeding, *Blood* **84**(10), 3363–3370.

de la Cuadra, J. L., Jeanneret, C., and Furlan, M., 1988, Modified *in vivo* platelet retention test for evaluating primary hemostasis, *Acta Haemat.* **80**:103–109.

De Nicola, P., 1962, Critical evaluation of methods for the study of blood coagulation, fibrinolysis and hemostasis, *Throm. Diath. Haemorrh.* **7**(Supp 1):325–333.

Duke, W. W., 1910, The relation of blood platelets to hemorrhagic diseases, *JAMA* **55**:1185.

du Nuoy, P. L., 1919, Cicatrization of wounds: X. A general equation for the law of cicatrization of surface wounds, *J. Exp. Med.* **29**:329–350.

Eaglestein, W. H., 1993, Occlusive dressings, *J. Dermatol. Surg. Oncol.* **19**:716–720.

Ehrlich, H. P., and Kelley, S. F., 1992, Hypertrophic scar: An interruption in the remodeling of repair; A laser Doppler blood flow study, *Plast. Reconstr. Surg.* **90**(6):993–998.

Feusner, J. H., 1980, Normal and abnormal bleeding times in neonates and young children utilizing a fully standardized template technic, *Am. J. Clin. Pathol.* **74**:73–77.

Fong, S. S., Hung, L. K., Cheng, J. C., 1997, The cutometer and ultrasonography in the assessment of postburn hypertrophic scar, *Burns* **23**(Suppl 1):S12–S18.

Garrel, D. R., Gaudreau, P., Zhang, L. M., Reeves, I., and Brazeau, P., 1991, Chronic administration of growth hormone-releasing factor increases wound strength and collagen maturation in granulation tissue, *J. Surg. Res.* **51**(4):297–302.

Ghamsari, S. M., Acorda, J. A., Taguchi, K., Abe, N., and Yamada, H., 1996, Evaluation of wound healing of the teat with and without low lever laser therapy in dairy cattle by laser Doppler flowmetry in comparison with histopathology, tensiometry and hydroxyproline analysis, *Br. Vet. J.* **152**(5):583–592.

Goss, R. J., 1972, Wound healing and antler regeneration, in: *Epidermal Wound Healing* (H. Maibach and D. Rovee, eds.), Yearbook Medical Publishers, Chicago.

Green, H. A., Bua, D., Anderson, R. R., and Nishioka, N. S., 1992, Burn depth estimation using indocyanine green fluorescence, *Arch. Dermatol.* **43**(7):128.

Grillo, H. C., Watts, G. T., and Gross, J., 1958, Studies in wound healing: I. Contraction and the wound contents, *Ann. Surg.* **148**:145–152.

Hansen, G. L., Sparrow, E. M., Kommamuri, N., and Iaizzo, P. A., 1996, Assessing wound severity with color and infrared imaging of reactive hyperemia, *Wound Rep. Reg.* **4**:386–392.

Hanisch, M. E., Baum, N., Beach, P. D., Griffith, D. P., and Tyler, M., 1975, A comparative evaluation of Avitene and Gelfoam for hemostasis in experimental canine prostatic wounds, *Invest. Urol.* **12**(5):333–336.

Hinman, C. D., and Maibach, H., 1963, Effect of air exposure and occlusion on experimental human skin wounds, *Nature* **200**:377–378.

Ivy, A. C., Nelson, D., and Bucher, G., 1941, The standardization of certain factors in the cutaneous "venostasis" bleeding time technique, *J. Lab. Clin. Med.* **26**:1812.

Janzarik, H., Remy, S., Morell, S., and Pabst, W., 1986, "Haemostasis time," a modified bleeding time test and its comparison with the Duke and IVY/template bleeding times. *BLUT* **52**(6):345–356.

Kennedy, D. F., and Cliff, W. J., 1979, A systematic study of wound contraction in mammalian skin, *Pathology* **11**:207–222.

Kirk, S. J., Hurson, M., Regan, M. C., Holt, D. R., Wasserkrug, B. A., and Barbul, A., 1993, Arginine stimulates would healing and immune function in elderly human beings, *Surgery* **144**:155–160.

Kjolseth, D., Frank, J. M., Barker, J. H., Anderson, G. L., Rosenthal, A., Acland, R. D., Schuschke, D., Campbell, F. R., Tobin, G. R., and Weiner, L. J., 1994, Comparison of the effects of commonly used wound agents on epithelialization and neovascularization, *J. Am. Coll. Surg.* **179**(3):305–312.

Ljungberg, B., and Johnsson, H., 1988, *In vivo* effects of low molecular weight heparin fragment on platelet aggregation and platelet dependent hemostasis in dogs, *Thromb. Haemost.* **60**(2):232–235.

Margolis, D. J., Gross, E. A., Wood, C. R., and Lazarus, G. S., 1993, Planimetric rate of healing in venous ulcers of the leg treated with pressure bandage and hydrocolloid dressing, *J. Am. Acad. Dermatol.* **28**:418–421.

Marshall, J. C., and Watson, R. W., 1997, Programmed cell death (apoptosis) and the resolution of systemic inflammation, *Can. J. Surg.* **40**(3):169–174.

Matthew, I. R., Browne, R. M., Frame, J. W., and Millar, B. G., 1994, Alginate fiber dressing for oral mucosal wounds, *Oral Surg. Oral Med. Oral Path.* **77**(5):456–460.

McHugh, A. A., Fowlkes, B. J., Maevsky, E. L., Smith, D. J., Jr., Rodriques, J. L., and Garner, W. L., 1997, Biomechanical alterations in normal skin and hypertrophic scar after thermal injury, *J. Burn Care Rehabil.* **18**(2):104–108.

Menger, M. D., Hammersen, F., and Messmer, K., 1992, *In vivo* assessment of neovascularization and incorporation of prosthetic biografts, *Thorac. Cardiovasc. Surg.* **40**(1):19–25.

Mielke, C. H., Kaneshiro, M. M., Maher, I. A., Weiner, J. M., and Rapaport, S. I., 1969, The standardized normal Ivy bleeding time and its prolongation by aspirin, *Blood* **34**:204.

Mollnes, T. E., and Fosse, E., 1994, The complement system in trauma-related and ischemic tissue damage: a brief review, *Shock* **2**(4):301–310.

Morris, O. E., Wu, L., Zhao, L. L., Bolton, L., Roth, S. I., Ladin, D. A., Mustoe, T. A., 1997, Acute and chronic animal models for excessive dermal scarring: Quantitative studies, *Plast. Reconstr. Surg.* **100**(3):674–681.

Netscher, D. T., Carlyle, T., Thornby, J., Bowen, D., Harris, S., and Clamon, J., 1996, Hemostasis at skin graft donor sites: Evaluation of topical agents, *Ann. Plast. Surg.* **36**(1):7–10.

Orbach, I., Stein, D., Palgi, Y., Asherov, J., Har-Even, D., and Elizur, A., 1996, Perception of physical pain in accident and suicide attempt patients: Self-preservation vs. self-destruction, *J. Psychiatr. Res.* **30**(4):307–320.

Padberg, F. T. Jr., Back, T. L., Hart, L. C., and Franco, C. D., 1992, Comparison of heated-probe laser Doppler and transcutaneous oxygen measurements for predicting outcome of ischemic wounds, *J. Cardiovasc. Surg.* **33**(6):715–722.

Paladini, R. D., Takahashi, K., Bravo, N. S., and Coulombe, P. A., 1996, Onset of re-epithelialization after skin injury correlates with a reorganization of keratin filaments in wound edge keratinocytes: Defining a potential role for keratin 16, *J. Cell Biol.* **132**(3):381–397.

Paul, R. G., Tarlton, J. F., Purslow, P. P., Sims, T. J., Watkins, P., Marshall, F., Ferguson, M. J., and Bailey, A. J., 1997, Biomechanical and biochemical study of a standardized wound healing model, *Int. J. Biochem. Cell Biol.* **29**(1):211–220.

Phillips, T. J., Gerstein, A. D., and Lordan, V., 1996, A randomized controlled trial of hydrocolloid dressing in the treatment of hypertrophic scars and keloids, *Dermatol. Surg.* **22**:775–778.

Pirone, L. A., Monte, K. A., Shannon, R. J., and Bolton, L. L., 1990, Wound healing under occlusion and non-occlusion in partial-thickness and full-thickness wounds in swine, *Wounds* **2**(2):74–81.

Rennie, J. M., and Gibson, T., 1985, Micromethod for bleeding time in the newborn, *Arch. Dis. Child.* **60**:51–53.

Riet, G., Kessels, A., and Knipschild, P., 1996, A randomized clinical trial of ultrasound in the treatment of pressure, *Phys. Ther.* **76**(12):1301–1312.

Rohrich, R. H., and Pittman, C. E., 1991, A clinical comparison of DuoDERM CGF and Op-Site donor site dressings, *Wounds* **3**(6):221–226.

Romanelli, M., 1997, Objective measurement of venous ulcer debridement and granulation with a skin color reflectance analyzer, *Wounds* **9**(4):122–126.

Sawada, D. N., Fass, J. A., Katzman, R. C., Bahn, R. C., and Bowie, E. J. W., 1986, Hemostatic plug formation in normal and von Willebrand pigs: The effect of the administration of cryoprecipitate and a monoclonal antibody to Willebrand factor, *Blood* **67**(5):1229–1239.

Sawada, Y., and Yotsuyanagi, T., 1992, A technique of hemostasis of the extremities after debridement of burn wounds, *Burns* **18**(5):412–415.

Silverstein, M. E., and Chvapil, M., 1981, Experimental and clinical experiences with collagen fleece as a hemostatic agent, *J. Trauma* **21**(5):388–393.

Smith, P. S., Baglini, R., and Meissner, G. F., 1985, The prolonged bleeding time in Hemophilia A: Comparison of two measuring technics and clinical associations, *Am. J. Clin. Pathol.* **83**:211–215.

Spann, K., Mileski, W. J., Atiles, L., Purdue, G., and Hunt, J., 1996, Use of pnematonometer in burn scar assessment, *J. Burn Care. Rehabil.* **17**(6):515–517.

Suetake, T., Sasai, S., Zhen, Y. X., Ohi, T., and Tagami, H., 1996, Functional analyses of the stratum corneum in scars: Sequential studies after injury and comparison among keloids, hypertrophic scars, and atrophic scars, *Arch. Dermatol.* **132**(12):1453–1458.

Sutor, A. H., and Bowie, E. J., 1971, Bleeding from standardized skin punctures: Automated technic for recording time, intensity and pattern of bleeding, *Am. J. Clin. Pathol.* **55**:541.

Thomas, A. C., and Wysocki, A. B., 1990, The healing wound: A comparison of three clinically useful methods of measurement, *Decubitus* **3**(1):18–25.

Thomas, R., Hessel, E. A., Dillard, D. H., and Haker, L. A., 1979, Standardized template bleeding time in dogs, *J. Surg. Res.* **27**:244–249.

Thomson, P. D., and Smith, D. J., 1994, What is infection?, *Am. J. Surg.* **167**(1A):7S–11S.

Todd, K. H., Funk, K. G., Funk, J. P., and Bonacci, R., 1996, Clinical significance of reported changes in pain severity, *Ann. Emerg. Med.* **27**(4):485–489.

Toomey, K. C., Hugh, K. C., Martin, K., and Parvin, S., 1986, Clinical trial of a new disposable bleeding-time device, *Am. J. Clin. Pathol.* **85**(5):610–613.

Uchiyama, S., Bach, M. L., Didisheim, P., and Bowie, E. J. W., 1984, Clinical evaluation of a new test of hemostasis: The filter bleeding time, *Thromb. Res.* **34**:397–405.

van der Does, A. J., 1989, Patients' and nurses' ratings of pain and anxiety during burn wound care, *Pain* **39**(1):95–101.

van Rijswijk, L., 1996, Wound assessment and documentation, *Wounds* **8**(2):57–69.

van Rijswijk, L., Multi-center leg ulcer study group, 1993, Full-thickness leg ulcers: Patient demographics and predictors of healing, *J. Fam. Pract.* **36**(6):625–632.

van Rijswijk, L., and Polansky, M., 1994, Predictors of time to healing deep pressure ulcers, *Ostomy/Wound Manage.* **40**(8):40–51.

Van Winkle, W. Jr., 1967, Wound contraction, *Surg. Gynecol. Obstet.* **125**:131–142.

Wallace, M. S., Dyck, J. B., Rossi, S. S., and Yaksh, T. L., 1996, Computer-controlled lidocaine infusion for the evaluation of neuropathic pain after peripheral nerve injury, *Pain* **66**(1):69–77.

Willoughby, M. L. N., and Arlington, M. J., 1961, The rate of blood loss from skin punctures during the Ivy bleeding time test, *J. Clin. Pathol.* **14**:381–384.

Wu, L., Siddiqui, A., Morris, D. E., Cox, D. A., Roth, S. I., and Mustoe, T. A., 1997, Transforming of growth factor $\beta 3$ (TGF$\beta 3$) accelerates wound healing without alteration of scar prominence histologic and competitive reverse transcription-polymerase chain reaction studies, *Arch. Surg.* **132**:753–760.

Yabon, S. A., Novick, E. S., Jain, S. S., Inhoffer, M., and Graves, D. E., 1995, Postoperative transcutaneous oxygen measurement in the prediction of delayed wound healing and prosthetic fitting among amputees during rehabilitation, *Am. J. Phys. Med. Rehabil.* **74**(3):193–198.

Young, E. K., Mann, R., Goldberg, M., Engrav. L., and Heimbach, D., 1996, Improved accuracy of burn wound assessment using laser Doppler, *J. Trauma* **40**(6):956–961.

Yosipovitch, G., Xiong, G. L., Haus, E., Sackett-Lundeen, L., Ashkenazi, I., and Maibach, H. I., 1998, Time-dependent variations of the skin barrier function in humans: transepidermal water loss, stratum corneum hydration, skin surface pH, and skin temperature, *J. Inv. Dermatol.* **110**:20–23.

Zitelli, J., 1984, Secondary intention healing: An alternative to surgical repair, *Clin. Dermatol.* **2**(3):92–106.

Chapter 18

Design of Clinical Trials to Evaluate Antimycotic Agents in the Treatment of Dermatophyte Infections

BEATRICE B. ABRAMS AND DENNIS E. BABEL

1. Introduction

Dermatophyte infections are among the most prevalent skin infections. They are caused by a group of filamentous fungi, the dermatophytes, which have evolved the ability to degrade keratin, the principal protein of the skin. Since the stratum corneum contains an abundance of this food, most dermatophytes "content themselves" with staying in the superficial regions of the skin. The immune and inflammatory defenses of the body further check the systemic invasion of these organisms. Systemic infections by dermatophytes are almost unknown in healthy hosts.

Infections caused by dermatophytes are referred to as tinea infections and are classified by the anatomical area involved. Thus, tinea pedis is an infection of the foot; tinea corporis an infection of the body; tinea capitis, an infection of the head. (One exception to the "tinea rule" is tinea versicolor, which is caused by a dimorphic fungus, variously named *Malassezia furfur*, *Pityrosporum obiculare*, *P. ovale*, but more correctly called *Pityriasis versicolor*.) The focus on body area may seem trivial; however, regional differences in the skin of the body may affect the clinical presentation of the infection. For example, the plantar surface of the foot often presents with a dry, scaly process. Maceration and secondary bacterial colonization/infection are common in the moist interdigital region of the foot. Infections of the

Beatrice B. Abrams • Clinical Research, Dermatology, Novartis Pharmaceuticals, Inc., East Hanover, New Jersey 07936-1080. *Dennis E. Babel* • Mycology Diagnostics Laboratory, Clinton Township, Michigan 48038.

Cutaneous Biometrics, edited by Doris A. Schwindt and Howard I. Maibach, Kluwer Academic/Plenum Publishers, New York, 2000.

hair may involve the dermal hair shaft, and thus be deeper than the epidermal infections characteristic of most dermatophytoses. Infections of the nails are initially defined by changes in the nail bed (hyperkeratosis) and later by changes in the nail plate (discoloration, onycholysis, and onychodystrophy).

Dermatophytes belong to the one of three genera: *Trichophyton*, *Microsporum*, or *Epidermophyton*. *T. rubrum* is the most common dermatophytic pathogen, being associated with tinea pedis, t. unguium, t. corporis, t. cruris, t. manuum, and t. barbae. *T. tonsurans* is the most prevalent agent associated with tinea capitis in North America and is becoming more prevalent in some European locations. Its prevalence has been increasing over time (Aly, 1994).

Dermatophytes can be classified further as: geophilic, those that live in the soil and infect animals/man when the chance presents itself; zoophilic, those that primarily infect animals but can also infect man; and anthropophilic, those that are obligate parasites of man. While these classifications may seem esoteric, the characteristics of the infections produced often reflect the "level of comfort" that has developed between pathogen and host. Thus, when a zoophilic organism, such as *Microsporum canis*, causes a human mycosis, the process tends to be highly acute and inflammatory. Anthropophilic species such as *Trichophyton rubrum* have developed a more intimate relationship with man over time, resulting in chronic, less inflammatory disease.

The following sections will review factors that bear on the evaluation of agents designed to treat dermatophyte infections. Many of the suggestions, observations, and opinions are based on the authors' experiences. Some useful general references have been added at the end of the paper. Before initiating clinical trials with antimycotic agents, clinical scientists should consult the Summary Basis of Approval and Medical Officer's Review of various antifungal agents approved by the United States Food and Drug Administration (FDA). These documents provide significant insight into the thinking of a highly critical regulatory agency (the US FDA) concerning clinical trial design and evaluation, and can be obtained from the Freedom of Information staff in Rockville, MD.

2. General Trial Design

Since many fungal infections are localized, the designation of target lesions for evaluation is usually feasible, facilitating both mycological and clinical assessments. Infected areas outside of the target should be treated and evaluated secondarily. Mycological assessments of nontarget lesions need not be made. Target lesions should include the most severe areas of involvement. For interdigital tinea pedis, a target foot may be chosen, incorporating assessments from all involved interdigital areas on that foot. Patients in tinea pedis trials should be stratified at enrollment into patients with plantar or moccasin involvement and those without, since the chronic plantar form of the disease is more difficult to treat. Labeling in the United States, in fact, must specify the type of tinea pedis studied, i.e., interdigital or plantar. For tinea unguium, the nail of one of the great toes should be designated as the target. The large size of this nail facilitates measurement of the area involved.

Clinical trials are usually divided into screening, treatment, and posttreatment phases. The duration of each phase is dependent on the mycosis treated, the pathogen (especially in tinea capitis), and the therapeutic agent used. With the "standard" antimycotic agents, tinea corporis generally requires a 2-week treatment period; tinea pedis 4 weeks; tinea capitis, approximately 8 weeks (or more) with griseofulvin. A 2- to 4-week posttreatment assessment is usually acceptable, during which the clinical assessments of patients who are successfully treated often improve.

As in any clinical trial a good history should be taken at baseline. In tinea studies, key data include: the duration of the current episode; an indication of how chronic the disease is; prior therapies for the infection, sex, age, and race of the patient. Concomitant fungal infections are common and should be noted. For example, many patients in tinea pedis studies have tinea cruris, and vice versa.

3. Elucidation of the Pathogenic Organism

One of the most important components in the diagnosis of a dermatophyte infection is the demonstration that the pathogenic organism is, indeed, a dermatophyte (Rippon, 1988; Rebell and Taplin, 1970; Baran and Aly, 1997). Two procedures are used: the KOH (potassium hydroxide) examination (or direct microscopy); and identification of organisms, ideally to genus and species, from cultured specimens. In the clinical setting, these procedures are often bypassed, and the diagnosis is made solely on the basis of clinical appearance and "gut feel." However, some fungal pathogens, which produce similar disease presentations, may require different antifungal therapies. Even the most experienced clinician will admit to cases where implementation of microbiological procedures early in the diagnostic process might have saved time and effort in the long run. In the more rigorous world of regulatory submissions and scientific investigation, microbiological confirmation of the diagnosis is fundamental.

3.1. Specimen Procurement

Both KOH and culture procedures rely on obtaining biological specimens from the patient by scraping, plucking, clipping, or probing. The superficial nature of the infections makes procurement relatively nontraumatic. However, if this step is trivialized, adequate viable fungal elements will not be obtained from the patient, and a complete microbiological diagnosis will be impossible.

In all infections, several sites within the target area should be sampled. Where possible, sites should be thoroughly cleaned with an isopropyl alcohol swab to remove contaminating debris, topical medications or cosmetics, natural hair oils, and the like. In general, the leading edge of a superficial infection (e.g., tinea corporis or tinea pedis) should be targeted, since it is here that the viable fungi can be found in relative abundance. Scaly, erythematous areas from the lesion's edge, tops of vesicles, and pustules are good sources of fungal material. Moist, macerated intertriginous skin should be avoided as it contains no viable fungus and in all likelihood is contaminated with bacteria that may interfere with the fungal culture.

Samples can be obtained with a sterile scalpel, Bard-Parker blade, or with the edge of a clean glass microscope slide. Cotton swabs should not be used since personnel not familiar with the technique can mistake cotton filaments for fungal hyphae during the KOH examination. While delineation of appropriate sites for sampling is relatively easy in the early part of a trial, as the clinical signs of infection resolve, the choice of the area to be sampled may become difficult. Baseline photographs of the target lesion may serve as a reminder of the original areas sampled.

For subungual nail infections, specimens should be obtained from the advancing onychomycotic border, where the greatest numbers of viable hyphae exist. For this purpose, material must be collected as proximally as possible from the nail bed with a 1-mm curette (a 1.5-mm Skeel curette also is useful for this purpose.) The nail plate itself need not be used since subungual onychomycosis is primarily an infection of the nail bed. However, if it is used, the nail fragments should be micronized. In tinea capitis, specimen collection can be made by vigorously scrubbing the prepared collection site with a new toothbrush. Scalp stratum corneum and hair fragments that become lodged within the hairbrush bristles can be transferred to microscope slides or culture plates. Alternatively, a scalpel blade or the edge of a glass microscope slide can be used to scrape the scalp within an area of alopecia. Fluorescing hair stubs, common to some tinea capitis infections, may be plucked with forceps.

3.2. KOH Examination

KOH examinations can be made in the office by trained personnel at the time of, or within a day of the patient's visit. A number of procedures and reagents are available. The strong basic reagent is used to clear the specimen of keratin and nonfungal material, allowing the resistant fungal elements to be more clearly visualized. The KOH examination permits rapid visualization of fungal elements; however it does not permit evaluation of the viability of the fungi. Classification of the pathogen to genus and species level is seldom feasible from a KOH slide preparation.

For most infections, specimens are placed on a clean glass microscope slide and a 20% KOH solution added. After a coverglass is applied, the slide should be gently pressed to expel air bubbles and heated without boiling. If dimethyl sulfoxide (DMSO) is used as the solvent for the KOH, heating is not required. Addition of the dye, chlorazol black-E, in the KOH/DMSO solution facilitates visualization of the hyphae even under low (10 ×) microscopic magnification.

The treated slides should be read immediately or within 5 min by scanning with the 10 × objective and confirming with the 40 × microscope objective. When material is obtained from nail specimens and cleared with aqueous KOH, the treated/heated slides may be kept hydrated for 24 hr before reading to permit dissolution of any hard keratin material. KOH/DMSO cleared nail specimens can be read within 5 to 10 min. A KOH examination is considered positive when hyaline hyphae are observed. For endothrix tinea capitis, a positive specimen will contain hair fragments filled with masses of barrel-shaped to round arthroconidia and/or loose hyphae and arthroconidia among the epithelial cells. In ectothrix infections, spores can be seen on the surface of the hair shaft.

3.3. Culture

At some point during the diagnostic procedure, specimens isolated from lesions are inoculated onto a culture medium and allowed to grow. While many clinicians have received training in the identification of fungi from cultures, and special media can make identification of certain dermatophytes less difficult, concerns for consistency in the clinical trial setting generally mandate use of a centralized mycology laboratory for identification of the pathogen from culture. The mycology laboratory, obviously, should have certified mycologists on its staff and should have a record of successful annual proficiency testing by a regulatory body. The laboratory should: be able to provide assurances of successful participation in other clinical trials; have facilities/personnel sufficient to provide investigational sites with culture media on an on-going basis during the trial and to receive and process large numbers of specimens; be equipped and happy to manage the mounds of paperwork required for clinical trials. References should be provided by the laboratory when possible.

Fungal specimens can be sent to the laboratory in two forms: as scale placed in glassine or cardboard envelopes, or as inoculated cultures. In the latter case, the laboratory must provide the clinical sites with appropriate culture plates and be able to resupply the sites as the medium is used up or becomes out of date. When culture plates are used, the clinical site should inoculate two: one is kept at the site as a backup, while the other is shipped to the laboratory. While inoculation of culture plates at the investigational site can be more difficult logistically than "scale in the mail," it can save time and effort in the long run. Patient specimens not directly inoculated onto culture media and sent dry in a mailer produce delays from culture setup by the mycology laboratory and lost time since the fungal growth starts only after incubation at the mycology laboratory. "Scale in the mail" also is associated with a diminished pathogen recovery rate.

The culture medium most often employed is Saboraud's dextrose agar with chloramphenicol and cyclohexamide. The antibiotics are added to control the overgrowth of contaminants; however, they may also interfere with growth of certain fungi, e.g., *Fusarium* sp., *Scopulariopsis* sp. (saprophytic molds), and yeast such as *Cryptococcus neoformans*, *Candida tropicalis*, and *Trichosporon beigelii*. Since these sensitive fungi are not often isolated in studies of routine dermatophyte infections, more is usually gained than lost by the addition of the antibiotics. Potato extract or peptone also can be added to the growth medium to enhance development of fruiting bodies by the fungi, which are important in determining the species of the organism. Cultures are generally grown at room temperature (25–28°C).

Growth of most dermatophytes is painstakingly slow, often requiring 3–4 weeks before material suitable for identification is available. *Trichophyton tonsurans*, the pathogen most often associated with tinea capitis in North America and some part of Europe, may grow faster (i.e., within 1 week). This frustratingly slow growth has repercussions during both the initial and final phases of a clinical study. During patient enrollment, therapy may be completed before culture confirmation of the diagnosis, especially when therapy is administered for only 1 week. Delay of treatment until culture results are available is not practical, especially in patients with acutely inflamed, pruritic disease. Patients, therefore, generally begin therapy with only a clinical diagnosis and positive KOH. Patients who are later found to be culture-negative are not included in the per-protocol analyses of the data. Enrollment of patients without positive culture results clearly facilitates enrollment. However, negative culture results in the presence of "clear" clinical presentations of dermatophyte infections and positive KOH results can run as high as 50% of enrolled patients. In a recently approved DNA for a product to treat tinea pedis, 26–30% of patients enrolled had negative baseline culture results.* These percentages are in line with our experience. The FDA reviewer commenting on the large percentage of patients excluded for negative baseline culture results stated that, "For future studies of topical antifungal drug products, it may be of interest to include those patients with a negative fungal culture (but who meet the other study criteria by having a positive KOH and clinical findings) in the study to determine their response, since, in practice, the diagnosis of tinea pedis is frequently made on clinical grounds alone." While the comment suggests a possible change in attitude, US approvals historically have required culture confirmation of the diagnosis. Sample size calculations for clinical trials should allow for the relatively high rate of unevaluable patients due to negative baseline mycological results.

During the latter stages of the study, when culture results are used to prove mycological cure, the designation of cure cannot be applied until enough time has elapsed to ensure that viable dermatophytes have had adequate time to grow. An allowance of 3 weeks is usually credible.

Interpretation of culture results from nail infections presents additional problems, since the subungual debris collected from infected nails often contains a wide variety of microorganisms. The role these organisms play in the actual infection is not easy to ascertain. Nail infections due solely to nondermatophyte infections, however, are rare. Since some of the fungi that have been associated with nail infections are sensitive to antibiotics used in the culture medium, the goals of a study in onychomycosis need to be clearly defined before the microbiological section of the protocol is finalized. If the study focuses primarily on the most prevalent form of the disease, i.e., that due to *Trichophyton*, an antibiotic-containing medium may be employed, since problems with overgrowth of dermatophytes by contaminants is of more concern than detection of the less prevalent organisms associated with nail infections. If a more general claim is being sought, antibiotic-free media should be used in addition to that containing antibiotics to ensure that sensitive organisms are not overlooked.

An additional issue, especially acute in trials involving topical therapies, involves possible spillover of the antifungal drug into the specimen being plated. If enough of the drug is present in the skin scrapings, a false negative evaluation could result. Thorough washing of the area to be sampled can reduce chances of spillover. Assessment of lesions after cessation of treatment—in a time frame consistent with

*Medical Officer's Review of NDA 20-524. Butenafine hydrochloride, 1% topical cream. Submission date, 4/3/95; Review date, 2/29/96. Obtained pursuant to the provisions of the Freedom of Information Act, through FOI Services, Inc., Gaithersburg, MD 20878.

disappearance of most of the test substance—can further strengthen arguments of a culture-proven cure. Validation studies to determine the impact of spillover can be developed; most regulatory authorities do not appear fixated on this issue, however, and the extra work may not be required.

4. Clinical Assessments

Clinical assessments of dermatophyte infections should focus on the key signs/symptoms of the disease. Overall or global evaluations are also employed. Careful assessment of clinical presentation is critical to the success of an antimycotic trial, for it is usually the clinical efficacy, not the mycological results, that limits the efficacy claims for an antifungal agent. While 60–80% of patients in various clinical trials may show negative culture/KOH results, far fewer generally show a complete clearance of signs/symptoms.

The key signs and symptoms may vary with the anatomical site/infection type and the patient. Scale is common to almost all tinea infections. Erythema can be pronounced in tinea corporis/cruris; it is common in interdigital tinea pedis, but is less evident in the plantar form of the disease. Vesicles are relatively infrequent in most presentations. Cracking/fissures and maceration are relatively common to interdigital tinea pedis. Subungual hyperkeratosis is the only sign (other than assessments of nail plate involvement) considered of any importance in the diagnosis of subungual onychomycosis. Hair breakage and areas of alopecia are common to tinea capitis. A minority of patients with tinea capitis may manifest highly inflamed kerion. Postcervical lymphadenopathy may be exhibited even by patients with noninflammatory tinea capitis. (Depending on the trial objectives, patients with kerion should be placed in special strata at enrollment or excluded.) Fluorescence under Wood's lamp (wavelength 365 nm) may provide a rapid, preliminary identification of tinea capitis due to *Microsporum canis* (positive fluorescence) and *Trichophyton tonsurans* (negative fluorescence).

Because of the variable presentations associated even with one tinea type, many clinical trial designs define primary and secondary clinical parameters for entry/enrollment. Minimal erythema, pruritus, and scaling scores often are used to define an acceptable baseline disease severity in patients with tinea pedis (not plantar form), tinea cruris, or corporis. As the final assessment of efficacy, some regulatory authorities may insist on complete clinical cure—the absence of all signs and symptoms. Since results using this stringent measure are often relatively unimpressive even with good antifungals, marketing departments—and the physicians and patients themselves—may find that a minimal sign/symptom score can represent a successful secondary outcome of treatment (Note: since some European regulatory agencies do accept this less stringent assessment as a primary outcome, regulatory agencies should be consulted before designating which variables are primary or secondary in multinational trials).

Whatever the final measure, the signs and symptoms are usually rated individually on a scale of 0 = none, 1 = mild (barely perceptible); 2 = moderate (definitely present); and 3 = severe (marked, intense), which has proven itself in numerous clinical trials. While more categories can be added, there is little to be gained from their inclusion. For example, since the category, "3 = severe," seems to describe a miserable existence quite well, little additional information is gained by inclusion of an additional indistinct category, "4 = very severe." The use of visual analog scales to describe symptom severity has been debated *ad nauseum*. In our opinion the categorical responses found in the simple four-point scale provide a patient/subject with enough adjectives to describe his/her level of discomfort.

Assessment of clinical efficacy in distal subungual onychomycosis is a more complex matter. The primary clinical assessment in this indication focuses on the evaluation of clear vs. abnormal nail plate. Subungual hyperkeratosis can be assessed as a secondary variable. The presence of infection extending from the edge of the nail to the lunula (spikes) may be used a criterion for patient exclusion or prospective stratification, since patients with this sign often do not respond well to antimycotic therapy.

Two methods are used to assess clear vs. abnormal nail plate. Zaias (1980) described a way of etching nails at the leading edge of involvement and following this mark as a measure of the appearance of clinically normal nail plate. The notch method provides a good first assessment of healing; however, it cannot take into account the asymmetric presentation of most subungual infections. The extent of disease on a nail can be measured best using tracings of the involved area, made with a fine-tipped marker pen, photography of the marked target nail, and planimetric assessment of the marked area. This procedure requires a high degree of attention to validation of the methods employed prior to initiation of the trial, meticulous standardization in the photography of the nails, and a relatively large clinical budget. Plans for archiving photographs should be made prior to trial initiation, since the quantities of photographs in a large clinical trial can overwhelm most standard archival systems. New computerized methods in photography may make this last issue less critical. Clinicians must be trained in how and what to trace on the infected nails. Equipment must be provided to allow clinicians to obtain consistent photographic representations of the infected nail. The quality and consistency of the film must be ensured throughout the relatively long course of trial in onychomycosis. However, the results from these efforts can be rewarding.

Tinea capitis is evaluated using common signs/symptoms. In addition, the presence of broken hairs (black dot) and alopecia should be assessed, but regrowth of hair to a normal state cannot be expected during the time interval studied in a tinea capitis study, and so should not be included in a summed sign/symptom score (FDA, 1998).

The global or overall assessment of clinical efficacy has had a number of representations. It often is comprised of seven categories given names suitable for marketing's promotional campaigns, along with more quantitative descriptions. For example: 0 = clear, no sign/symptoms of disease; 2 = excellent, 80–90% improvement in the clinical signs/symptoms compared with baseline; 3 = good, 50–79% improvement; 4 = fair, 25–49% improvement; 5 = poor, <25% improvement; 6 = unchanged; 7 = worsened. Use of quartiles throughout the evaluations might be considered by some to be frivolous since improvement of 25 to 49% includes levels, that are of little clinical value. A lower limit of 50% improvement is more indicative of clinically relevant improvement (i.e., improvement less than 50% is considered clinically irrelevant). The comparison with baseline in the measurement in these definitions allows improvement to be assessed in the context of the patient's original condition. A 90% improvement in a severely affected patient certainly is more dramatic than a 90% improvement in a patient with limited disease. On the other hand, the residual 10% unimproved status in these two patients could represent substantial differences in the disease activity present at the time of assessment.

Because some regulators may feel that the critical criterion for success is normal, unaffected skin, a global score rating normalcy, not improvement may be required. Thus the global score would include categories such as 1 = excellent, 90–99% normal skin, etc. However, the definition or interpretation of these categories seems problematic. Because of the difficulties surrounding global evaluations, many trials have begun to rely on sign/symptom evaluations as the primary clinical variable. Interestingly, overall evaluations and minimal sign/symptom scores may not always give the same outcome. In the Medical Officer's Review of the butenafine hydrochloride cream submission to FDA,* a number of cases were identified where the signs/symptoms score was zero, while the overall evaluation was indicative of only 80–99% improvement. The author has also dealt with instances in which global evaluations of "clear" had been given to patients still rated as showing minimal signs or symptoms. The investigator stated in these instances that the signs/symptoms were not attributed to an active fungal infection (e.g., erythema due to friction from a shoe in a tinea pedis patient). Discrepancies between sign/symptom scores and global evaluations should be discussed with the investigators during monitoring visits. The rationale for the discrepancies should be noted in the case report form.

The final measure of efficacy in a clinical antimycotic trial is a variable comprised of results from KOH, culture, and clinical assessments. If both KOH and culture results are negative (no fungal elements on KOH examination and no hyphal growth after 3 weeks of culture), the patient is said to exhibit a mycological cure. KOH and culture results are generally close, although there are usually more negative culture than KOH results. In our unpublished clinical trials, KOH and clinical results often were closer than culture and clinical results, with the former indicative of less efficacy than the latter. In early phases of the trial, nonviable hyphae may still be present producing results that truly are KOH-positive and culture-negative. The results also suggest a notable presence of false negative culture results.

As discussed above, the clinical component of the composite variable can be expressed in a number of ways. No matter which expression is chosen, regulatory authorities, especially the FDA, may insist that only a variable representing complete absence of any sign or symptom of disease is acceptable. Since this rigorous measure of success is often disappointingly low, companies usually prefer a variable allowing the presence of some minimal degree of signs/symptoms. "Effective treatment," as defined in the butenafine NDA,* is comprised of mycological cure (negative culture and KOH) and investigator's global assessment of cleared (100% remission of clinical signs and symptoms compared to baseline) or excellent (80–99% improvement of clinical signs and symptoms compared to baseline). "Successful outcome,"

*Medical Officer's Review of NDA 20-524. Butenafine hydrochloride, 1% topical cream. Submission date, 4/3/95; Review date, 2/29/96. Obtained pursuant to the provisions of the Freedom of Information Act, through FOI Services, Inc., Gaithersburg, MD 20878.

TABLE 1. Comparison of Tinea Pedis Results Using Mycological Cure, Effective Treatment and Overall Cure[a,b]

	Week 4		Week 8	
	Butenafine	Vehicle	Butenafine	Vehicle
Mycological cure	48/53 (91%)	33/52 (63%)	44/53 (83%)	20/52 (38%)
Effective treatment	31/53 (58%)	16/52 (31%)	36/53 (68%)	13/52 (25%)
Overall cure	7/53 (13%)	5/52 (10%)	11/53 (21%)	4/52 (8%)

[a]Medical Officer's Review of NDA 20-524. Butenafine hydrochloride, 1% topical cream. Submission date, 4/3/95; Review date, 2/29/96. Obtained pursuant to the provisions of the Freedom of Information Act, through FOI Services, Inc., Gaithersburg, MD 20878.

[b]Week 8 is 4 weeks posttreatment; results are from modified intent-to-treat (MITT) population. Mycological cure = negative KOH and negative culture. Effective treatment = mycological cure (negative culture and KOH) and investigator's global assessment of cleared (100% remission of clinical signs and symptoms compared to baseline) or excellent (80–99%) improvement of clinical signs and symptoms compared to baseline). Overall cure = mycological cure (negative culture and KOH) and investigator's global assessment of cleared (100% remission of clinical signs and symptoms compared to baseline).

presented in the US terbinafine cream label, is defined as negative mycology and a total clinical (sign/symptom) score of less than 2 (sign/symptom severity graded on a scale of 0–3).†

In the composite score, the mycological scores generally are far better than the clinical, and the allowance of a minimal sign/symptom or global score usually enhances results considerably over use of the clinically clear designation (Table 1).

5. Alternative Designs

For most purposes clinical trials should be randomized, parallel group designs. However, during early evaluations of new drugs or dosage forms, a within-patient design can be used. The method described by Aly *et al.* (1992) using experimentally induced dermatophyte infections requires an expert clinical center to conduct it. However, the results can provide a relatively rapid assessment of efficacy in a limited number of subjects.

6. Conclusions

Many antifungal agents are available in the market, representing many different degrees of efficacy. Unfortunately, the myriad of variables used to assess and describe efficacy often make determination of the true value of the test agent impossible. As

†Lamisil Cream 1% label in Physicians' Desk Reference (ed. 50), 1996, Medical Economics Co., Montvale NJ., p2265.

regulatory reviews become more rigorous, consistency in outcomes should become more evident in labeling. Marketing claims and data presentations in the literature should be based on the same strict requirements for data presentation. Because of the high rate of negative mycological cures, claims based on this variable should never be given credence. Criteria for clinical success need to be carefully specified: do they depend on minimal sign/symptom scores, global assessments, or total absence of clinical signs/symptoms. As more antifungal agents move over the counter, the ability of the consumer to understand the basis of claims diminishes even further. The industry should develop clear, consistent, and meaningful criteria, and then use them.

ACKNOWLEDGMENT: The authors wish to thank Dr. Raza Aly for his help in the preparation of this manuscript.

References

Aly, R, 1994, Ecology and epidemiology of dermatophyte infections, *J. Am. Acad. Dermatol.* **31:**S21–S25.

Aly, R., Maibach, H., Ho, I., and Abrams, B., 1992, Simplified experimental human dermatophytosis model, *Int. J. Dermatol.* **31:**122–125.

Baran, R., and Aly, R., 1997, Diagnosis and new treatment in the management of onychomycosis, in: *Cutaneous Infection and Therapy* (R. Aly, K. Beutner, and H. I. Maibach, eds.). Marcel Dekker, New York, pp. 149–156.

FDA, 1998, U.S. Department of Health and Human Services, Public Health Service, Food and Drug Administration, Tinea capita (sic) clinical trials issues, Transcripts of the Ophthalmic and Dermatologic Drugs Advisory Committee #50. Open Session, Vol. 2, Oct. 22, 1998.

Rebell, G., and Taplin, D., 1970, *Dermatophytes: Their Recognition and Identification*, 2nd Ed., University of Miami Press, Coral Gables.

Rippon, J. W., 1988, Cutaneous infections: Dermatophytosis and dermatomycosis, in: *Medical Mycology: The Pathogenic Fungi and the Pathogenic Actinomycetes*, 3rd Ed., Saunders, Philadelphia, pp. 169–275.

Zaias, N, 1980, Onychomycosis, in The Nail in Health and Disease (N. Zaias, ed.), SP Medical and Scientific Books, New York, pp. 91–113.

Chapter 19

Sites of Contact Dermatitis in Relationship to Particular Allergens: A Follow-Up Study

BJÖRN EDMAN

1. Introduction

In two previous studies I reported on some relationships between allergens and eczema sites, e.g., balsam of Peru was found to be an overrepresented allergen in patients with eczema of the forehead, cheeks, and the lower legs, using the dermatology software system DALUK (Edman, 1988). The first study (Edman, 1985) included 1016 patients tested during 1982–1983, the second (Edman, 1988) was an extended survey involving the previous patients as well as new ones, for a total of 3652 patients tested during the years 1982–1987.

Since then new allergens have been introduced and added to the standard series and some have been removed. Therefore, it is of interest to look again for relationships between today's allergen and eczema sites. Furthermore, it would be interesting to compare the old findings with the new data. We hope that the correlations found between contact allergens and eczema sites will help to guide dermatologists in their efforts to find explanations and relevance of the most common allergens.

2. Material and Methods

During January 1995 and August 1997, we tested 931 patients (288 male and 643 females). Age and sex distribution is presented in Table 1. Patch-testing was performed with the Swedish standard series, which consists of the European standard series, including some extra allergens, using the Finn chamber on Scanpor.

Björn Edman • Department of Dermatology, University Hospital, Malmö, S-20502 Sweden.
Cutaneous Biometrics, edited by Doris A. Schwindt and Howard I. Maibach, Kluwer Academic/Plenum Publishers, New York, 2000.

TABLE 1. Sex and Age Distribution

Age	Males	Females	Total
0–9	10	4	14
10–19	26	58	84
20–29	42	120	162
30–39	48	76	124
40–49	49	117	166
50–59	43	111	154
60–69	35	75	110
70–79	28	57	85
80–89	7	23	30
90–99	0	2	2
Total	288	643	931

The Swedish standard series including the frequency of positive patch tests is shown in Table 2. Owing to the fact that some patients were also tested with other series, e.g., dental screening series, acrylate series, etc., significant correlations to allergens outside the standard series were also found and will be presented here.

All test results as well as information about, e.g., eczema sites, occupation, duration, and drugs being taken at the time of patch application were recorded by the QSDALUK computer system (from Chemotechnique Diagnostics). The patches were left for 48 hr and read 3 and 7 days after application. An inflammatory reaction with papulo-vesicles and/or infiltration was required for the test to be regarded as positive.

3. Statistics

Information about positive patch tests and eczema sites was then extracted. Each combination of allergen and eczema site was statistically tested using the Fisher test. Only allergens with a positive patch test frequency of 1% or more were included in order to minimize the number of analysis. The test was designed as:

	Eczema site X	Other sites
Positive allergen Y	a	b
Negative allergen Y	c	d

In this way the frequency of positive reactions to a certain allergen in combination with a certain eczema site (a) will be compared to all other combinations. Fourteen hundred tests were performed and 64 statistically significant correlations were found with $p < 0.01$. This level of significance was chosen in order to avoid the phenomenon of mass significance. It could be calculated that with the given number of tests and level of significance ($P < 0.01$), 22%, i.e., 14 correlations might be random findings.

TABLE 2. Frequency of Tested Allergens Included in Standard Series

Allergen	Conc.	Males		Females		Total	
		Number	Percent	Number	Percent	Number	Percent
Amercholl 101	As is	21	7.3	18	2.8	39	4.2
Bensocaine	5% pet.	4	1.4	3	0.5	7	0.8
Black rubber mix	0.6% pet.	1	0.3	5	0.8	6	0.6
Bronopoll (5-bromo-5-nitro-1.3-dioxane)	2.5% pet.	15	5.2	21	3.3	36	3.9
Budesonide	0.02% alc.	19	6.6	17	2.6	36	3.9
Butylphenol formaldehyde resin (PTBP)	1% pet.	4	1.4	6	0.9	10	1.1
Caine mix II	10% pet.	13	4.5	14	2.2	27	2.9
Desonide		4	1.4	0	0.0	4	0.4
Diklofluanid	0.05% alc.	4	1.4	4	0.6	8	0.9
Epoxy resin	1% pet.	20	6.9	27	4.2	47	5.0
Ethylene diamine dihydrochloride	1% pet.	17	5.9	33	5.1	50	5.4
Ethyl meruric chloride	0.066% pet.	11	3.8	32	5.0	43	4.6
Euxyl K 400	1% pet.	12	4.2	24	3.7	36	3.9
Phenol formaldehyde resin	1% pet.	3	1.0	17	2.6	20	2.1
Phenylene diamine (P-)	1% pet.	5	1.7	20	3.1	25	2.7
Formaldehyde	1% aq.	13	4.5	22	3.4	35	3.8
Fragrance-mix	16% pet.	0	0.0	1	0.2	1	0.1
Germal-115	2% aq.	7	2.4	3	0.5	10	1.1
Germal-2	2% aq.	7	2.4	8	1.2	15	1.6
Gold sodium thiosulfate	0.5% pet.	109	37.8	234	36.4	343	36.8
Hydrocortisone-17-butyrate	1% pet.	0	0.0	3	0.5	3	0.3
Hydrocortisone alcohol	0.1% pet.	6	2.1	9	1.4	15	1.6
Kathon CG	0.02% aq.	14	4.9	16	2.5	30	3.2
Quinoline mix	6% pet.	6	2.1	8	1.2	14	1.5
Cobalt	0.5% pet.	23	8.0	58	9.0	81	8.7
Colophony	20% pet.	19	6.6	39	6.1	58	6.2
Chromium	0.5% pet.	14	4.9	21	3.3	35	3.8
Lichen mix	0.3% pet.	6	2.1	9	1.4	15	1.6
Mercapto mix	2% pet.	1	0.3	7	1.1	8	0.9
Neomycin	20% pet.	14	4.9	15	2.3	29	3.1
Nickel sulfate	5% pet.	77	26.7	242	37.6	319	34.3
Palladium chloride	2% pet.	36	12.5	95	14.8	131	14.1
Parabens	15% pet.	7	2.4	11	1.7	18	1.9
Balsam of Peru	25% pet.	21	7.3	47	7.3	68	7.3
Primin	0.01% pet.	1	0.3	10	1.6	11	1.2
Propyl gallate	1% pet.	2	0.7	2	0.3	4	0.4
Quaternium-15	1% pet.	4	1.4	13	2.0	17	1.8
Sesquiterpene lacton-mix	0.1% pet.	3	1.0	6	0.9	9	1.0
Thimerosal (merthiolate)	0.1% pet.	28	9.7	54	8.4	82	8.8
Thiosalicylic acid	0.039% pet.	2	0.7	6	0.9	8	0.9
Thiuram mix	1% pet.	8	2.8	15	2.3	23	2.5
Tixocortol pivalate	0.1% pet.	7	2.4	10	1.6	17	1.8
Toluensulfonamid-formaldehyde resin	10% pet.	4	1.4	4	0.6	8	0.9
Triamcinolon acetonide		3	1.0	0	0.0	3	0.3

TABLE 3. Significant Relationships by Allergens

Allergen	Localization	P-Value
Bis-GMA	Neck	0.0085
Amerchol® L 101	Popliteal fossa	0.0053
Bronopol	Hands	0.0069
	Finger	0.0027
Budesonide	Back of hand	0.0055
Caine mix III	Popliteal fossa	0.0070
Epoxy resin	Chest	0.0001
	Sole	0.0001
	Scalp central	0.0001
	Back	0.0069
Ethylendiamine dihydrochloride	Neck	0.0080
	Eye lids	0.0027
Ethyl mercuric chloride	Elbow flexure	0.0074
	General	0.0014
	Cheek	0.0002
Euxyl K 400	Lower leg	0.0094
	Cheek	0.0023
Phenylene diamine (P)	Cheek	
Formaldehyde	Chin	
Fragrance mix	Palm edge	0.0096
Germal 115	Chin	0.0067
	Periorbital	0.0004
Germal 2	Chin	0.0009
	Periorbital	0.0002
Glycidyl methacrylate	Eyelids	0.0053
Gold sodium thiosulfate	Ano-genitals	0.0003
	Hands	0.0032
	Ears	0.0018
Hydroxypropyl methacrylat (2-)	Cuticles	0.0088
Isobutoxi meth acrylamide	Joints	0.0048
Cobalt	Belly	0.0006
	Wrist	0.0050
Chromium	Hands	0.0004
	Lower arm	0.0062
Mercury	Elbow flexure	0.0023
	Wrist	0.0034
	Oral mucous	0.0022
Mercapto mix	Feet	0.0028
	Groin	0.0035
Mercapto benzotiazol	Shoulders	0.0024
	Feet	0.0003
	Groin	0.0042
Sodium laurylsulfate	Palm center	0.0002
Neomycin	Lower leg	0.0084
Nickel sulfate	Belly	0.0004
	Nose	0.0047
	Perioralt	0.0056
Nitro-4-phenylendiamine (2-)	Scalp	0.0097
Octylgallate	Cheek	0.0004

TABLE 3. Continued

Allergen	Localization	P-Value
Palladium chloride	Perioral	0.0012
Parabens	General	0.0066
Primin	Lower arm	0.0100
Propylene glykol (methylglykol)	Hands	0.0090
Thimerosal (merthiolat)	Finger	0.0037
	Elbow flexure	0.0001
	Oral mucous	0.0001
	Lips	0.0029
Thiosalicylic acid	Finger tip	0.0001
	Sole	0.0001
	Palm	0.0002
Thiuram mix	Ankle	0.0075
Tixocortol pivalate	Arms	0.0081
	Axillas	0.0004
	Genitals	0.0006

4. Results

We found 64 statistically significant correlations and these are listed in Table 3 by allergens and in Table 4 by sites. Among the substances included in the standard series, 45 correlations to various sites were found for: *preservatives*: Euxyl K 400, formaldehyde, Germal 115, Germal 2, parabens, thimerosal; *pharmaceutical drugs*: Budesonide, neomycin, tixocortal pivalate; *ointment/cosmetic bases*: Amerchol L 101, ethylene diamine; *metals*: Cobalt, chromium, gold sodium tiosulfate, palladium chloride; *rubber chemicals*: Mercapto mix, thiuram mix; *plastic chemicals*: Epoxy resin; *dyes*: Phenylene diamine; *other*: Fragrance mix, primin.

Among substances *not* included in the standard series, 19 correlations to various sites were found: *plastic chemicals*: BIS-GMA, glycidyl metacrylate, hydroxypropyl metacrylate(2-), isobutoxi metacrylateamide; *preservatives*: bronopol, ethyl mercury chloride, mercury, octyl gallate, thiosalicylic acid; *ointment/cosmetic bases*: sodium laurylsulfate, propylene glycol; *dyes*: nitro-4-phenylenediamine (2-)

5. Discussion

First, let us analyze the correlations of allergens used in the standard series found in this study.

- Four substances were correlated with externally applied pharmaceutical products: Budesonide and back of hands; Tixocortal pivalate and arms and axillas. The correlations of corticosteroids reflect the use pattern of each one, the stronger budesoside is often used on the hands, while tixocortol pivalate, a well-known hydrocortisone allergy marker, represents the weaker corticosteroid group used on thinner and more sensitive skin areas such as axillas and arms (Rietschel and

TABLE 4. Significant Relationships by Sites

Ankle	Thiuram mix
Ano-genitals	Gold sodiumtiosulfate
Arms	Tixocortol pivalate
Axillas	Tixocortol pivalate
Back	Epoxy resin 340
Back of hand	Budesonide
Belly	Cobalt
	Nickel sulfate
Cheek	Ethyl mercury chloride
	Euxyl K 400
	Phenylene diamine (P)
	Octylgallat
Chest	Epoxy resin 340
Chin	Formaldehyde
	Germal 115
	Germal 2
Cuticles	Hydroxypropyl metacrylat (2-)
Ears	Gold sodium tiosulfate
Elbow flexure	Ethyl mercury chloride
	Mercury
	Thimerosal (merthiolat)
Eyelids	Ethylendiamine dihydrochloride
	Glycidyl metacrylate
Feet	Mercapto mix
	Mercapto benzotiazol
Finger	Bronopol
	Thimerosal (merthiolat)
Finger tip	Thiosalicylic acid
General	Parabens
	Ethyl mercury chloride
Genitals	Tixocortol pivalate
Groin	Mercapto mix
	Mercapto benzotiazol
Hands	Bronopol
	Gold sodium tiosulfate
	Chromium
	Propylene glycol (methylglycol)
Joints	Isobutoxi metacrylamide
Lips	Thimerosal (merthiolat)
Lower arm	Chromium
	Primin
Lower leg	Euxyl K 400
	Neomycin
Neck	BIS-GMA
	Ethylendiamine dihydrochloride
Nose	Nickel sulfate
Oral mucous	Mercury
	Thimerosal (merthiolat)
Palm	Thiosalicylic acid
Palm center	Sodium laurylsulfate
Palm edge	Fragrance mix
Perioral	Nickel sulfate
	Palladium chloride
Periorbital	Germal 115
	Germal 2 (diazolidinylurea)
Popliteal fossa	Amerchol L 101
	Caine mix III
Scalp	Nitro-4-phenylendiamine (2-)
Scalp center	Epoxy resin 340
Shoulders	Mercapto benzotiazol
Sole	Epoxy resin 340
	Thiosalicylic acid
Wrist	Cobalt
	Mercury

Fowler, 1995). Neomycin and lower leg. Neomycin allergy in the lower legs is a well-known condition owing to the use of neomycin-containing pharmaceutical products on that site (Edman, 1985). Caine mix and popliteal fossa. The main allergen of this mix is usually benzocaine and dibucaine. One possible explanation for this correlation might be that atopic patients use O.T.C. products containing caines, in order to stop the itching derived from their atopic dermatitis.

- Two mixes of rubber were found to have correlations to different sites: Mercapto mix and feet and groins; and Thiuram mix and ankle; and Mercaptocbenzothiazole and shoulders, feet and groin. Contact allergy to rubber chemicals is often correlated to sites where rubber in clothing touches the skin such as in shoes, stockings, underwear, and bras (Ryctroft *et al.*, 1992).

- There were four metals associated reactions. Nickel and the belly, nose, and perioral. Nickel allergy is well known to cause eczema under jeans buttons, but, the other locations might reflect the new use of nickel items, such as in the nose and on the lips (Edman, 1985). Cobalt and the belly and wrist. Cobalt is also a well-known allergen causing dermatitis under jeans buttons and under wrist watches (Edman, 1985). Chromium and hands and lower arm, gold sodium tiosulfate and ano-genitals, hands and ears. Chromium allergy is also a common cause of eczema on the hands and forearms owing to the use of leather gloves. The sites correlated to gold contact allergy were described earlier (Möller *et al.*, 1996) Hands and ears are common eczema sites for this allergy, because of the use of gold jewelery. However, a connection to ano-genital sites is yet to be explained—maybe the excretion of gold salts from gold dental fillings.
- Epoxy resin allergy was found to have four correlations: the chest, sole, scalp, and the back. This allergen is sometimes found in markers in clothing, which could explain the eczemas of the back and neck (Ryctroft *et al.*, 1992). Epoxy-based glues are normally not used in shoes, but, it is easy to imagine people mending their shoes using epoxy glues. However, an explanation of the correlation to the scalp is harder to find.
- Ethylene diamine had two correlations to the neck and eyelids. The main source of the allergen is known to be Kenacombin®, and owing to its broad indications it can be used almost anywhere on the body (Edman and Möller, 1986).
- Six substances used as preservatives had various relationships: Formaldehyde and chin; the two formaldehyde releasers Germal 115 and Germal 2 to both the chin and periorbital. The similarities between eczema sites of these three allergens strongly suggest that cosmetics and shampoos are the main source (Rietschel and Fowler, 1995). Contact allergy to shampoo very seldom cause eczema in the scalp, but rather causes more often causes a reaction on the face. Euxyl K 400 and lower leg and cheek. The sites correlated with allergy to Euxyl K 400 may reflect the fact that it is used in both cosmetics as well as in externally applied pharmaceutical products. Parabens and the whole body. Parabens, being among the most frequently used preservatives, causes dermatitis everywhere on the body (Ryctroft *et al.*, 1992). Thimerosal and fingers, elbow flexure, oral mucous, and lips. Thimerosal is used mostly as a preservative in eye drops and ointments and in contact lense solutions (Ryctroft *et al.*, 1992). However, allergy to mercury and mercury salts may cause dermatitis on several different sites, probably owing to allergy to amalgam dental fillings (Laine *et al.*, 1997).
- Fragrance mix was correlated to only the edge of the *palms*. The hands are usually exposed to a variety of fragrance-containing products, as they are used to apply these products on different skin sites.
- Primin was related only to the *lower arm*. This is a well-known correlation, which occurs when patients with a contact allergy to primin handle their primula obconica (Rietschel and Fowler, 1995).
- Phenylene diamine was correlated to the *cheek*. Since this allergen is found in hair dyes as well as in dyes for mascara (Ryctroft *et al.*, 1992), its seems that it still is

a problem in Sweden despite the fact that it is forbidden. It also cross-reacts with other permitted chemically similar dyes.

- Amerchol L 101 and the popliteal fossa were found to have a significant relationship. This lanolin derivate is found in some corticosteroid-containing externally applied pharmaceutical products in Sweden, and this finding implies that it is a problem mostly for atopic patients (Edman, 1985).

Other allergens outside the standard series were found to have correlations to various sites:

- First four acrylates: BIS-GMA and neck, glycidyl metacrylate and eye lids, hydroxypropyl metacrylate and joints, iso-butoxy methacrylamid and joints. All of these sites are on the fingers or on sites often touched by fingers, which strongly suggests that artificial nails are the main cause (Freeman *et al.*, 1995). Different acrylates are used as well to build up the nails and to attach them to the natural nail.
- Bronopol and hand and finger, ethyl mercury chloride and elbow flexure, eye lids and cheek. Bronopol is a preservative widely used in cosmetics such as hair conditioners, shampoos, and cleansing preparations (Rietschel and Fowler, 1995), and correlations to hand and fingers must be almost unavoidable.
- Mercury and elbow flexure, wrist, and oral mucous. These findings follow the correlations of thimerosal noted above, a fact that stresses even more that these correlations are caused by allergy to amalgam in dental fillings.
- Nitro-4-phenylenediamine and the scalp. As discussed in connection with phenylene diamine allergy, this allergen is found in hair dyes (Ryctroft *et al.*, 1992). It also cross-reacts with phenylene diamine, another hair dye that now is forbidden in Sweden.
- Octyl gallat and cheek. This allergen is known to cause lip dermatitis (Ryctroft *et al.*, 1992), which suggests that the causes are cosmetic lip or face products.
- Propylene glycol and hands. This allergen is widely used in cosmetics as well as in externally applied pharmaceutical products (Ryctroft *et al.*, 1992), so the hands are very likely to be exposed.
- Thiosalicylic acid and finger tip, sole, and palm. This allergen is said to be the responsible agent for thimerosal allergy (Osawa *et al.*, 1991). The sites correlated to this allergen are very much alike endogenously. One source of thimerosal is vaccine (Osawa *et al.*, 1991), which might be the explanation as there are no other known sources connected with these sites.

Finally, I will compare these new findings with the findings of the previous study (Edman, 1985). Contact allergy to Amerchol L 101 and lanolin were both previously found to be correlated to eczema on the lower leg, and in the present study Amerchol L 101 is found in patients with eczema on the popliteal fossa, so one could say that this correlation holds.

The same is true for contact allergy to caines and neomycin, both previously correlated to lower legs and today to popliteal fossa.

The earlier findings of balsam of Peru were not in evidence at all; perhaps it is an allergen that is not used in the same amount nowadays.

The old finding of chromium and the back did not appear this time, and in the previous study it was discussed whether it was a random finding or not, and so its seems today.

The old finding of cobalt and wrist still stands, as do some of the correlations of nickel allergy. However, nickel contact allergy and eczema of the nose has been added.

Originally I found that fragrance mix allergy had a correlation of the axilla; today this site has been exchanged with eczema of the hands, so it appears that the source of fragrance contact allergy has moved from deodorants to cosmetics applied by the hands onto the skin.

References

Edman, B., 1985, Sites of contact allergy in relationship to particular allergens, *Contact Dermatitis*, **13**(3):129–135.

Edman, B., 1988, Computerized patch test data in contact allergy, Thesis, Malmö, Sweden.

Edman, B., and Möller, H., 1986, Medicament contact allergy, *Derm. Beruf. Umwelt.* **34**(5):139–143.

Freeman, S., Lee, M.-J., and Gudmundsen, K., 1995, Adverse contact reactions to sculptured acrylic nails: 4 case reports and a literature review, *Contact Dermatitis* **33**(6):381–385.

Laine, J., Kalimo, K., Happonen, R. P., 1997, Contact allergy to dental restorative materials in patients with oral lichenoid lesions, *Contact Dermatitis* **36**(3):141–146.

Möller, H., Björkner, B., and Bruze, M., 1996. Clinical reactions to systemic provocation with gold sodium thiomalate in patients with contact allergy to gold. *Br. J. Dermatol.* **135**(3):423–427.

Osawa, J., Kitamura, K., Ikezawa, Z., and Nakajima, H., 1991, A probable role for vaccines containing thimersal in thimerosal hypersensitivity, *Contact Dermatitis* **24**(3):178–182.

Rietschel, R. L., and Fowler, J. F., 1995, *Fisher's Contact Dermatitis*, Williams & Wilkins, Baltimore.

Ryctroft, R. J. G., Menné, T., Frosch, P. J., Benezra, C. (eds.), 1992, *Textbook of Contact Dermatitis*, Springer-Verlag, Berlin.

Chapter 20

Evaluating Surveillance Systems in Contact Dermatitis

AXEL SCHNUCH

1. Introduction

Contact dermatitis may be due to exposure to irritants resulting in "irritant contact dermatitis" or to exposure to allergenic substances causing sensitization and subsequent "allergic contact dermatitis." Both irritant and allergic contact dermatitis are major occupational and nonoccupational diseases.

Allergic sensitization diagnosed in contact dermatitis patients may be related to the disease, but unrelated sensitizations also exist. In these cases, as well as in the "healthy" population without a manifest eczema, sensitization can be understood as an immunological "scar," irrelevant for the disease. Either allergic contact dermatitis or allergic sensitization or both may be a matter for epidemiologic observation.

Prevention can be considered to be an efficient method of medical intervention (Wall, 1995). In order to identify the probable target of prevention as early as possible, the causes of contact dermatitis must be identified (Elsner *et al.*, 1996). Preventive methods may generate costs and may be difficult from a technical (e.g., removal of an irritant or allergenic compound) as well as from a psychological (motivational) point of view. They need, therefore, to be substantiated by a quantification of the disease to be prevented and by a quantification of the identified cause (the impact of a cause and a disease) and not just by the simple diagnoses of a cause as such, as in single case reports. One quantitative method to decide whether (the *numerical* aspect) and how (the *causal* aspect) to enter into preventive measures is epidemiologic observation of which there are several methods.

Axel Schnuch • IVDK-Zentrale, Universitätshautklinik, D-37075 Göttingen, Germany.

Cutaneous Biometrics, edited by Doris A. Schwindt and Howard I. Maibach, Kluwer Academic/Plenum Publishers, New York, 2000.

2. Epidemiologic Methods

Surveillance is the "ongoing scrutiny, generally using methods distinguished by their practicability, uniformity, and frequently their rapidity, rather than by complete accuracy. Its main purpose is to detect changes in trend or distribution in order to initiate investigative or control measures" (Last, 1995). On the other hand, a *register* is "a file of data, collected for a specific public health purpose, containing *all* (identifiable) cases of a particular disease or other health-relevant condition in a defined population such that the cases can be related to a population base. With this information incidence rates can be calculated" (Last, 1995; Solomon *et al.*, 1991). Data from registries may be useful in a surveillance system (Thacker *et al.*, 1988).

Epidemiology may be defined as "the study of the distribution and determinants of health-related states or events in specified populations and the application of this study to control of health problems" (Last, 1995). From an historical point of view epidemiology studied infectious diseases, but today chronic diseases comprise a second subject matter area. Sensitization generally persists, whereas contact dermatitis may or may not be a chronic condition. Therefore, one may distinguish between the "epidemiology of allergic contact dermatitis" and the "epidemiology of allergic sensitizations."

The term *clinical epidemiology*, introduced by J. R. Paul in 1938, has not yet been accepted as unequivocally as the concept of (traditional or classical) epidemiology (Kramer, 1988; Last, 1995). The two epidemiologic approaches have in common that they may employ the same methods, and that they are concerned with the causality of diseases. The most important distinction between the two approaches is that in clinical epidemiology the denominator contains a clinical group, determined by studies of patients who have a particular condition or disease, whereas in public health epidemiology it is a general population, or in occupational health epidemiology, it is an occupational population. In studies of skin allergy the clinical group (the denominator) is made up of those patients with contact dermatitis who seek medical advice and treatment. In those patients, patch-testing is done in order to reveal the underlying sensitizations. Subgroups may be defined by sex and age, atopic diseases, or occupation. The question to be answered may be: what are the most important sensitizations in a subgroup of patients belonging to a specific occupation.

3. Measures of Frequency: The Time Factor

Incidence and *prevalence* are measures of disease frequency. The frequency is expressed as a rate, i.e., number of cases per number of persons in the group within whom the disease occurred. The cumulative incidence is the number of new cases during a given period of time divided by the total population at risk (Hennekens and Buring, 1987). The prevalence rate measures numbers of existing cases of a disease in a population at one point in time ["a snapshot of the frequency that prevails at a given moment" (Kramer, 1988)]. In diseases with a short duration (e.g., due to successful treatment), the frequency may be 0% at a given point in time, whereas, in the periods before and after, cases accumulate. For such diseases, the appropriate measure may be the 1-year prevalence, the number of all cases (new and recurrent) that are observed during a 1-year period (the lens of the camera would be open for 1 year).

There are different reasons to favor prevalence rates over incidence rates in studies of contact dermatitis. Allergic contact dermatitis is a nonfatal, chronic disease. Even with low incidence rates the prevalence may be high, the new cases being added to an already high basic rate. Prevalence is more useful to describe the "burden" of the disease; incidence rates would underestimate the extent of the disease. Prevalence rates can be applied to both the sensitization rate to a certain allergen and to the manifest disease, allergic contact dermatitis. Incidence rates can be calculated for the disease only, not for sensitizations, as the exposure that led to sensitization may be different from the exposure that led to elicitation, i.e., this case being a new one only with regard to the disease, sensitization having been acquired long before. Finally, the incidence rate is the frequency measure of choice when interest focuses on the hitherto unknown cause of the disease. On principle, the cause of allergic contact dermatitis, the allergen, is diagnosed by a laboratory procedure (patch-testing). The calculation of incidence rates only for this purpose would be useless. On the other hand, increasing incidence rates in defined populations may serve as a warning signal followed by studies on the causal exposure (Elsner *et al.*, 1996).

4. Data Generation: Active or Passive

There is one fundamental difference in the nature of data generation in epidemiology and surveillance (see also Coenraads and Smit, 1992). We would propose the following distinction: (a) Active data generation would mean that people (workers) in the source population are scrutinized (active case ascertainment) and the results are evaluated. Ideally, all cases would be identified and registered. (b) Passive data generation would mean that people consult a physician because they suffer from contact dermatitis. There is sufficient concordance between the diagnoses of the patient affected and the diagnoses of the dermatologist. Of those considering themselves to have hand eczema, the diagnoses were confirmed in 89.4% (Meding and Swanbeck, 1987), but only 15 to 36% of patients with contact dermatitis seek medical advice (Smit *et al.*, 1993). It could be concluded that the difference between active and passive data generation is, in the final analyses, not due to cases underdiagnosed (information bias) but to "self-underreporting" (selection bias).

5. Study Population (Problem: Selection Bias)

Basic demographical data are generally used in epidemiology to describe the population studied. Characteristics such as age and sex are of utmost importance (Kwangsukstith and Maibach, 1995). Although, from a theoretical point of view, everybody may suffer from contact dermatitis, the chance of acquiring this exogenous disease is unevenly distributed among different groups of people. One reason is a different susceptibility. People with an atopic diathesis or even atopic dermatitis are at greater risk of irritant contact dermatits (Gollhausen, 1991). The susceptibility to become sensitized and to develop allergic contact dermatitis is also different but the responsible endogenous traits are not yet clearly defined (Moss *et al.*, 1985).

The second reason is the uneven occurrence of noxious agents in different subpopulations. Different exposure to noxious agents is the most important reason for the different distribution of contact dermatitis in different subpopulations. The exposure may be known (e.g., from questionnaires), may be suspected (e.g., in certain occupations with defined occupational activities), or may be unknown (e.g., several products are suspected without further information on the ingredients) or completely unknown (no cause of contact dermatitis would have been suspected or even identified).

As the precise exposure is often difficult to assess (Elsner *et al.*, 1996), further information may be taken from population characteristics as surrogate markers of exposure. Not only age and sex were shown to be of importance, but also the body site affected by the dermatoses (e.g., leg dermatitis and face dermatitis), and the occupational background. Studying the frequency of contact dermatitis in different populations (differing with regard to one or more of the above-mentioned characteristics) would yield different results (Schnuch *et al.*, 1997). Studies not considering these factors are highly vulnerable to a selection bias.

To adjust for the more general factors, age and sex, an instrument of standardization (PAFS) was proposed (Schnuch, 1996), and was used in the evaluation of multicenter patch test data (Schnuch *et al.*, 1997). Other factors are long since known to influence the results of patch test studies, i.e., the percentages of occupational skin disease, atopic dermatitis, hand dermatitis, and leg dermatitis. To describe a patch test population, the MOHL index (M = male; O = occupational; H = hand dermatitis; L = leg dermatitis) was introduced (Wilkinson *et al.*, 1980), later atopy was added as a further descriptive item. Based on recent studies it was felt necessary to extend the original MOAHL index, and to add face dermatitis and age above 40 (the acronym would then be MOAHLFA) (Schnuch *et al.*, 1997). Face dermatitis can be regarded as a surrogate marker of an exposure to cosmetics. The items of the MOAHLFA index are not yet used routinely for further standardization of data, but as descriptive items they should at least be communicated in order to allow a discussion of results. However, one important effect of standardization should be mentioned: it may become impossible to pinpoint specific subgroups at risk, as extreme sensitization rates in certain groups are leveled by the procedure of standardization (Chan *et al.*, 1988; Schnuch *et al.*, 1997). To analyze and identify groups at risk, standardization must be supplemented by an analysis of the differences between standardized and crude rates induced by specific patterns of the determinants.

6. Diagnostic Procedure (Problem: Information Bias)

It is of utmost importance that the data being entered into a surveillance system are valid from a diagnostic point of view. Occupational physicians or general practitioners often are the first to be consulted, but they may fail to consider the differential diagnoses of irritant contact dermatitis, allergic contact dermatitis, or atopic dermatitis. The instrument of patch-testing may be used inappropriately. The dermatologist involved in the further elucidation of a case should meet the demands of international defined guidelines for patch-testing (Fregert, 1981). For the diagnoses of atopic dermatitis, standards were proposed (Hanifin and Rajka, 1980), and it is advisable to indicate which of them were used to diagnose atopic dermatitis or atopic diathesis (Diepgen, 1991).

It is therefore decisive that the skin disease be diagnosed (or verified) by a dermatologist (Aberer *et al.*, 1993). On the other hand, it may be helpful in specific cases to consult an occupational physician in order to learn more about possible causal exposures in the workplace. (One could conceive of surveillance systems that integrate collaboration with occupational physicians on a regular basis). Last but not least, the diagnostic process should be concluded by the assessment of relevance (Ale and Maibrach, 1995). The most important feature is to know the (possibly or unequivocally causal) exposure, completed by data on occupation or occupational activities.

Data gathered during the diagnostic process and compiled in a data bank may become subject to quality control. It should audit the completeness and plausibility of data, the fitting to established rules (of diagnoses and registration), and further should identify results (e.g., sensitization rates) that seriously deviate from the means of the surveillance system, thus pointing at possible systematic errors. Finally, the participants in a surveillance system may be involved in a feedback process. Ongoing discussions of results may help regulate the data structure (new items to be introduced, others to be omitted) or the criteria of diagnoses and patch test practices (e.g., by changing the concentration or vehicle).

7. General Aspects of Evaluating a Surveillance System

Apart from the above-mentioned topics specific for the surveillance of contact dermatitis, further characteristics, discussed in comprehensive reviews by Klaucke (1992) and Klaucke *et al.* (1988), should be considered and checked more or less thoroughly if a surveillance system is to be organized:

1. Public health importance of the disease to be surveyed.
2. System description: objectives, health event(s) under surveillance, component(s) (e.g., population under surveillance, information collected, data processing, external data usage).
3. Usefulness: for what purposes (e.g., preventive action) the output of the system was used or is to be used in the future.
4. System attributes: simplicity, flexibility, acceptability, sensitivity, predictive value positive, representativeness, timeliness.
5. Resources for the system: personal and financial resources applied to the collection, processing, analysis, and dissemination of the surveillance data.

A multicenter epidemiologic surveillance system on contact dermatitis was evaluated using the above criteria (Schnuch and Lehmacher, 1992), and a proposal for establishing guidelines for evaluating surveillance systems on contact dermatitis is presented in the appendix.

8. Discussion of Several Methods

8.1. Active Data Generation

1. If the objective is continuously updated information about the health status of workers and about the offending exposure, then a continuous (institutionalized) examination (patch-testing included), as well as the continuous registration of the results, would be necessary. This ideal approach would be time-consuming and expensive. From an economic point of view, case ascertainment by questionnaires, distributed regularly over time, would be preferable. Questionnaires have been used to identify cases in many epidemiological studies (Smit *et al.*, 1993). The questions themselves are important to the validity of the questionnaire, but the questionnaires are rarely evaluated (Smit *et al.*, 1992*a*), although such evaluation should be a prerequisite for their use. The cases thus identified would have to be interviewed about their working conditions. In certain cases, the exposure would have to be analyzed by the physician. In any case the health status would have to be examined (patch-testing included). Data generated actively should be stored and evaluated regularly. The system would be a register and could be used for surveillance as well as epidemiologic purposes, i.e., calculation of prevalence, incidence, risk factors, identification of causal exposures, etc.

2. If the objective is to know prevalence rates, incidence rates, risks linked to certain exposures, or the natural course of allergic contact dermatitis, then observational studies in clearly defined populations are indicated. Publications on epidemiological studies on contact dermatitis have been reviewed recently (Schnuch, 1994; Smit and Coenraads, 1993). In some studies, only hand dermatitis was examined in a cross-sectional approach (Smit *et al.*, 1993) or as a prospective cohort study (Uter *et al.*, 1995), but without further diagnosis by patch-testing. In other studies, the diagnosis of contact dermatitis was followed by patch-testing, in cross-sectional studies (Meding, 1990) as well as in prospective studies (Smit *et al.*, 1992*b*). Others were (only) interested in the prevalence rates of sensitization in healthy populations [general population (Nielsen and Menné, 1992), or nondermatologic patients (Magnusson and Möller, 1979)].

These epidemiological studies yield static "facts": The prevalence rate of contact dermatitis or allergic sensitization in the general population or the identification of risk factors (wet work, history of hand dermatitis, smoking, or atopic diathesis) in specific occupations (Uter *et al.*, 1995). To quantify the risk to become sensitized by exposure to a certain allergen, the exposure (e.g., time of exposure, concentration of the allergen) would have to be measured, and after a given time, patch-testing would have to be done in persons exposed and not exposed. The ideal measure would be incidence density. These studies have never been done. Given the complex work environments, with exposure to many allergens as well as exposure to various cofactors, quantification of exposure to a certain allergen will be very difficult if not impossible. Nevertheless, some information may be drawn from semiquantitative studies. For example, it has been shown by logistic regression analysis that exposure ($>2\times$/day) to acidic permanent wave (glycerolmonothioglycolate) has a greater risk for hand dermatitis than exposure to alkaline permanent wave (ammonium thioglycolate) (OR 1.61 vs. OR 1.11) (Uter *et al.*, 1995). Unfortunately, patch-testing has not been done in the study population. In a prospective cohort study in Germany (North Bavaria) more than 2000 apprentices were examined for predisposing factors to develop hand dermatitis before the start of the apprenticeship, and later followed up. The most important predisposing factors were atopic eczema and dyshidrosis (Elsner *et al.*, 1996). Based on this information the targets for prevention can be identified and intervention can take place (Elsner *et al.*, 1996).

8.2. Passive Data Generation

Passive data generation occurs when people consult a physician because they suffer from contact dermatitis. Clinical examination and patch-testing would yield the diagnosis (allergic contact dermatitis) and the cause of the illness (sensitization to an allergen). Consultation with a physician may be initiated by the observation of cases by employers, health safety officers, or, in the case of a dermatitis owing to consumer products, by the cosmetic industry involved. In any event, passive data generation is the basis of morbidity statistics. Special issues such as contact dermatitis owing to cosmetics and to external drugs are not considered further. For cosmetics, surveillance systems have been built up in collaboration with dermatologists (Berne *et al.*, 1996; Vigan, 1997). Reactions to external drugs are often reported to the national drug safety agencies by practitioners without further diagnoses (with respect to irritation or allergy).

Two main sources of data can be distinguished: (a) official authorities of occupational health care; and (b) dermatological departments or dermatological practitioners.

(a) Official Authorities. Registers of occupational skin diseases are kept in several European countries and in the United States (Kanerva *et al.*, 1994; Mathias *et al.*, 1990; Rothe and Bräunlich, 1991). Most of the registers concern all types of skin disease and no distinction is made with regard to the type of eczema (irritant or allergic contact dermatitis). Data are designed to assess economic rather than occupational safety and health factors. Although some registers provide some epidemiological data (incidence of occupational skin disease 0.6 to 1.9 per 1000 workers (Mathias *et al.*, 1990), they provide no data on causation or for setting priorities for development of occupational health standards. The data published by the German *Berufsgenossenschaften* (insurance companies for all employees) indicate that the total number of cases with suspected work-related dermatological disease were 20,584 in 1991 and increased to 21,224 in 1992, and reported only the financial burden of the different industrial sectors. The annually published report of the *Berufsgenossenschaft für Gesundheitsdienst und Wohlfahrtspflege* (insurance company for health care/personal care workers in appropriate services) showed that the total costs for compensation due to occupational skin diseases were 105 million DM, of which 59% (DM 63 million) were directed to hairdressers. A detailed evaluation of the offending agents and the affected occupational branches was done sporadically (Plinske *et al.*, 1989). On the other hand, there are (or have been) at least two register systems in Europe differentiating the type of skin disease (e.g., allergic contact dermatitis), the occupation, and the causal allergens. In Finland, the most important

sensitizer was colophony, followed by rubber chemicals (Kanerva *et al.*, 1994). In the former (East)German Democratic Republic (GDR), the most important allergens were chromate and formaldehyde. Incidence rates of occupational skin disease were calculated, e.g., for electroplaters (58.8 per 10,000 per year) or for hairdressers (18.4 per 10,000 per year) (Rothe and Bräunlich, 1991). Since the occupational health care system in the former GDR had to be abandoned, the statistics are no longer collected. An epidemiological study in Germany (North Bavaria) was started in 1990 (Diepgen *et al.*, 1994). All reports on suspected occupational skin diseases observed in this region are evaluated, i.e., 1200 to 1400 cases (6% of all suspected occupational cases in Germany). The notification and evaluation system in Germany, regulated by law, contains on principle most of the relevant information (clinical examination, patch and prick tests, and, occasionally, reports of work site visits), which converge within the institution of an official government physician. The evaluation of these data in North Bavaria permits the calculation of incidence rates in specific industrial branches and occuaptions, the identification of specific occupational or personal risk factors, type and course of the dermatological disease, and of the offending allergens (Elsner *et al.*, 1996).

The data source type in all these cases is one of a register. Data drawn from registers can be used for epidemiological studies as well as for surveillance purposes.

(b) Dermatological Departments or Dermatological Practitioners. The most important data source on occupational skin diseases is the statistical information generated by the dermatologist. He/she takes the patient's history (e.g., occupation), examines the case clinically (constitutional risk factors included), and establishes the allergic diagnosis by patch-testing. Most of the information reported in the literature and in contact dermatitis textbooks is based upon this data source. From statistics of the clinical eczema population we learn that: (a) nickel and fragrance mix are the most important causes of allergic contact dermatitis, (b) the typical allergens found with hairdressers with contact dermatitis are glycerylmonothioglycolate and hair coloring substances, (c) the typical allergens found with dental technicians with allergic contact dermatitis are acrylates, and (d) the occupation affected most by glutardialdehyde is the (hospital) health service. Data generated from this source are best suited for allergen surveillance, as the dermatologist is the first and often the only one to diagnose a skin allergy.

Until recently, this data source was used sporadically for evaluation. In the United Kingdom, dermatologists from 16 centers started a pilot study in order to assess the viability of maintaining a surveillance register for occupational dermatoses through a simple card reporting system. In the first 12 months, this group identified 1466 new cases of occupational dermatitis (Beck, 1992). In the United States, several dermatologists joined to pool their results and to evaluate patch test data and data from patients' histories. Reports from different periods of observation were published (Marks *et al.*, 1995). In 1989, a multicenter project for registering the most important data from patch-test patients was developed in Germany [the information network of departments of dermatology (IVDK)]. In 30 participating departments, standardized information (demographical data, occupational data, clinical diagnosis, patch test results, relevance of the sensitization) is continuously stored and regularly transferred to the data center in Goettingen for statistical evaluation (Schnuch *et al.*, 1997).

The convergence of large numbers of data batches within a short period reduces the time necessary to detect new trends in sensitization or to identify (new) occupational risk exposures regarded as harmless until now. The system is conceived as a surveillance system where evaluations are done within clinical epidemiology. The Center for Disease Control of the United States has published "Guidelines for Evaluating Surveillance Systems" (Klaucke *et al.*, 1988), and the IVDK has been described as following these guidelines (Schnuch and Lehmacher, 1992). Annual reports on the frequency of sensitizations of the most important allergens have been published, as well as reports on several occupations and data analyses to improve diagnostic procedures (Uter *et al.*, 1998).

9. Conclusions

The objective of surveillance systems is the rapid identification of problems defined as sentinel health events. Registers based on continuous active data generation do not yet exist. Incidence and prevalence rates could be provided and could function as a surveillance system, but the high costs involved would be a disadvantage. Incidence and prevalence rates (with regard to the general population) can be calculated on the basis of epidemiological studies with active data generation, but this does not fulfill the criteria of continuous surveillance.

Data from "official sources" (e.g., worker insurance companies) have the character of a register. All "occupational cases" (operationally defined) are reported continuously. If, rarely, the data structure contains the relevant information about demographic, medical, and allergy data (as, e.g., the Finnish or North Bavarian registers), this register can be used for epidemiological studies (incidence, prevalence, risk factors). The structure of the register would, on principle, enable its use as a surveillance system. One problem would be slow entry of data into the system (e.g., the entry of a case in the system depends on notification and sometimes long-lasting evaluation). A further problem is the operationally defined occupational case. The true figure of work-related dermatological cases is estimated to be 10 to 50 times higher than the number of reported cases (Emmett, 1986). Third, the occupational case is a problem not only with regard to its legal restrictions but also in terms of its medical evaluation (no standardized occupational history, no standardized occupation related patch-testing, no standardized exposure assessment, etc.). It may be concluded that all statistics from official sources, although yielding apparently "exact" incidence and prevalence figures, suffer essentially from the "numerator" problem. Nevertheless, the increased risks that have been identified by these registries are valuable and must be taken seriously, as they may even underestimate the real situation.

The incidence and prevalence rates cannot be calculated on the basis of clinical data. If the problem of the official registries (and their epidemiological studies) is the numerator, the problem in clinical surveillance systems is the denominator. However, these rates are not necessarily essential to meet the objectives of a surveillance system on skin allergies. On the other hand, rate changes within the clinical population can be identified without any delay by a clinical surveillance system, especially if it is an "active" surveillance system, e.g., IVDK [passive surveillance systems depend on reporting (e.g., Beck, 1992), active systems collect data regularly and do not depend on the "good will" of the reporter].

The earliest signal coming from a (new) allergen is a diseased person. As this person is first examined and identified in the clinical practice, the evaluation and statistical analysis of such cases yield the most rapid information. Based on this information, the structure of data generation in "official systems" can be changed (e.g., introduction of new allergens in occupational patch-testing) or further studies can be undertaken (e.g., epidemiological studies with an numerical objective).

Appendix: Proposal to Establish Guidelines for Evaluating Surveillance Systems in Contact Dermatitis

Apart from the following questions, the general aspects of a surveillance system should be also considered (see Section 7).

1. What are the diseases under surveillance? irritant contact dermatitis, allergic contact dermatitis, atopic dermatitis. On which criteria is the differential diagnosis based?
2. Who "diagnoses" contact dermatitis (what is the diagnostic evidence of the signal to be reported): dermatologist, occupational physician, general practitioner, or only the patient himself (with or without a questionnaire), or a consumer, or workers, or others (pharmacists, nurses etc.)?
3. How is the population under surveillance defined: general population, workers from a certain occupation or a certain plant, subpopulations, e.g., children, patients, consumers of certain products, e.g., cosmetics?
4. Epidemiologic system: Surveillance system, register, epidemiological study (type, e.g., cross-sectional), clinical epidemiology (dermatological practitioner, dermatological department, multicenter).
5. Is followup of cases possible?
6. Duration of the study/the system: limited, repeated, or continuous?
7. Measure of frequency: incidence or prevalence?
8. Data acquisition: active or passive (see Section 4). Further information on the methods used (e.g., questionnaires—evaluated for sensitivity and specificity), diagnoses by medical experts?
9. Is confidentiality or anonymity in connection with personal data guaranteed?
10. If allergic contact dermatitis is addressed: is sensitization confirmed and specified by patch-testing? Are the methods of patch-testing (material used, procedure) communicated? Do they comply with internationally accepted rules?
11. If atopic dermatitis is addressed: which criteria are used to diagnose atopic dermatitis or atopic diathesis?
12. Are possible causative exposures assessed and registered (e.g., occupation, occupational activity, nonoccupational exposure such as to consumer products, leisure activities)?
13. Is the relevance of a single contact allergy assessed and registered? What set of criteria is used for its assessment?

14. Are basic demographic data (age, sex, geographic origin) registered?
15. Which methods and instruments are used for (a) data exchange, storage, processing, and communication between participants, and (b) communication to institutions to be advised or informed?
16. Are the results partially (e.g., for age and sex) or extensively adjusted (standardized) by considering other factors such as occupation?
17. Is the study population described by the MOAHL index or the recently extended MOAHL index (MOAHLFA)? Are the results discussed with respect to these items?
18. Are measures of quality control (completeness, plausibilty, comparison among different participants of the system, feedback and discussion of the results) part of the system? Are diagnostic materials (i.e., patch test preparations) subject to quality control?
19. What is the ultimate goal of the system—simple counting of the "signal" for the purpose of public (or occupational) health statistics or contributing to prevention by analyzing data with regard to causative factors? To whom are the results adressed? On a regular basis or sporadically? Are there confidential elements (not to be published) reserved for certain groups of interest (industry, regulatory institutions, workers' insurance companies)?
20. Is external evaluation possible? To what degree did the information delivered contribute to successful preventive decisions/actions? Does the economic benefit of the system exceed the costs of the system?

References

Aberer, W., Andersen, K. E., and White, I. R., 1993, Should patch testing be restricted to dermatologists only? *Contact Dermatitis* **28**:1–2.

Ale, S. I., and Maibach, H. I., 1995, Clinical relevance in allergic contact dermatitis: An algorithmic approach, *Dermatosen* **43**:119–121.

Beck, M. H., 1992, Occupational dermatoses surveillance in the United Kingdom, *Br. J. Dermatol.* **127**:16–17.

Berne, B., Bostrom, A., Grahnen, A. F., and Tammela, M., 1996, Adverse effects of cosmetics and toiletries reported to the Swedish Medical Products Agency 1989–1994, *Contact Dermatitis* **34**:359–362.

Chan, C. K., Feinstein, A. R., Jekel, J. F., and Wells, C. K., 1988, The value and hazards of standardization in clinical epidemiologic research, *J. Clin. Epidemiol.* **41**:1125–1134.

Coenraads, P.-J., and Smit, J., 1992, Methodological aspects of epidemiologic studies of contact dermatitis, in: *Textbook of Contact Dermatitis* (R. J. G. Rycroft, T. Menné, P. J. Frosch, and C. Benezra, eds.), Springer-Verlag, Berlin, Heidelberg, New York, pp. 133–150.

Diepgen, T. L., 1991, *Die atopische Hautdiathese*, Gentner Verlag, Stuttgart.

Diepgen, T. L., Schmidt, A. M., and Fartasch, M., 1994, Berufsekzeme und Berufskrankheitsverfahren—epidemiologische Aspekte, *Allergologie* **17**:84–89.

Elsner, P., Lachapelle, J. M., Wahlberg, J. E., and Maibach, H. I. (eds.), 1996, *Prevention of Contact Dermatitis*, Karger, Basel.

Emmett, E. A., 1986, Dermatological screening, *J. Occup. Med.* **28**:1045–1050.

Fregert, S., 1981, *Manual of Contact Dermatitis*, 2nd Ed., On behalf of the International Contact Dermatitis Research Group and the North American Contact Dermatitis Group, Munksgaard Publishers, Copenhagen.

Gollhausen, R., 1991, The phenomenon of irritable skin in atopic eczema. in: *Handbook of Atopic Eczema* (T. Ruzicka, J. Ring, and B. Przybilla, eds.), Springer-Verlag, Berlin, Heidelberg, New York, pp. 306–318.

Hanifin, J. M., and Rajka, G., 1980, Diagnostic features of atopic dermatitis, *Acta Derm. Venereol. Suppl. Stockh.* **92**:44–47.

Hennekens, C. H., and Buring, J. E., 1987, *Epidemiology in Medicine*, Little, Brown and Company, Boston, Toronto.

Kanerva, L., Jolanki, R., and Toikkanen, J., 1994, Frequencies of occupational allergic diseases and gender differences in Finland, *Int. Arch. Occup. Environ. Health* **66**:111–116.

Klaucke, D. N., 1992, Evaluating public health surveillance systems, in: *Public Health Surveillance* (W. Halperin and E. L. Baker, and R. R. Monson, eds.) Van Nostrand Reinhold, New York.

Klaucke, D. N., Buehler, J. W., Thacker, S. B., Parrish, R. G., Trowbridge, F. L., and Berkelman, R. L., 1988, Guidelines for evaluating surveillance systems, *MMWR* **37** (Suppl 5):1–18.

Kramer, M. S., 1988, *Clinical Epidemiology and Biostatistics*, Springer-Verlag, Berlin, Heidelberg, New York.

Kwangsukstith, C., and Maibach, H. I., 1995, Effect of age and sex on the induction and elicitation of allergic contact dermatitis, *Contact Dermatitis* **33**:289–298.

Last, J. M. (ed.), 1995, *A Dictionary of Epidemiology*, 3rd Ed., Oxford University Press, New York, Oxford, Toronto.

Magnusson, B., and Möller, H., Contact allergy without skin disease, *Acta Derm. Venereol. Stockh.* **59** (Suppl. 85):113–115.

Marks, J. G., Belsito, D. V., DeLeo, V. A., Fowler, J. F., Fransway, A. F., Maibach, H. I. Mathias, C. G. T., Nethercott, J. R., Rietschel, R. L., Rosenthal, L. E., Sheretz, E. F., Storrs, F. J., Taylor, J. S., 1995, North American Contact Dermatitis Group standard tray patch test results, (1992 to 1994), *Am. J. Contact Dermatitis* **6**:160–165.

Mathias, C. G. T., Sinks, T. H., Seligman, P. J., and Halperin, W. E., 1990, Surveillance of occupational skin disease: A method utilizing worker's compensation claims, *Am. J. Ind. Med.* **17**:363–370.

Meding, B., 1990, Epidemiology of hand eczema in an industrial city, *Acta Derm. Venereol. Stockh. Suppl.* **153**:2–43.

Meding, B., and Swanbeck, G., 1987, Prevalence of hand eczema in an industrial city, *Br. J. Dermatol.* **116**:627–634.

Moss, C., Friedmann, P. S., Shuster, S., and Simpson, J. M., 1985, Susceptibility and amplification of sensitivity in contact dermatitis, *Clin. Exp. Immunol.* **61**:232–241.

Nielsen, N. H., and Menné, T., 1992, Allergic contact sensitization in an unselected Danish population: The Glostrup allergy study, Denmark, *Acta Derm. Venereol. (Stockh.)* **72**:456–460.

Plinske, W., Drexel, G., Francks, H. P., and Butz, M., 1989, *BK-DOK—Dokumentation des Berufskrankheitengeschehens in der Bundesrepublik Deutschland*, Hauptverband der gewerblichen Berufsgenossenschaften, Sankt Augustin.

Rothe, A., and Bräunlich, A., 1991, Zur Epidemiologie der Berufsdermatosen in der ehemaligen DDR. in: *Epidemiologie allergischer Erkrankungen* (J. Ring, ed.), MMV Medizin Verlag München, pp. 83–97.

Schnuch, A., 1994, Die Verbreitung des Kontaktekzems in der Allgemeinbevölkerung und in verschiedenen Berufen, in: *Manuale allergologicum* (E. Fuchs and K.-H. Schulz, eds.), Dustri Verlag, Deisenhofen, V/16.2 pp. 1–42.

Schnuch, A., 1996, PAFS: Population adjusted frequency of sensitization (I)—Influence of sex and age, *Contact Dermatitis* **34**:377–382.

Schnuch, A., and Lehmacher, W., 1992, Epidemiologische Überwachung des Kontaktekzems—Darstellung des Projektes "Informationsverbund Dermatologischer Kliniken" (IVDK) in Anlehnung an die "Guidelines for Evaluating Surveillance Systems" des Center of Disease Control (CDC), *Dermatosen* **40**:177–189.

Schnuch, A., Geier, J., Uter, W., Frosch, P. J., Lehmacher, W., Aberer, W., Agathos, M., Arnold, R., Fuchs, T., Laubstein, B., Lischka, G., Pietrzyk, P. M., Rakoski, J., Richter, G., Rueff, F., 1997, National rates and regional differences in sensitization to allergens of the standard series: Population adjusted frequencies of sensitization (PAFS) in 40,000 patients from a multicenter study (IVDK), *Contact Dermatitis* **37**:200–209.

Smit, H. A., and Coenraads, P. J., 1993, Epidemiology of contact dermatitis, in: *Epidemiology of Clinical Allergy* (M. L. Burr, ed.), Monographs in Allergy, Vol. 31, Karger, Basel, pp. 29–48.

Smit, H. A., Coenraads, P. J., Lavrijsen, A. P. M., and Nater, J. P., 1992*a*, Evaluation of a self-administered questionnaire on hand dermatitis, *Contact Dermatitis* **26**:11–16.

Smit, H. A., van Rijssen, A., Vandenbroucke, J. P., and Coenraads, P. J., 1992*b*, Individual susceptibility and the incidence of hand dermatitis in a cohort of apprentice hairdressers and nurses, in: *Work Related Hand Dermatitis—Epidemiological Studies in Occupational Dermatology* (H. A. Smit, ed.), Proefschrift Rijksuniversiteit, Groningen, pp. 97–120.

Smit, H. A. Burdorf, A., Coenraads, P. J., 1993, The prevalence of hand dermatitis in different occupations, *Int. J. Epidemiol.* **22**:288–293.

Solomon, D. J., Henry, R. C., Hogan, J. G., van Amburg, G. H., and Taylor, J., 1991, Evaluation and implementation of public health registries, *Public Health Rep.* **106**:142–150.

Thacker, S. B., Parrish, R. G., Trowbridge, F. L., 1988, Surveillance coordination group: A method for evaluating systems of epidemiological surveillance, *World Health Stat* **Q41**:11–18.

Uter, W., Gefeller, O., Schwanitz, H. J., 1995, Frühe irritative Hautschäden bei Friseurlehrlingen, *Hautarzt* **46**:771–778.

Uter, W., Schnuch, A., Geier, J., Frosch, P. J., Epidemiology of contact dermatitis: The information network of the departments of dermatology (IVDK) in Germany—a surveillance system on contact allergies, *Eur. J. Dermatol.* **8**:36–40.

Vigan, M., 1997, Les nouveaux allergènes des cosmétiques: La cosmétovigilance, *Ann. Dermatol. Venereol.* **124**:571–575.

Wall, S., 1995, Epidemiology for prevention, *Int. J. Epidemiol.* **24**:655–664.

Wilkinson, J. D., Hambly, E. M., and Wilkinson, D. S., 1980, Comparison of patch test results in two adjacent areas of England: II. Medicaments, *Acta. Derm. Venereol. Stockh.* **60**:245–249.

Chapter 21

Sodium Lauryl Sulfate: Water Soluble Irritant Dermatitis Model

CHEOL HEON LEE AND HOWARD I. MAIBACH

1. Sodium Lauryl Sulfate

Sodium lauryl sulfate (SLS) is an anionic surface active agent used as an emulsifier in pharmaceutical vehicles, cosmetics, foaming dentifrices, and foods; it is the sodium salt of lauryl sulfate that conforms to the formula: $CH_3(CH_2)_{10}CH_2OSO_3Na$ (Nikitakis *et al.*, 1991). The action of SLS on surface tension is putatively the cause of its irritancy, and its great capacity for altering the stratum corneum makes it useful in enhancing penetration of other substances in patch tests and animal assays. Kligman (1966) found no sensitization to SLS in a hundred volunteers in whom SLS was employed in provocative or prophetic patch test procedures.

Some important characteristics have been proposed for irritants used experimentally: no systemic toxicity, noncarcinogenic, not a sensitizer, chemically well defined, no extreme pH value, and not a cause of cosmetic inconvenience to exposed subjects (Wahlberg and Maibach, 1980). SLS fulfills these criteria.

2. Application Methods

Many studies concerned with cutaneous irritation utilize a 24-hr patch application. A 7-hr patch (Loden and Andersson, 1996) and 4-hr patch (Basketter *et al.*, 1996) with high concentrations of SLS have been developed. In real life surfactant exposure is usually of short duration, open application, and cumulative. A single challenge to the skin with an irritant insult is a momentary reflection of the skin's

Cheol Heon Lee • Department of Dermatology, Kangnam Sacred Heart Hospital, College of Medicine, Hallym University, Seoul 150-071, Korea. *Howard I. Maibach* • Department of Dermatology, School of Medicine, University of California, San Francisco, California 94143-0989.

Cutaneous Biometrics, edited by Doris A. Schwindt and Howard I. Maibach, Kluwer Academic/Plenum Publishers, New York, 2000.

susceptibility, which does not consider the cumulative effect of irritation or the skin's repair mechanisms. Repetitive challenges allow for these effects. Assay methods such as the repeated short-duration chamber test (Frosch and Kligman, 1979*a*; Tupker *et al.*, 1989*a, b*), the repeated open application test (Lammintausta *et al.*, 1987*a*; Algood *et al.*, 1990; Wilhelm *et al.*, 1990; Lee and Maibach, 1994), the plastic occlusion stress test (POST) (van der Valk and Maibach, 1989b; Berardesca and Maibach, 1990), and the soak or wash test (Lukacovic *et al.*, 1988; Klein *et al.*, 1992) may mimic consumer use.

A correlation coefficient of 0.63 between a single exposure and a four-day repetitive exposure to patch-testing with SLS was found (Pinnagoda *et al.*, 1989). With repeated open application of SLS for 5 days as well as a single 24-hr patch test with SLS using small (8-mm) patch test chambers, only the degree of skin damage caused by the repeated open test was associated with prior skin complaints (Lammintausta *et al.*, 1988). Lammintausta *et al.* (1987a) observed the decrease in patch test reactivity secondary to the cumulative open SLS application using small (8-mm) patch test chambers and suggested that the induced hyporeactivity might be one of false negative diagnostic patch tests.

There are variations in skin responses to identical patch tests and standardization of patch test procedure is necessary to minimize the variations in patch test responses.

2.1. Purity and Carbon Length of SLS

There were significant differences in the irritant potential *in vivo* for different qualities of SLS, and there were cases in which some of the C_{12} chains had been substituted by longer and less irritating carbon chains (Agner *et al.*, 1989). The presence of C_{12} chains of SLS is known to elicit a maximum irritant reaction (Kligman and Wooding, 1967; Stillman *et al.*, 1975; Wilhelm *et al.*, 1993). Agner *et al.* (1989) suggested that only SLS qualities of high purity ($>99\%$) should be used for irritant patch-testing and that the quality and the purity of SLS should be stated.

2.2. Quantity and Concentration of the Test Solution

Quantity of test solution is important and larger quantities of test solution give more intense skin reactions, although the concentration of the irritant is kept constant (Magnusson and Hersle, 1965; Frosch and Kligman, 1979*b*) and Agner (1992) suggested that the Duhring chamber, the 12-mm Finn chamber, or even large chambers having bigger test areas are more effective in eliciting a response. Mikulowska and Andersson (1996) observed that the effect of 8-mm chambers could result in increased, unchanged, or decreased numbers of Langerhans cells (LC), while 12-mm chambers always produced a decrease in LCs. Lee *et al.* (1997*a*) also compared the effect of chamber size on SLS irritation on the volar forearm using three different size (8-mm, 12-mm, 18-mm) Finn chambers. The increase in skin response [visual score and transepidermal water loss (TEWL)] with the large (12-mm) Finn chamber was larger than that with the small (8-mm) Finn chamber.

However there were no significant differences between the large and extra-large (18-mm) Finn chambers.

2.3. Evaporation and Temperature of the Test Solution

Berardesca *et al.* (1995) reported significantly different skin responses to the temperature of test solution (4°C, 20°C, and 40°C). Skin damage was more severe in sites treated with warmer temperatures: there was a highly significant correlation between irritation and the temperature of the test solution. The evaporation rate of aqueous solutions from Finn chambers was reported as 1 mg/3 min (Fisher and Maibach, 1984). Evaporation from the patch before application inhibits the inflammatory response, even though the relative concentration of the irritant is increased by the process (Dahl and Roering, 1984). This inhibition of skin irritation could be the result of decreased amount or lowered temperature owing to evaporation of test solution.

2.4. Time of Evaluation

When noninvasive measurements of the skin response are made, the interval between removal of the patch and the measurements should allow for a period of increased evaporation following occlusion. In most papers, for measurements of TEWL the interval was reported to be 30 min (Beradesca and Maibach, 1988*b*; Freeman and Maibach, 1988; Goh and Chia, 1988). The time course of TEWL after SLS patch-testing demonstrated significant reduction in TEWL values from 30 to 60 min after removal of the patch, but not from 60 to 180 min (Agner and Serup, 1993). Others have argued that a minimum waiting period of 2 or 3 hr should be allowed for evaporation of excessive water due to occlusion (Baker and Kligman, 1967; Pinnagoda *et al.*, 1989).

3. Biologic Endpoints

Several noninvasive bioengineering methods to obtain and quantify information that is not detectable clinically have been developed (Table 1) (Lee and Maibach, 1995). Many surfactants including SLS disrupt the skin barrier function, resulting in increased TEWL (Scheuplein and Ross, 1970; Elias, 1983) and increased blood flow, clinically visible as erythema (van der Valk *et al.*, 1984). A number of hypotheses on the mechanism of SLS-induced skin irritation have been proposed. Leveque *et al.* (1993) suggested that an increase in TEWL did not necessarily imply the alteration of stratum corneum and SLS-induced dry skin could hardly be interpreted in terms of lipid removal (Froebe *et al.*, 1990), and the most likely explanation of SLS-induced increase in TEWL lies in the hyperhydration of stratum corneum and a possible disorganization of lipid bilayers.

Measurement of TEWL as a technique to evaluate skin barrier function is widely used (Berardesca and Maibach, 1988*a*; Wilhelm *et al.*, 1989) and a positive dose–response relationship for skin response to SLS as measured by TEWL has

TABLE 1. Noninvasive Bioengineering Techniques Used in the Evalution of Cutaneous Irritation

Technique	Measured skin function	Information obtained
Evaporimeter	TEWL	Positive dose–response relationship for skin response to SLS. Most sensitive method for SLS-induced irritation
Laser-Doppler flowmeter	Blood flow	Positive relationship between applied dose of SLS and blood flows. Wide fluctuations in response to SLS due to spotty erythema
Ultrasound	Skin thickness	No preconditioning is necessary. Good correlation to SLS concentrations, but minimal correlation with erythema or epidermal damage
Impedance, conductance, capacitance	Skin hydration	Correlation with epidermal damage, but intraindividual variation is so high that this method is unhelpful
Colorimeter	Skin colors	Positive correlation between changes in the a^* color coordinates and doses of SLS, but not with epidermal damage

been demonstrated (Agner and Serup, 1990*a*). When attempting to quantify irritant patch test reactions by electrical conductance measurement, the intraindividual variation in the results was so high that the method was found unhelpful for this purpose (Agner and Serup, 1990*b*). A positive relationship was found between the dose of SLS and blood flow values recorded by laser Doppler flowmetry (Nilsson *et al.*, 1982; Agner and Serup, 1990*a*). However, wide fluctuations in laser Doppler blood flow values in response to SLS patches were found owing to spotty erythema (Freeman and Maibach, 1988). The skin color is expressed in a three-dimensional coordinate system: a^* (from green to red), b^* (from blue to yellow), and L^* (from black to white) values (Robertson, 1977). Color a^* coordinates have been demonstrated to correlate well with visual scoring of erythema in inflammatory reactions caused by soap or SLS (Babulak *et al.*, 1986; Wilhelm *et al.*, 1989; Serup and Agner, 1990). Ultrasound examination has the advantage that no preconditioning of the subjects is necessary before measurement. Ultrasound A-scan has been found suitable for quantification of patch test reactions (Serup *et al.*, 1984; Serup and Staberg, 1987) and also a promising method for quantification of SLS-induced inflammatory response, being consistently more sensitive than measurement of skin color (Agner and Serup, 1990*a*), and Seidenari and di Nardo (1992) demonstrated that B-scanning evaluation showed a good correlation with TEWL values in assessing superficial skin damage induced by SLS.

In a comparison among evaporimetry, laser Doppler flowmetry, ultrasound A-scan, and measurement of skin color, evaporimetry was found to be the best method for evaluation of SLS-induced skin damage (Wilhelm *et al.*, 1989; Agner and Serup, 1990*b*). Lee *et al.* (1997*b*) observed that measurement of erythema index using

Dermaspectrometer was less sensitive than TEWL measurement when comparing the cutaneous irritation to two different size (8-mm and 12-mm) Finn chambers. Wilhelm *et al.* (1989) suggested that although TEWL measurements may constitute an accurate and sensitive method of evaluating skin irritation, color reflectance measurements may be a helpful complementary tool for the clinician because of its convenience. Serup (1995) suggested that TEWL is sensitive and useful in the study of corrosive irritants, such as SLS, especially in the induction phase of irritant reaction, but does not have direct clinical relevance, and the results need be backed up with other relevant measures.

Tupker *et al.* (1990) found that the time course of TEWL after a 24-hr SLS patch varied among different subjects. Using SLS in varying concentrations, Serup and Staberg (1987) found a delayed response only for reactions clinically scored as 1+, but not for more intense reactions, indicating that the kinetics of the response may depend on the severity of the reaction (Dahl and Trancik, 1977; Agner and Serup, 1989).

4. Host-Related Factors

There are many host-related factors in cutaneous irritation: those that are considered as skin diseases and those that represent variations from normal skin predisposed to irritation (Table 2).

4.1. Age

Increased susceptibility to SLS was reported in young compared to older females, when assessed by visual scoring and TEWL, and the increase in TEWL values was found to be more persistent in the older group (Cua *et al.*, 1990; Elsner *et al.*, 1990). These findings imply less reaction to an irritant stimulus but a prolonged healing period in older people. There is no significant influence on skin susceptibility between 18 and 50 years of age (Agner, 1991*a*).

4.2. Sex

Hand eczema occurs more frequently among women than men. However, many investigators have found no sex correlation in skin susceptibility (Goh and Chia, 1988; Bjornberg, 1975; Lammintausta *et al.*, 1987*b*; Tupker *et al.*, 1989*c*). Reactivity to SLS at day 1 increased during the menstrual cycle compared to day 9–11, when

TABLE 2. Host-Related Factors in Cutaneous Irritation

Age	Skin hydration
Sex	Sensitive skin
Anatomic region	Hyperirritable skin
Race and skin color	Skin disease (atopic dermatitis, hand eczema, seborrheic dermatitis)

tested on opposite arms in healthy women (Agner *et al.*, 1991). Since no cyclical variation was found in baseline TEWL, the increased reactivity of the skin at day 1 in the menstrual cycle probably reflects an increased inflammatory reactivity rather than changes in the barrier function.

4.3. Anatomic Region

Variation in skin responses within the same individual to identical irritant patch tests may be considerable. Van der Valk and Maibach (1989) studied the differences in sensitivity of the volar surface of the forearm to SLS and demonstrated that the potential for irritation increases from the wrist to the cubital fossa, and Panisset *et al.* (1992) showed that TEWL values next to the wrist were greater than on the other sites of the volar forearm. Cua *et al.* (1990) reported that the thigh had the highest reactivity and the palm the lowest. Dahl *et al.* (1984) found that for simultaneous Al-patch testing with SLS, the corresponding sites on the right and the left side were scored identically in only 53% of cases. With large (12-mm) Finn chambers, 84% of SLS patches showed identical visual scores when tested simultaneously on right and left arms (Agner and Serup, 1990*b*). Rogiers (1995) suggested that measurement of TEWL should be carried out on identical anatomic sites for all subjects; the volar forearm is a good measurement site and corresponding places on the right and left forearms exhibit the same TEWL.

4.4. Skin Color

Bjornberg *et al.* (1979) reported that subjects with fair skin and blue eyes showed a high intensity of the inflammatory response to a mechanical irritant. By determination of MED in Caucasians, the cutaneous sensitivity to UV light and to seven different chemical irritants was found to correlate positively, while skin phototype based on complexion and history of sunburn proved less reliable (Frosch and Wissing, 1982). In contrast to these reports, an inclination to increased susceptibility to SLS in Black and Hispanic skin types as compared to white skin types was found when evaluated by measurement of TEWL (Berardesca and Maibach, 1988*b*, *c*). When skin color was assessed by a tristimulus colorimeter, an association between light reflection (L^*) from the skin surface and susceptibility to SLS was found (Agner, 1991). Tanning may influence the susceptibility to irritants. A diminished reaction to SLS after UVB exposure was reported (Larmi *et al.*, 1989).

4.5. Skin Hydration

In repetitive exposure to SLS, higher susceptibility was reported in dry skin than in clinically normal skin in eczematous subjects and controls (Tupker *et al.*, 1990). Comparing winter and summer skin, decreased skin hydration was found in winter, when a higher reactivity to SLS was also found (Agner and Serup, 1989). Low outdoor temperature and low relative humidity in the winter lead to decreased ability of the stratum corneum to retain water (Spencer *et al.*, 1975). Thus, these studies indicate that a decreased skin hydration state may be associated with impaired barrier function and increased skin susceptibility. In contrast, Lammintausta *et al.* (1988) found no relationship between clinically dry skin and the response to repeated SLS exposure.

4.6. Sensitive Skin

Frosch and Kligman (1977) reported a significant correlation between the skin response to particular irritants in healthy volunteers and patients with skin diseases. A 24-hr forearm chamber exposure to 5% SLS was used for preselection of hyperreactors (Frosch and Kligman, 1979*a*). Murahata *et al.* (1986) suggested a relationship between skin susceptibility to detergents and high-baseline TEWL, and a highly significant correlation between baseline TEWL and TEWL after a single or repeated exposure to SLS was reported (Pinnagoda *et al.*, 1989; Tupker *et al.* 1989*c*; Agner *et al.*, 1991). However, other studies reported an absent or poor correlation between baseline TEWL and TEWL after SLS exposure (Berardesca and Maibach, 1988*b*, *c*; Freeman and Maibach, 1988; Wilhelm and Maibach, 1990).

4.7. Hyperirritable Skin (Excited Skin Syndrome)

Mitchell (1977) introduced the term "angry back," to describe the phenomenon of a single strong positive patch test reaction creating a back that is hyperreactive to other patch test applications. The excited skin syndrome was illustrated experimentally in guinea pigs and increased susceptibility to an ointment containing 1% SLS was observed in animals stressed by inflammatory reactions in the neck area (Andersen and Maibach, 1980). An increased susceptibility to SLS in patients with acute hand eczema, as compared to patients with chronic or healed eczema, was reported (Agner, 1991*b*). Bruynzeel *et al.* (1983) attempted to use SLS patches as markers of hyperirritability, and Agner (1991*b*) found SLS to be useful as a marker of hyperirritable skin. Shahidullah *et al.* (1969) reported increased TEWL values in the clinically normal skin of patients with eczema, but there was no significant difference in baseline TEWL values between patients with eczema and controls (van der Valk *et al.*, 1985; Agner, 1991*b*).

4.8. Atopic Dermatitis

Several studies demonstrated a high risk for atopic persons to develop irritant contact dermatitis. Agner (1990) reported that the increase in TEWL was not higher in atopics than in controls, but TEWL values before and after SLS were increased in atopics. Patients with atopic dermatitis in a quiescent phase reacted more severely to SLS than controls, as assessed by measurement of TEWL (van der Valk *et al.*, 1985; Tupker *et al.*, 1990), and there was an enhanced skin reactivity to SLS in patients with current atopic dermatitis compared to controls, when measured by visual scoring and by increase in skin thickness (Agner, 1990) and by laser Doppler flowmetry (Cowley and Farr, 1992).

4.9. Hand Eczema

Baseline TEWL values in patients with localized, inactive, or healed eczema were not significantly higher than in controls (van der Valk *et al.*, 1989*b*; Agner, 1991*b*). Agner (1991*b*) observed no increased skin reactivity to SLS in patients with chronic or healed eczema compared to controls, while hand eczema patients with acute eczema showed an increased skin reactivity to SLS compared to controls.

4.10. Seborrheic Dermatitis

There were several reports indicating that patients with seborrheic dermatitis could be easily irritated by exposure to some chemicals including SLS (Lammintausta and Maibach, 1988; Cowley and Farr, 1992). Tollesson and Frithz (1993) observed increased TEWL values and abnormality in essential fatty acids in infantile seborrheic dermatitis, and they normalized TEWL values by applying the borage oil containing γ-linoleic acid.

5. Conclusion

SLS data does not provide for a unanimous opinion on all points. Yet, the preponderence of the observations suggests that we are beginning to understand some of the parameters, such as purity, dose, patch, anatomic site, single-vs.-multiple application, occluded-vs.-open application, that influence diverse responses of skin irritation.

References

Agner, T., 1990, Susceptibility of atopic dermatitis patients to irritant dermatitis caused by sodium lauryl sulphate, *Acta Derm. Venereol. (Stockh.)* **71**:296–300.

Agner, T., 1991*a*, Basal transepidermal water loss, skin thickness, skin blood flow and skin colour in relation to sodium-lauryl-sulphate-induced irritation in normal skin, *Contact Dermatitis* **25**:108–114.

Agner, T., 1991*b*, Skin susceptibility in uninvolved skin of hand eczema patients and healthy controls, *Br. J. Dermatol.* **125**:140–146.

Agner, T., 1992, Noninvasive measuring methods for the investigation of irritant patch test reactions: A study of patients with hand eczema, atopic dermatitis, and controls, *Acta Derm. Venereol. (Stockhl.) Suppl* **173**:1–26.

Agner, T., and Serup, J., 1989, Seasonal variation of skin resistance to irritants, *Br. J. Dermatol.* **121**:323–328.

Agner, T., and Serup, J., 1990*a*, Sodium lauryl sulphate for irritant patch testing—A dose–response study using bioengineering methods for determination of skin irritation, *J. Invest. Dermatol.* **95**:543–547.

Agner, T., and Serup, J., 1990*b*, Individual and instrumental variations in irritant patch-test reactions-clinical evaluation and quantification by bioengineering methods, *Clin. Exp. Dermatol.* **15**:29–33.

Agner, T., and Serup, J., 1993, Time course of occlusive effects on skin evaluated by measurement of transepidermal water loss (TEWL): Including patch tests with sodium lauryl sulphate and water, *Contact Dermatitis* **28**:6–9.

Agner, T., Serup, J., Handlos, V., and Batsberg, W., 1989, Different skin irritation abilities of different qualities of sodium lauryl sulphate, *Contact Dermatitis* **21**:184–188.

Agner, T., Damm, P., and Skouby, S. O., 1991, Menstrual cycle and skin reactivity, *J. Am. Acad. Dermatol.* **24**:566–570.

Algood, G. S., Altringer, L. A., and Kraus, A. L., 1990, Development of 14 day axillary irritation test, *J. Toxicol. Cut. Ocular Toxicol.* **9**:67–75.

Andersen, K. E., and Maibach, H. I., 1980, Cumulative irritancy in the guinea pig from low grade irritant vehicles and the angry skin syndrome, *Contact Dermatitis* **6**:430–434.

Babulak, S. W., Rhein, L. D., Scala, D. D., Simion, F. A., and Grove G. L., 1986, Quantification of erythema in a soap chamber test using the Minolta Chroma (reflectance) Meter: Comparison of instrumental results with visual assessment, *J. Soc. Cosmet. Chem.* **37**:475–479.

Baker, H., and Kligman, A. M., 1967, Measurement of transepidermal water loss by electrical hygrometry, *Arch. Dermatol.* **96**:441–452.

Basketter, D. A., Griffiths, H. A., Wang, X. M., Wilhelm, K. P., and McFadden, J., 1996, Individual, ethnic and seasonal variability in irritant susceptibility of skin: The implications for a predictive human patch test, *Contact Dermatitis* **35**:208–213.

Berardesca, E., and Maibach, H. I., 1988*a*, Bioengineering and the patch test, *Contact Dermatitis* **18**:3–9.

Berardesca, E., and Maibach, H. I., 1988*b*, Racial differences in sodium lauryl sulphate induced cutaneous irritation: Black and white, *Contact Dermatitis* **18**:65–70.

Berardesca, E., and Maibach, H. I., 1988*c*, Sodium-lauryl-sulphate-induced cutaneous irritation: Comparison of white and hispanic subjects, *Contact Dermatitis* **19**:136–140.

Berardesca, E., and Maibach, H. I., 1990, Monitoring the water-holding capacity in visually non-irritated skin by plastic occlusion stress test (POST), *Clin. Exp. Dermatol.* **15**:107–110.

Berardesca, E., Vignoli, G. P., Distante, F., Brizzi, P., and Rabbiosi, G., 1995, Effect of water temperature on surfactant-induced skin irritation, *Contact Dermatitis* **32**:83–87.

Bjornberg, A., 1975, Skin reactions to primary irritants in men and women, *Acta Derm. Venereol. (Stockh.)* **55**:191–194.

Bjornberg, A., Lowhagen. G., and Tengberg, J., 1979, Relationship between intensities of skin test reactions to glass-fibres and chemical irritants, *Contact Dermatitis* **5**:171–174.

Bruynzeel, D. P., van Ketel, W. G., von Blomberg-van der Flier, M., and Scheper, R. J., 1983, Angry back or the excited skin syndrome, *J. Am. Acad. Dermatol.* **8**:392–397.

Cowley, N. C., and Farr, P. M., 1992, A dose–response study of irritant reactions to sodium lauryl sulphate in patients with seborrheic dermatitis and atopic eczema, *Acta Derm. Venereol. (Stockh.)* **72**:432–435.

Cua, A. B., Wilhelm, K. P., and Maibach, H. I., 1990, Cutaneous sodium lauryl sulphate irritation potential: Age and regional variability, *Br. J. Dermatol.* **123**:607–613.

Dahl, M. V., and Roering, M. J., 1984, Sodium lauryl sulphate irritant patch tests: III. Evaporation of aqueous vehicle influences inflammatory response, *J. Am. Acad. Dermatol.* **11**:477–479.

Dahl, M. V., and Trancik, R. J., 1977, Sodium lauryl sulphate irritant patch tests: Degree of inflammation at various times, *Contact Dermatitis* **3**:263–266.

Dahl, M. V., Pass, F., and Trancik, R. J., 1984, Sodium lauryl sulphate irritant patch tests: II. Variation of test responses among subjects and comparison to variations of allergic responses elicited by Toxicodendron extract, *J. Am. Acad. Dermatol.* **11**:474–477.

Elias, P. M., 1983, Epidermal lipids, barrier function, and desquamation, *J. Invest. Dermatol.* **80**:44S–49S.

Elsner, P., Wilhelm, D., and Maibach H. I., 1990, Sodium lauryl sulfate-induced irritant contact dermatitis in vulvar and forearm skin of premenopausal and postmenopausal women, *J. Am. Acad. Dermatol.* **23**:648–652.

Fischer, T., and Maibach, H. I., 1984, Finn chamber patch test technique, *Contact Dermatitis* **11**:137–140.

Freeman, S., and Maibach, H. I., 1988, Study of irritant contact dermatitis produced by repeat patch testing with sodium lauryl sulphate and assessed by visual methods, transepidermal water loss and laser Doppler velocimetry, *J. Am. Acad. Dermatol.* **19**:496–502.

Froebe, C. L., Simion, F. A., Rhein, L. D., Cagan, L. H., and Kligman, A., 1990, Stratum corneum lipid removal by surfactants: Relation to *in vivo* irritation, *Dermatologica* **181**:277–283.

Frosch, P. J., and Kligman, A. M., 1977, Rapid blister formation in human skin with ammonium hydroxide, *Br. J. Dermatol.* **96**:461–473.

Frosch, P. J., and Kligman, A. M., 1979*a*, The soap chamber test: A new method for assessing the irritancy of soaps, *J. Am. Acad. Dermatol.* **1**:35–41.

Frosch, P. J., and Kligman, A. M., 1979*b*, The Duhring chamber test, *Contact Dermatitis* **5**:73–81.

Frosch, P. J., and Wissing, C., 1982, Cutaneous sensitivity to ultraviolet light and chemical irritants. *Arch. Dermatol. Res.* **272**:269–278.

Goh, C. L., and Chia, S. E., 1988, Skin irritability to sodium lauryl sulphate as measured by skin vapour loss by sex and race, *Clin. Exp. Dermatol.* **13**:16–19.

Klein, G., Grubauer, G., and Fritsch, P., 1992, The influence of daily dish-washing with synthetic detergent on human skin, *Br. J. Dermatol.* **127**:131–137.

Kligman, A. M., 1966, The SLS provocative patch test in allergic contact sensitization, *J. Invest. Dermatol.* **36**:573–583.

Kligman, A. M., and Wooding, W. M., 1967, A method for the measurement and evaluation of irritants on human skin, *J. Invest. Dermatol.* **49**:78–94.

Lammintausta, K., and Maibach, H. I., 1988, Exogenous and endogenous factors in skin irritation, *Int. J. Dermatol.* **27**:213–222.

Lammintausta, K., Maibach, H. I., and Wilson, D., 1987*a*, Human cutaneous irritation: Induced hyporeactivity, *Contact Dermatitis* **17**:193–198.

Lammintausta, K., Maibach, H. I., and Wilson, D., 1987*b*, Irritant reactivity in males and females, *Contact Dermatitis* **17**:276–280.

Lammintausta, K., Maibach, H. I., and Wilson, D., 1988, Susceptibility to cumulative and acute irritant dermatitis, *Contact Dermatitis* **19**:84–90.

Larmi, E., Lahti, A., and Hannuksela, M., 1989, Effect of ultraviolet B on nonimmunologic contact reactions induced by dimethyl sulfoxide, phenol and sodium lauryl sulphate, *Photodermatology* **6**:258–262.

Lee, C. H., and Maibach, H. I., 1994, Study of cumulative irritant contact dermatitis in man utilizing open application on subclinically irritated skin, *Contact Dermatitis* **30**:271–275.

Lee, C. H., and Maibach, H. I., 1995, The sodium lauryl sulfate model: An overview, *Contact Dermatitis* **33**:1–7.

Lee, K., Y., Park, C., W., and Lee, C., H., 1997*a*, The effect of chamber size and volume of test solution on cutaneous irritation, *Kor. J. Dermatol.* **35**:424–430.

Lee, K. Y., Shin, K. Y., Park, C. W., and Lee, C. H., 1997*b*, Cutaneous irritation to sodium lauryl sulfate and sodium lauroyl glutamate, *Kor. J. Dermatol.* **35**:491–498.

Leveque, J. L., de Rigal, J., Saint-Leger, D., Billy, D., 1993, How does sodium lauryl sulfate alter the skin barrier function in man? A multiparametric approach, *Skin Pharmacol.* **6**:111–115.

Loden, M., and Andersson, A. C., 1996, Effect of topically applied lipids on surfactant-irritated skin, *Br. J. Dermatol.* **134**:215–220.

Lukacovic, M. F., Dunlap, F. E., Michaels, S. E., Visscher, M. O., and Watson, D. D., 1988, Forearm wash test to evaluate the clinical mildness of cleansing products, *J. Soc. Cosmet. Chem.* **39**:355–366.

Magnusson, B., and Hersle, K., 1965, Patch test methods: I. A comparative study of six different types of patch tests, *Acta Derm. Venereol. (Stockh.)* **45**:123–128.

Mikulowska, A., and Andersson, A., 1996, Sodium lauryl sulfate effect on the density of epidermal Langerhans cells: Evaluation of different test models, *Contact Dermatitis* **34**:397–401.

Mitchell, J. C., 1977, Multiple concomitant positive patch test reactions, *Contact Dermatitis* **3**:315–320.

Murahata, R., Crove, D. M., and Roheim, J. R., 1986, The use of transepidermal water loss to measure and predict the irritation response to surfactants, *Int. J. Cosmet. Sci.* **8**:225–231.

Nikitakis, J. M., McEwen, G. N., and Wenninger, J. A., 1991, *CTFA International Cosmetic Ingredient Dictionary*, 4th Ed., The Cosmetic, Toiletry, and Fragrance Association Inc., Washington DC.

Nilsson, G. E., Otto, U., and Wahlberg, J. E., 1982, Assessment of skin irritancy in man by laser Doppler flowmetry, *Contact Dermatitis* **8**:401–406.

Panisset, F., Treffel, P., Faivre, B., Lecomte, P. B., and Agache, P., 1992, Transepidermal water loss related to volar forearm sites in humans, *Acta Derm. Venereol. (Stockh.)* **72**:4–5.

Pinnagoda, J., Tupker, R. A., Coenraads, P. J., and Nater, J. P., 1989, Prediction of susceptibility to an irritant response by transepidermal water loss, *Contact Dermatitis* **20**:341–346.

Robertson, A. R., 1977, The CIE 1976 color difference formulas, *Color Res. Appl.* **2**:7–11.

Rogiers, V., 1995, Transepidermal water loss measurements in patch test assessment: The need for standardization, in: *Irritant Dermatitis. New Clinical and Experimental Aspects* (P. Elsner and H. I. Maibach, eds.), Karger, Basel, pp. 152–158.

Scheuplein, R. J., and Ross, L., 1970, Effects of surfactants and solvents on the permeability of epidermis, *J. Soc. Cosmet. Chem.* **21**:853–873.

Seidenari, S., and di Nardo A., 1992, B-scanning evaluation of irritant reactions with binary transformation and image analysis, *Acta Derm. Venereol. (Stockh.) Suppl.* **175**:9–13.

Serup, J, 1995, The spectrum of irritancy and application of bioengineering techniques, in: *Irritant Dermatitis: New Clinical and Experimental Aspects* (P. Elsner and H. I. Maibach, eds.), Karger, Basel, pp. 131–143.

Serup, J., and Agner, T., 1990, Colorimetric quantification of erythema—a comparison of two colorimeters (Lange Micro Color and Minolta Chroma Meter CR-200) with a clinical scoring scheme and laser Doppler flowmetry, *Clin. Exp. Dermatol.*. **15**:267–272.

Serup, J., and Staberg, B., 1987, Ultrasound for assessment of allergic and irritant patch test reactions, *Contact Dermatitis* **17**:80–84.

Serup, J., Staberg, B., and Klemp, P., 1984, Quantification of cutaneous oedema in patch test reactions by measurement of skin thickness with high-frequency pulsed ultrasound, *Contact Dermatitis* **10**:88–93.

Shahidullah, M., Raffle, E. J., Rimmer, A. R., and Frain-Bell, W., 1969, Transepidermal water loss in patients with dermatitis, *Br. J. Dermatol.* **81**:722–730.

Spencer, T. S., Linamen, C. E., Akers, W. A., and Jones, H. E., 1975, Temperature dependence of water content of the stratum corneum, *Br. J. Dermatol.* **93**:159–164.

Stillman, M. A., Maibach, H. I., and Shalita, A. R., 1975, Relative irritancy of free fatty acids of different chain length, *Contact Dermatitis* **1**:65–69.

Tollesson, A., and Frithz, A., 1993, Transepidermal water loss and water content in stratum corneum in infantile seborrheic dermatitis, *Acta Derm. Venereol.* (*Stockh.*) **73**:18–20.

Tupker, R. A., Pinnagoda, J., Coenraads, P. J., Kerstholt, H., and Nater, J. P., 1989*a*, Evaluation of hand cleansers: Assessment of composition, skin compatibility by transepidermal water loss measurements, and cleansing power, *J. Soc. Cosmet. Chem.* **40**:33–39.

Tupker, R. A., Pinnagoda, J., Coenraads, P. J., and Nater, J. P., 1989*b*, The influence of repeated exposure to surfactants on human skin as determined by transepidermal water loss and visual scoring, *Contact Dermatitis* **20**:108–114.

Tupker, R. A., Coenraads, P. J., Pinnagoda, J., and Nater, J. P., 1989*c*, Baseline transepidermal water loss (TEWL) as a prediction of susceptibility to sodium lauryl sulphate, *Contact Dermatitis* **20**:265–269.

Tupker, R. A., Pinnagoda, J., Coenraads, P. J., and Nater J. P., 1990, Susceptibility to irritants: Role of barrier function, skin dryness and history of atopic dermatitis, *Br. J. Dermatol.* **123**:199–205.

van der Valk, P. G. M., and Maibach, H. I., 1989*a*, Potential for irritation increases from the wrist to the cubital fossa, *Br. J. Dermatol.* **121**:709–712.

van der Valk, P. G. M., and Maibach, H. I., 1989*b*, Post-application occlusion substantially increases the irritant response of the skin to repeated short-term sodium lauryl sulphate (SLS) exposure, *Contact Dermatitis* **21**:335–338.

van der Valk, P. G. M., Nater, J. P., and Bleumink, E., 1984, Skin irritancy of surfactants as assessed by water vapor loss measurements, *J. Invest. Dermatol.* **82**:291–293.

van der Valk, P. G. M., Nater, J. P., and Bleumink E., 1985, Vulnerability of the skin to surfactants in different groups of eczema patients and controls as measured by water vapour loss, *Clin. Exp. Dermatol.* **10**:98–103.

Wahlberg, J. E., and Maibach, H. I., 1980, Nonanoic acid irritation: A positive control at routine patch testing? *Contact Dermatitis* **6**:128–130.

Wilhelm, K. P., and Maibach, H. I., 1990, Susceptibility to irritant dermatitis induced by sodium lauryl sulphate, *J. Am. Acad. Dermatol.* **23**:122–124.

Wilhelm, K. P., Saunders, J. C., and Maibach, H. I., 1989, Quantification of sodium lauryl sulphate dermatitis in man: Comparison of four techniques: skin color reflectance, transepidermal water loss, laser Doppler flow measurement and visual scores, *Arch. Dermatol. Res.* **281**:293–295.

Wilhelm, K. P., Saunders, J. C., and Maibach, H, I., 1990, Increased stratum corneum turnover induced by subclinical irritant dermatitis, *Br. J. Dermatol.* **122**:793–798.

Wilhelm, K. P., Cua, A. B., Wolf H., H., and Maibach, H. I., 1993, Surfactant-induced stratum corneum hydration *in vivo*: Prediction of the irritation potential of anionic surfactants, *J. Invest. Dermatol.* **101**:310–315.

Chapter 22

Quaternium-18 Bentonite in the Prevention of Allergic Contact Dermatitis

ROBERT L. RIETSCHEL

1. Introduction

The prevention of poison ivy/oak dermatitis by application of topical pretreatment is an effort to achieve an effective barrier to the penetration of an allergenic material. The system to be discussed has been shown to achieve meaningful protection from experimentally induced poison ivy/oak dermatitis and, as such, can serve as a model for similar barrier studies. Quaternium-18 bentonite lotion can be applied before exposure to the urushiol allergen from poison ivy/oak in an experimental system and prevent about 70% of sensitized subjects from reacting (Marks *et al.*, 1995). In order for the study of barrier function to be meaningful, attention to several specific details is necessary: time of occlusion, timing of patch test readings, and grading scale.

2. Time of Occlusion

Barriers can be topical creams, lotions, gels, pastes, ointments, or solid objects such as gloves. Patch tests can be placed over the applied barrier for varying periods of time, removed, and compared with control (untreated) sites. In traditional diagnostic patch-testing of patients with suspected allergic contact dermatitis, a 2-day occlusive period is customary (Rietschel and Fowler, 1995). However, this 2 days may be a totally inappropriate duration of contact between a barrier and an allergen. An example of this can be most clearly seen with the evaluation of gloves. Gloves are not generally worn for 2 days of continuous contact with a potentially noxious substance. An 8-hr work shift is likely to represent a maximal exposure. The duration of contact between the allergen and glove should reflect actual conditions

Robert L. Rietschel • Department of Dermatology, Ochsner Clinic and Alton Ochsner Medical Foundation, New Orleans, Louisiana 70121.

Cutaneous Biometrics, edited by Doris A. Schwindt and Howard I. Maibach, Kluwer Academic/Plenum Publishers, New York, 2000.

as much as possible. Knowledge of the breakthrough time for the allergen vs. the glove in question can be useful in designing the proper experiment along with the nature of the expected working conditions (Rietschel *et al.*, 1984).

In naturally occurring allergic contact dermatitis, the concentration of the allergenic exposure is rarely known, but it is controlled in experimental circumstances. An overwhelming concentration occluded for an extended period may overwhelm a barrier that may be more than adequate in real settings. One way of adjusting the concentration of allergen and duration of occlusion is to test normal skin with the planned system in order to find a combination of dose and time of exposure that mimics a typical clinical reaction seen in the naturally occurring disease state. In the urushiol study that is the paradigm for this chapter (Marks *et al.*, 1995), the commercially obtained allergen was diluted with an equal amount of 95% ethanol and occluded for 4 hr, then washed off. This gave a vesiculo-bullous response in more than half of the sensitive subjects tested and represented a pattern of dermatitis typical of poison ivy/oak eruptions. This type of information is necessary to determine if the barrier to be evaluated is performing in a meaningful way.

3. When to Read Patch Tests

The standard diagnostic patch tests are read at day 2 or 3 after application and a second time between day 4 and day 7, depending on the investigator (Rietschel *et al.*, 1988). While this is somewhat standardized, it should be intuitively obvious that the magnitude of the challenge could affect the dynamics of how rapidly or slowly the response occurs. In our paradigm, readings were taken at three points: days 2, 5, and 8. The peak response to urushiol on normal skin plateaued at day 5, but there was a slight increase in reactivity on skin treated with quaternium-18 bentonite between day 5 and day 8, albeit quite small (Marks *et al.*, 1995). Since the barrier may delay and/or retard reactions, a reading beyond the normal time frame for clinical work may be useful. It is possible that a barrier will fail and conversely accelerate or intensify a reaction. Preliminary work with the model system used experimentally will be necessary with observation earlier and later than the customary clinical parameters to be certain that the experiment is appropriately designed.

4. What to Grade When Reading Patch Tests

It is traditional to grade patch tests in clinical practice on a three-point scale. The most widely used grading scale is seen in Table 1.

The morphologic qualities used to evaluate patch tests are erythema, edema (infiltration), papulovesicles, bullae, and surface area. The arbitrary three-point scale of positive allergic responses does not provide for subtle distinctions and, while utilitarian in the practice of diagnostic dermatology, this grading system does not capture all the varied responses seen even in diagnostic work. Some patch tests produce very compact, papulovesicular reactions, while others cause large erythematous, edematous responses. It is tempting to take each of the above morphology observations—erythema, edema, vesicles, and surface area—and code

TABLE 1. Clinician's Patch Test Grading Scale

+	Weak reaction; erythema, infiltration, possibly microvesicles (also referred to as papules)
++	Strong reaction; erythema, infiltration, papulovesicles
+++	Extreme reaction; bullous or spreading reaction
Ir	Irritant reaction, not felt to represent true allergy
?	Doubtful reaction; faint erythema of macular quality

each component separately as mild, moderate, or severe and then summate the scores. This would be unacceptable as it could result in equal weight being given to moderate erythema and moderate vesiculation. While a continuum from erythema to edema to vesicles is implied by the grading system, vesiculation is a better guide to allergic reactions and carries more weight. In our paradigm, the scale was expanded to seven points for positive reactions (Table 2).

The system used to evaluate poison ivy/oak reactions acknowledges the importance of vesiculation and is weighted to assess this variable with more discriminating power than the clinician's three-point scale. There is integration of the other variables of erythema and edema. Surface area of vesiculation is considered, while this variable is silent in the clinician's scale (Table 1). In actual practice, an integration of surface area, degree of edema, and erythema occurs and is summated as a patch test reading. This practice would be expected to lead to some variability in the interpretation of patch tests, which, in fact, has been reported even when all the observers are experts (Bruze *et al.*, 1995). Five experts read 400 patch test reactions identically 144 times (36%), but at least similarly 82% of the time (Bruze *et al.*, 1995).

The grading system chosen should integrate the key patch test morphology of the actual dermatitis being studied and provide sufficient discriminating power to make judgments that would be appropriate to the natural setting. Well-spelled-out definitions and preliminary testing of chosen concentrations, duration of exposure, and time table for data collection (patch test readings) will allow for a meaningful experimental design.

TABLE 2. An Expanded Patch Test Grading Scale[a]

1+	Erythema with itching
2+	Erythema with edema and itching
3+	Erythema, edema, and beginning vesiculation involving less than 25% of the test site
4+	As in 3+, but vesicles involving 25 to 50% of the treatment site
5+	As in 3+, but vesicles involving 50 to 75% of the treatment site
6+	As in 3+, but vesicles confluent in a circular pattern on the test site
7+	Erythema, edema, vesiculation, and evidence of ulcerative breakdown
±	Questionable reaction (erythema, with or without edema, without itching)

[a]Used with permission from Marks, J. G., Jr., Fowler, J. F., Jr., Sheretz, E. F., and Rietschel, R. L., 1995, Prevention of poison ivy and oak allergic contact dermatitis by quaternium-18 bentonite, *J. Am. Acad. Dermatol.* **33**:212–216.

References

Bruze, M., Isaksson, M., Edman, B., Bjoerkner, B., Fregert, S., and Moeller, H., 1995, A study on expert reading of patch test reactions: Inter-individual accordance, *Contact Dermatitis* **32**:331–337.

Marks, J. G., Jr., Fowler, J. F., Jr., Sheretz, E. F., and Rietschel, R. L., 1995, Prevention of poison ivy and oak allergic contact dermatitis by quaternium-18 bentonite, *J. Am. Acad. Dermatol.* **33**:212–216.

Rietschel, R. L., and Fowler, J. F. (eds.), *Fisher's Contact Dermatitis*, 4th Ed., Williams and Wilkins, Baltimore, pp. 11–32.

Rietschel, R. L., Huggins, R., Levy, N., and Pruitt, P. M., 1984, *In vivo* and *in vitro* testing of gloves for protection against UV-curable acrylate resin systems, *Contact Dermatitis.* **11**:279–282.

Rietschel, R. L., Adams, R. M., Maibach, H. I., Storrs, F. J., and Rosenthal, L. E., 1988, The case for patch test readings beyond day 2, *J. Am. Acad. Dermatol.* **18**:42–45.

Chapter 23

Personal Cleansers: Irritation and Moisturization

KEITH D. ERTEL

1. Introduction

Soaps are among the most widely used personal care products, accounting for approximately half of the personal hygiene market in 1996 with a value of over $2 billion (Mcardle, 1997). Soap has been used for personal cleansing since antiquity. The past few decades have witnessed the introduction of cleansing bars based in part or in toto on synthetic surfactant systems. More recently, the industry has seen a rise in the popularity of body washes. Many of these body washes are based entirely on synthetic surfactant systems to offer "ultramild" cleansing. Additionally, a number of bar and body wash products now claim to offer consumers additional convenience by not only cleansing, but also moisturizing the skin.

Personal cleansing products are the subject of a sizable body of scientific literature. Despite this attention, or perhaps because of it, the effects on skin reported for these products have been inconsistent (Frosch and Kligman, 1979). Most consumers can use personal cleansers without developing overt clinical symptoms or excessive irritation (Blank, 1969; Rieger, 1995). Yet, results from numerous patch and exaggerated washing tests have appeared in the literature demonstrating a potential for personal cleansers to cause skin damage. The proliferation of test methods focusing on discriminating ever smaller product differences has led some to question whether the focus on personal cleansing products is misplaced, emphasizing mildness rather than cleansing ability (Wolf, 1994, 1996).

Keith D. Ertel • The Procter & Gamble Company, Sharon Woods Technical Center, Cincinnati, Ohio 45241.

Cutaneous Biometrics, edited by Doris A. Schwindt and Howard I. Maibach, Kluwer Academic/Plenum Publishers, New York, 2000.

2. Irritancy of Personal Cleansing Products

2.1. Surfactant Interactions with Skin

(a) Theoretical Considerations. The absence of overt clinical symptoms or irritation from personal cleansing products does not mean that such cleansers have no effect on skin. Indeed, a range of interactions between surfactants and the skin has been postulated and explored. Because soaps are intended to clean the skin of dirt and exogenous oils, the most obvious interaction is the removal of natural lipids from the skin. Lipids play an important role in maintaining stratum corneum barrier integrity (Elias *et al.*, 1981; Grubauer *et al.*, 1989; Abrams *et al.*, 1993); however, the effect of soap washing on lipids and skin condition remains unclear. Kawai and Imokawa (1984) proposed lipid removal from the skin as a prerequisite for the development of tightness sensation. Dugard and Scheuplein (1973) reported increased stratum corneum permeability following *in vitro* surfactant exposure. However, this effect was reversible to various degrees after surfactant removal, indicating that lipid extraction alone was not responsible for the permeability increase. Middleton (1968) found a markedly decreased water-binding capacity following lipid extraction and water washing of corneum and proposed a role for lipids in retaining the skin's water-binding substances. Fulmer and Kramer (1986) noted decreased stratum corneum hydration and increased scaling after washing with sodium dodecyl sulfate (SDS) for 2 weeks. These authors found that the total amount of lipid was conserved, but alterations in ceramide, cholesterol, and free fatty acid profiles were noted. Abnormal stratum corneum lipid compositions may be a predisposing factor to surfactant irritation (di Nardo *et al.*, 1996). Froebe *et al.* (1990) also noted slight alterations in lipid levels following treatment of stratum corneum with sodium lauryl sulfate (SLS) and ammonium lauryl sulfonate. Coupling these results with soap chamber test data, the authors proposed that surfactant interactions with keratin, rather than lipid, are responsible for erythema development.

SLS, often used as a model surfactant, has the ability to penetrate the stratum corneum and alter its barrier properties (Lodén, 1990; Patil *et al.*, 1995). Surfactants will also bind to surfaces, and surfactant binding to skin and stratum corneum proteins has been implicated as a factor governing harshness and tightness sensation (Ananthapadmanabhan *et al.*, 1996; Wortzman *et al.*, 1986; Kawai and Imokawa, 1984). Surfactants may also cause morphologic changes in isolated stratum corneum. Rhein *et al.* (1986) observed differing stratum corneum swelling responses to a series of surfactants and noted that these responses corresponded to the surfactants' expected *in vivo* irritancy. Shukuwa *et al.* (1997) reported stratum corneum responses ranging from slight swelling to cell destruction following *in vitro* exposure to solutions prepared from a series of commercial soaps. The relative responses to these products paralleled irritation responses seen in a soap chamber test. Muhkerjee *et al.* (1995) noted responses ranging from little damage to corneocyte cell wall damage and loss of intercellular cohesion following *in vitro* washing with commercial soap solutions. Again, the relative changes were reported to parallel irritancy rankings obtained under *in vivo* test conditions.

The above provides only a sampling of the body of work that has examined the interactions between surfactants and skin. The reader is referred to a recent review article for a more detailed discussion of this topic (Rieger, 1995).

(b) Practical Considerations. Many of the studies examining surfactant interactions with skin have been conducted with harsh surfactants or under exaggerated exposure conditions. It is possible to demonstrate clinical differences in personal cleanser effects on skin without resorting to such extremes. For example, we conducted a double-blind, 4-week ad lib usage study using a variety of commercial and prototype cleansing products as test articles. Subjects with clinically normal skin were randomly assigned to use a single test product in place of their usual personal cleanser for a period of 4 weeks. No other changes in personal cleansing habits were required. Some 75 subjects were assigned to use each test product. Subjects' skin condition was evaluated at baseline and after 2 and 4 weeks of product use. Expert forearm dryness scores for representative synthetic detergent (syndet), combination (synthetic and soap surfactants), and soap bars included in this study are summarized in Fig. 1. After 2 weeks of use, the syndet bar produced significantly ($P < 0.01$) less dryness than the combination and soap bars; this difference persisted after 4 weeks of product use. While this study does not identify a mechanism for cleanser interactions with skin, it does demonstrate a practical and important outcome of these interactions, namely, the potential for significant differential skin effects to develop from personal cleansing products, even under normal use conditions. Subjects in this study completed an exit questionnaire about their test product.

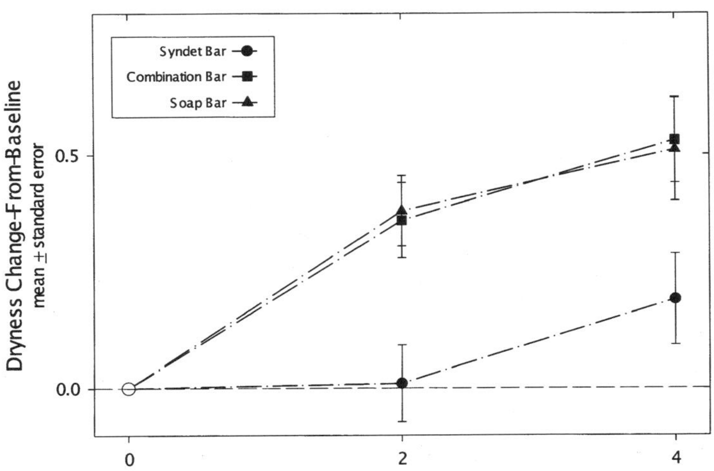

FIGURE 1. Visual dryness results obtained with synthetic detergent (syndet), combination, and soap bars in a 4-week home use study. The syndet bar produced significantly less dryness than the combination and soap bars after only 2 weeks of product use; this difference persisted at the final evaluation. These results demonstrate the potential for personal cleansing products to induce differential changes in skin condition, even under normal use conditions.

The soap bar received a significantly ($P \leqslant 0.10$) poorer overall rating and a greater number of unfavorable comments for making the skin itch and look dry than the other two bars. These results provide additional evidence of cleansing products' potential to produce differential skin effects under normal use conditions and demonstrate consumers' ability to discern these effects. This was also shown in a retrospective study conducted among female clients of beauticians by de Groot *et al.* (1988), which found a high incidence of subjective (e.g., itchiness, dry feeling) and objective (e.g., irritation, dryness) complaints attributed to personal cleansing products.

Certain skin conditions may increase the likelihood of developing adverse reactions to personal cleansers. For example, the de Groot study found that individuals with an atopic diathesis are predisposed to adverse reactions to cosmetics and toiletries products. Increased irritant susceptibility has been demonstrated in persons with atopy (Tupker *et al.*, 1990; Cowley and Farr, 1992; Agner, 1990, 1991; van der Valk *et al.*, 1985; Nassif *et al.*, 1994), although this may be dependent on the test material and exposure conditions (Hannuksela and Hannuksela, 1995). Indeed, soaps and detergents are often identified by clinicians as aggravating factors in atopic disease (Rajka, 1986).

Frank skin disease is not a prerequisite for increased irritant susceptibility. For example, susceptibility to irritant response is known to vary with weather, becoming more acute under cold, dry conditions (Agner and Serup, 1989). Table 1 presents data showing the seasonal variation in the irritant response to two commercial cleansing bars under a mildly exaggerated wash protocol (Ertel *et al.*, 1995). It should be noted that although the absolute product response varies, the relative difference between the two products remains reasonably constant, indicating

TABLE 1. Visual Dryness and Redness Scores for a Syndet Bar and a Combination Bar Tested in a Series of FCAT Studies[a]

Study	Mean dew point (°F)	Mean endpoint dryness		Mean endpoint redness	
		Syndet bar	Combination bar	Syndet bar	Combination bar
1	27.0	1.10	1.81	1.55	2.02
2	42.4	0.47	1.09	0.70	1.24
3	11.6	1.25	1.90	1.15	1.73
4	27.0	1.32	2.05	1.08	1.61
5	30.0	1.40	2.09	1.28	1.71
6	71.0	0.43	0.97	0.70	1.14
7	72.9	0.65	1.25	0.53	0.85
8	35.9	0.99	1.68	0.97	1.35
9	28.7	1.37	2.12	1.46	2.13
10	20.5	0.80	1.53	0.79	1.25

[a]Sample sizes ranged from 38 to 48 subjects. All within-study comparisons of these two products are significant at $P \leqslant 0.05$. The average relative dryness difference for this study series is 0.67 ± 0.021 (mean $\pm$ SEM); the average relative redness difference is 0.48 ± 0.032.

that the relative difference in the products' irritation potential is unaffected by weather.

Product use habits may also increase irritant susceptibility. Most consumers use a variety of cosmetics and toiletry products, each containing numerous potential irritants or sensitizers. While a given compound may not induce a strong irritant reaction on its own, it may predispose the skin to irritation from another ingredient or act in combination with other ingredients to produce an irritant response (Malten, 1981). Allenby *et al.* (1993) demonstrated enhanced susceptibility to a number of irritants after the skin was presoaked in SDS solution. The amount of irritation resulting from this presoak was estimated to be within the normal clinical range expected after a morning of housework. Soaps have been shown to elicit a sequential response in skin, with subclinical barrier damage preceding erythema development (Simion *et al.*, 1991). The response to a cleansing product need not progress beyond the subclinical stage to place the skin in a hyperirritable state. For example, Lahti *et al.* (1995) reported an enhanced irritant response for benzoic acid applied to skin that had been washed repeatedly with dilute dishwashing liquid.

2.2. Estimating Personal Cleanser Irritation Potential

There is little doubt that surfactants in personal cleansing products have the potential to interact with the stratum corneum and elicit noticeable changes, even under normal use conditions. Further, this potential may be enhanced under certain use conditions or in certain populations. Given these considerations, it is prudent to test personal cleansing products to help dimension their potential to cause untoward skin effects under in-use conditions. The soap chamber test was one of the first methods introduced to differentiate products on the basis of their relative irritancy potential (Frosch and Kligman, 1979). A variety of test protocols employing different exposure models and levels of exaggeration has since been described. Not surprisingly, the relative irritancy predicted for a material or product is highly dependent on the test conditions used to make the evaluation (e.g., Frosch, 1982; Hannuksela and Hannuksela, 1995). For example, we have shown that it is possible to generate irritation far in excess of what would be expected in use, even for a lotion with therapeutic skin effects, under highly exaggerated exposure conditions (Ertel *et al.*, 1997*a*). This highlights a need to avoid highly exaggerated exposure conditions when testing personal cleansers to help ensure that the information generated is relevant and to minimize the risk of overestimating or falsely predicting a product's in-use irritancy potential.

The following presents two methods we have developed to dimension the relative irritancy potential of personal cleansing products. The mini-immersion method provides a tool for prototype screening during the early stages of formulation (Ertel and Hartwig, 1994). The forearm controlled application technique is used to obtain a more definitive measure of products' relative irritancy potentials (Ertel *et al.*, 1995).

(a) Mini-Immersion Method. The primary goal in developing the mini-immersion method was a protocol to rapidly screen the relative irritancy potentials of prototype personal cleanser formulas. The protocol developed incorporates some

aspects of the original soap chamber test; however, the mini-immersion method avoids the exaggerated exposures and concomitantly high level of irritation often associated with the chamber test. The mini-immersion method also provides more rapid turnover, with treatments completed in one day. General guidelines for conducting a mini-immersion study are presented below.

Subject Population: Fair-skinned females, aged 18–65, with clinically normal skin are eligible to participate. A typical panel includes 16–24 subjects. Subjects are given a mild cleansing bar to use for bathing and showering in place of their usual personal cleansing product for a 1-week period preceding treatment.

Test Material Preparation: A 10% (w/v) solution of each test product is prepared in sterile water. This concentration is consistent with the soap concentration present on skin during normal washing (Frosch, 1982). Equal doses of the prepared solutions are measured into test tubes and refrigerated. At the time of use, refrigerated solutions are warmed to 35–40°C in a dry bath before being applied to subjects' arms.

Study Design and Data Analysis: The studies use a Latin square design, with three test materials applied to each volar forearm. Treatment assignments are randomized and balanced. Test data are analyzed using an analysis of variance (ANOVA) model that accounts for baseline, subject, site, side, and treatment effects.

Treatment Procedure: All procedures are conducted in a controlled environment room (70 ± 2°F, 30 ± 2% RH). Subjects remain in this room throughout the study. Three application sites are marked on each volar forearm surface with the aid of a template; the spacing of these sites corresponds to cutouts in an insulating foam block that holds the test solution cylinders. The foam blocks are positioned on subjects' forearms and held in place with Velcro™ straps. The warmed test solutions are transferred into the appropriate solution cylinders, which are then covered with foam plugs. After 15 min, the test solutions are removed from the cylinders, the foam blocks are removed from the arms, and the application sites are rinsed with tap water and patted dry. The application procedure is repeated four times at 1-hr intervals.

Evaluations: Evaluations are conducted at baseline and 40 min after each treatment application. Pilot work conducted with water alone showed that stratum corneum hydration returned to baseline after this time. Parameters evaluated during initial method development work included visual dryness and redness scores (Lukacovic *et al.*, 1988), changes in TEWL rate (Servomed EP1), and changes in skin capacitance (IBS Skin Surface Hydrometer with a Measurement Technologies MT-8C probe). Only skin capacitance changes are evaluated in the current protocol.

Pilot work showed that visual scores were elevated as a result of treatment. However, the visual changes resulting from a range of products from syndet to full soap bars were not distinct from each other or the change produced by a water blank. Similarly, TEWL rates for these treatments were increased uniformly and returned to baseline within 2 hr after the last exposure. The transient nature of this increase suggests that the mini-immersion method induces minimal stratum corneum damage, even when full soap bars are tested. Treffel and Gabbard (1996) also found no change in skin color and TEWL rate after a 15-min patch exposure to SLS. These authors observed that this exposure caused changes in the stratum corneum's hygroscopic properties, but not in its hydration level.

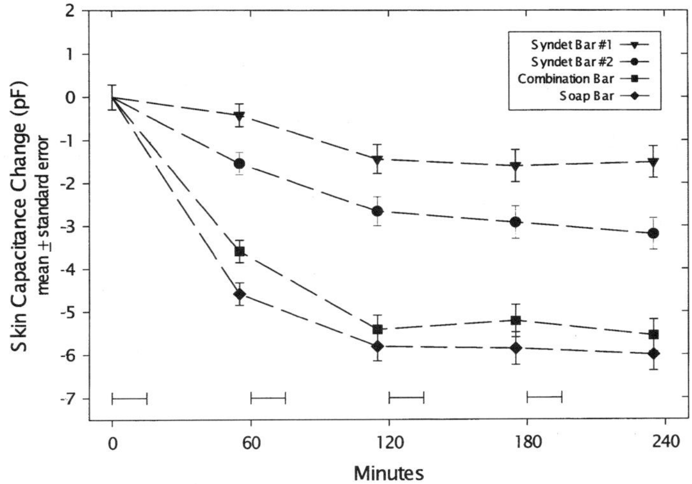

FIGURE 2. Results from a mini-immersion study conducted with synthetic detergent (syndet), combination, and soap bars. Horizontal bars indicate exposure periods; evaluations were conducted 40 min after each exposure. The difference between the syndet bars and the combination and soap bars is significant after a single exposure ($n = 24$).

Mini-immersion results for a series of commercial cleansing bar products covering a compositional range from syndet to full soap bars are shown in Fig. 2. Product differentiation is achieved rapidly—the syndet bars are significantly discriminated from the combination and full soap bars after a single, 15-min exposure. Note that syndet bar #1 and the combination bar were also discriminated in home use (Fig. 1) and FCAT (Table 1) testing, but over much longer exposure periods. There is generally good agreement between the relative irritancy rankings predicted by mini-immersion and FCAT testing. A potential limitation of the mini-immersion method, however, is that it may not detect mildness benefits afforded by special technologies. For example, polymers may be added to cleansing products to provide positive skin effects (Pugliese *et al.*, 1990). We have tested cleansing bars that are identical except for the presence or lack of polymer in their formula. The beneficial effect of polymer addition on mildness is readily shown in FCAT testing, but not in mini-immersion testing.

The mini-immersion method was developed from a practical, not a theoretical, standpoint. However, some additional insight into factors influencing product performance under this protocol is provided by the following. The skin capacitance data shown in Fig. 2 appear to follow an exponential decay. Since water in the stratum corneum is in equilibrium with the surrounding environment, the process should be described by a first-order equation of the form: $\ln[(C_t - C_\infty)/(C_0 - C_\infty)] = -kt$, where C_0, C_t, and C_∞ refer to the skin capacitance values at baseline, time $= t$, and time $= \infty$ (i.e., a long exposure time), respectively. Values for C_∞ were estimated by iteratively fitting each set of capacitance data to an asymptotic regression equation; the calculated values were then used to fit the capacitance data to the

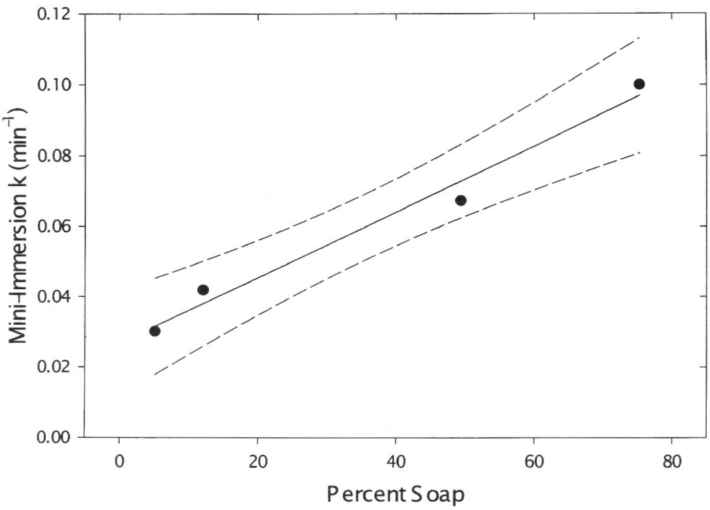

FIGURE 3. Mini-immersion "rate" constants calculated from the skin capacitance data shown in Fig. 2 plotted as a function of percent soap in the bars. Dashed lines indicate 95% confidence limits of the fitted least-squares line. The results indicate that mini-immersion results for the tested bars are driven by their soap content.

first-order equation shown above. An excellent fit ($r \geqslant 0.91$) was obtained in all cases (a much poorer fit was obtained when a simple first-order model was used). The calculated "rate" constants k are plotted vs. the percent soap in the test bars in Fig. 3. The fit is again excellent ($r > 0.98$). These results clearly show that the skin capacitance change for these bars is driven by the products' soap content. This outcome is consistent with the mini-immersion method's inability to demonstrate polymer mildness benefits that was noted previously.

(b) Forearm Controlled Application Technique (FCAT). The FCAT was developed to provide a more definitive measure of prototype and finished formulas' relative irritancy potentials. The FCAT uses an exposure model that is based on consumer washing habits and practices (Lukacovic *et al.*, 1988), and clinical results generated under this protocol have been shown to be predictive of consumer experience with personal cleansers (Keswick *et al.*, 1992). In fact, the syndet and combination bars reported in Table 1 were also differentiated under normal use conditions (Fig. 1). Table 1 comprises FCAT data generated at three test sites using six different expert graders, and thus demonstrates the robustness of the procedure. Because the FCAT method tests up to eight products on each subject, it also offers greater efficiency than paired tests. A detailed description of the FCAT procedure was given earlier (Ertel *et al.*, 1995).

One factor that influences personal cleanser skin effects, but that is often ignored, is water hardness. We recently conducted a study to examine the impact of water hardness on the mildness of various personal cleansing products, including a body wash based on a completely synthetic surfactant system, a syndet bar, a combination bar, and a soap-based bar (Ertel *et al.*, 1997*b*). Due to the mildness of

TABLE 2. Results from a Pooled Data Analysis Showing the Effect of Water Hardness on the Skin Drying Potential of the Different Treatments Used in this Study Series[a]

Treatment	Visual dryness *P*-value	Skin capacitance *P*-value
Synthetic body wash	NS	NS
Syndet bar	0.0808	NS
Combination bar	0.0004	0.0001
Soap-based bar	0.0001	0.0001

[a]A small *P*-value indicates that a product's performance was significantly impacted by water hardness over the range from 0 to 23 gpg, while NS indicates that a product's performance was not significantly impacted by water hardness, $P > 0.10$.

some of the products tested, the published FCAT procedure was modified slightly to provide greater washing exposure. Thus, products were lathered on the forearm for 20 sec and the lather residence time was decreased to 60 sec. Product skin effects were assessed at baseline and 3–6 hr after the final wash by visual scoring (Lukacovic *et al.*, 1988) and skin capacitance measurements (IBS Skin Surface Hydrometer with a Measurement Technologies MT-8C probe). Subjects acclimated in a controlled environmental room for 30 min before each evaluation.

Water with a total hardness of 0 grains per gallon (gpg), 3 gpg, 7 gpg, 15 gpg, or 23 gpg (approx. 3:1 calcium:magnesium) was used in the wash procedure. Logistical constraints required that the experiment be conducted as a series of four studies, each comparing two water hardness conditions. A 7 gpg water hardness condition was common to each study. Individual studies were analyzed using an analysis of covariance (ANCOVA) technique. Pooled data were analyzed using an ANCOVA technique with the 7 gpg data from each study as a covariant in a factorial design that accounted for water hardness and product effects.

The skin drying potential at a given water hardness paralleled the products' soap content, with the body wash producing significantly less dryness than the other products at each water hardness tested. Soap deposition increases as water hardness increases (Fujiwara *et al.*, 1992), thus, it is not surprising that a factorial analysis of the pooled data showed that the products' skin drying potential was related to water hardness, the dependence increasing as the soap content increased (Table 2). Since many areas have hard water, these results may have practical significance for consumers who are prone to experience negative skin effects from personal cleansing products.

3. Moisturization from Personal Cleansing Products

The term "moisturizing personal cleanser" seems, at first, to be an oxymoron. As discussed previously, personal cleansers are intended to remove dirt and exogenous oils from the skin. Delivering moisturizing agents, many of which are lipids, to the skin during the cleansing process seems implausible. Indeed, Blank (1969) observed that, "Theoretically, it is not feasible to add an emollient to the skin in the same process as that in which soil is removed from the skin."

The concept of delivering a moisturization benefit from a personal cleansing product is not new, although it has enjoyed increased attention in recent years with the spate of body wash introductions. Some personal cleansing products have, for years, been claimed to moisturize on the basis of ingredients they contain. Although many of these ingredients may be classified as having moisturizing or humectant properties and may provide these actions in the context of a leave-on product, it is not clear how this translates to moisturization in the context of a rinse-off product.

3.1. Aspects of Moisturization from Personal Cleansers

Applied materials may act as moisturizers in several ways. Some may cause a transient improvement in dryness, an action Prall *et al.* (1986) defined as cosmetic moisturization. Others may act on more fundamental skin processes, such as desquamation, to provide a more protracted improvement in dry skin appearance. This was termed functional moisturization. These same mechanisms should apply to materials if deposited on the skin by a personal cleansing product. Further, consumers will presumably rely on the same signals to judge moisturizing personal cleansers' performance as they do to judge the efficacy of moisturizing lotions.

Skin dryness and flaking is readily apparent to consumers and appears to be a key signal prompting moisturizer use. Prall *et al.* (1986) reported that their consumer research showed that concerns about skin dryness were sufficient to prompt regular application of moisturizing creams or lotions in over 70% of female consumers. Jemec and Serup (1992) also reported results showing that 67% of subjects used a moisturizer regularly to relieve dry skin, although clinical examination revealed little or no dryness in a majority of cases. This suggests that consumer perceptions of skin dryness and the need to apply moisturizer have more than a visual component.

Personal cleansers may provide a moisturization benefit without acting explicitly as a cosmetic or functional moisturizer. Dahlgren *et al.* (1986) reported results from a controlled wash study that included a soap bar formulated with 10% glycerin and a soap bar with an identical surfactant base but no glycerin. The bars were not significantly different for any of the clinical parameters measured, including visual dryness scores and skin electrical impedance measurements. When placed in a consumer test, however, the bar containing glycerin was rated significantly better for leaving the skin feeling moisturized after use and for leaving the skin feeling soft and smooth after use. Thus, both objective and subjective factors may contribute to a consumer's perception of a personal cleansing product's ability to provide a moisturization benefit.

3.2. Protocol Considerations for Demonstrating Moisturization from Personal Cleansers

A number of test protocols for dimensioning the efficacy of leave-on moisturizer products have been published. These protocols are not strictly applicable to personal cleansers since use habits for these two product classes differ markedly. As is the case with irritation protocols, appropriate design of protocols intended to show a moisturization benefit from personal cleansers is critical to assure reliable, consumer-relevant results. For example, protocols should include controls to account for factors such as weather and employ relevant exposure conditions to avoid overestimating a product's benefit. Another important consideration is the state of the skin on which products are tested. Moisturizer test protocols are usually conducted on the legs, which generally exhibit a higher level of dryness than other body sites. Many of these protocols, however, use a preconditioning period that incorporates both moisturizer withdrawal and controlled soap washing to help induce dry skin. The latter may not be appropriate in the context of testing personal cleansers.

We conducted a paired study to compare the effects of this type of pretreatment regimen to one involving a more natural dry skin induction process. Subjects were instructed to withdraw from moisturizer use and to wash one leg twice daily in a prescribed manner with a soap bar while using a mild cleansing bar to cleanse the remainder of their body. Subjects were not specifically instructed to wash either leg with the mild bar (i.e., they were instructed to follow their normal bathing and showering habits). After 1 week the soap-treated leg showed a greater dryness increase than did the untreated leg, although in both cases there was a sufficient level of dryness to meet study requirements. More importantly, there was a greater TEWL rate increase on the soap-treated leg than on the untreated leg, indicating that the soap washing induced more barrier damage. This is consistent with expectations based on numerous published results showing damage to skin following controlled washing with soap. As discussed previously, this may predispose the skin to further irritation. If the test products are milder than the pretreatment regimen the skin will begin to heal spontaneously during treatment, which may result in an overestimation of a treatment's beneficial effect. Thus, protocols intended to assess the moisturizing potential of personal cleansers should be conducted on natural, rather than surfactant-induced dry skin to avoid confounding clinical endpoints. Our experience has shown that allowing dry skin to develop naturally by withdrawing lotion does not significantly increase the difficulty of recruiting panels during most of the year.

As discussed previously, visual dryness is a key signal prompting moisturizer use. Further, dry skin is often an unwanted effect associated with personal cleanser use. Thus, protocols intended to show a moisturization or dry skin improvement benefit from personal cleansing products should employ a visual dryness endpoint. The study by Dahlgren *et al.* (1986) demonstrates that perception is also an important component of the moisturization experience, and self-assessment may prove a useful endpoint in the context of testing moisturizing personal cleansers. Objective endpoints are often included in protocols used to test lotions. For example, a decrease in stratum corneum hydration has been shown to accompany xerosis (Horri *et al.*, 1989; Long and Marks, 1992), and many lotion protocols incorporate skin impedance measurements to provide an objective measure of stratum corneum hydration. However, the stratum corneum hydration state may not correlate with "dry skin" perceptions. For example, Laufer and Dikstein (1996) found no significant correlation between women's' complaints of dry facial skin and stratum corneum moisture content. This suggests that impedance measurements alone may not provide sufficient evidence for a consumer-relevant moisturization benefit.

3.3. Examples of Moisturizing from Personal Cleansing Products

To date, relatively little work has appeared that is specific to moisturizing personal cleansers. Dick (1970) demonstrated emollient deposition on skin after washing with a soapless cleansing bar formulation that contained 25% unsaponified, partially hydrogenated fat. A high percentage of subjects (86%) who used the formulation over a period of weeks reported that their skin felt softer and more supple compared to the use of regular soaps or the avoidance of soaps. As noted earlier, Dahlgren *et al.* (1986) reported positive perceptions of leaving skin feeling moisturized and leaving skin soft and smooth from a bar containing 10% glycerin. Boccanfuso *et al.* (1978) described two controlled wash studies in which subjects' legs were washed twice daily with a moisturizing soap bar and a control soap bar. The moisturizing soap bar produced a lower level of dryness than the control bars, although the level of dryness was decreased from baseline in only one of the studies. Oddo *et al.* (1997) presented a protocol employing a single wash exposure for testing moisturizing personal cleansers. These authors included a water control in their design to correct for changes in baseline skin condition due to factors such as weather. The data presented indicate a moisturizing benefit for one of the products tested (a body wash). The authors cite several advantages for their single exposure design; however, because of the protocol's short duration, a single evaluation parameter, skin impedance measurement, was used to assess treatment effects.

These examples demonstrate the potential to deliver a moisturization or dry skin improvement benefit from personal cleansing products. However, there is obviously considerable variation in the methodology used to show these effects. Given the large number of personal cleansers currently claiming to provide a moisturization benefit, a key challenge facing the industry is the development of more standardized protocols that will reliably detect a consumer-relevant benefit from these types of products.

References

Abrams, K., Harvell, J D., Shriner, D., Wertz, P., Maibach, H., Maibach, H. I., and Rehfeld, S. J., 1993, Effect of organic solvents on *in vitro* human skin water barrier function, *J. Invest. Dermatol.* **101**:609–613.

Agner, T., 1990, Susceptibility of atopic dermatitis patients to irritant dermatitis caused by sodium lauryl sulphate, *Acta Derm. Venereol. (Stockh.)* **71**:296–300.

Agner, T., 1991, Skin susceptibility in uninvolved skin of hand eczema patients and healthy controls, *Br. J. Dermatol.* **125**:140–146.

Agner, T., and Serup, J., 1989, Seasonal variation of skin resistance to irritants, *Br. J. Dermatol.* **121**:323–328.

Allenby, C. F., Basketter, D. A., Dickens, A., Barnes, E. G., and Brough, H. C., 1993, An arm immersion model of compromised skin: (I) Influence on irritation reactions, *Contact Dermatitis* **28**:84–88.

Ananthapadmanabhan, K. P., Yu, K. K., Meyers, C. L., and Aronson, M. P., 1996, Binding of surfactants to stratum corneum, *J. Soc. Cosmet. Chem.* **47**:185–200.

Blank, I. H., 1969, Action of soaps and detergents on the skin, *Practitioner* **202**:147–151.

Boccanfuso, S. M., Volpe, A. R., and Bensel, A., 1978, Skin xerosis: Clinical report on the effect of a moisturizing soap bar, *Cutis* **21**:703–707.

Cowley, N. C., Farr, P. M., 1992, A dose-response study of irritant reactions to sodium lauryl sulphate in patients with seborrhoeic dermatitis and atopic eczema, *Acta Derm. Venereol. (Stockh.)* **72**:432–435.

Dahlgren, R. M., Lukacovic, M. F., Michaels, S. F., Visscher, M. O., 1986, Effects of bar soap constituents on product mildness, Second World Conference on Detergents, Montreaux, Switzerland.

de Groot, A. C., Beverdam, E. G. A., Ayong, C. T., Coenraads, P. J., Nater, J. P., 1988, The role of contact allergy in the spectrum of adverse effects caused by cosmetics and toiletries, *Contact Dermatitis* **19**:195–201.

Dick, L. A., 1970, A new emollient cleanser for dry skin conditions, *J. Am. Geriatr. Soc.* **18**:425–427.

di Nardo, A., Sugino, K., Wertz, P., Ademola, J., and Maibach, H.I., 1996, Sodium lauryl sulfate (SLS) induced irritant contact dermatitis: A correlation study between ceramides and *in vivo* parameters of irritation, *Contact Dermatitis* **35**:86–91.

Dugard, P. H., and Scheuplein, R. J., 1973, Effects of ionic surfactants on the permeability of human epidermis: An electrometric study, *J. Invest. Dermatol.* **60**:263–269.

Elias, P. M., Cooper, E. R., Korc, A., and Brown, B. E., 1981, Percutaneous transport in relation to stratum corneum structure and lipid composition, *J. Invest. Dermatol.* **76**:297–301.

Ertel, K. D., and Hartwig, P. M., 1994, A mini-immersion method for screening personal cleanser mildness using electrical impedance, 10th International Symposium on Bioengineering and the Skin, Cincinnati, OH.

Ertel, K. D., Keswick, B. H., and Bryant, P. B., 1995, A forearm controlled application technique for estimating the relative mildness of personal cleansing products, *J. Soc. Cosmet. Chem.* **46**:67–76.

Ertel, K. D., Neumann, P. B., Keswick, B. H., Kligman, A. M., and Stoudemayer, T., 1997*a*, A comparison of two antecubital fossa tests with personal care products, *J. Toxicol.: Cut. & Ocular Toxicol.* **16**:19–30.

Ertel, K. D., Knight, E. A., Rains, G. Y., Neumann, P. B., and Keswick, B. H., 1997*b*, A body wash personal cleanser option to provide mild cleansing under a wide range of water hardness conditions. American Academy of Dermatology *Academy '97* Meeting, New York, NY.

Froebe, C. L., Simion, F. A., Rhein, L. D., Cagan, R. H., and Kligman, A., 1990, Stratum corneum lipid removal by surfactants: relation to *in vivo* irritation, *Dermatologica* **181**:277–283.

Frosch, P. J., 1982, Irritancy of soaps and detergent bars, in: *Principles of Cosmetics for the Dermatologist* (P. Frost and S. N. Horwitz, eds.), C. V. Mosby, St. Louis, pp. 5–12.

Frosch, P. J., and Kligman, A. M., 1979, The soap chamber test: A new method for assessing the irritancy of soaps, *J. Am. Acad. Dermatol.* **1**:35–41.

Fujiwara, N., Toyooka, I., Ohnishi, K., and Onohara, E., 1992, Absorbed residue of fatty acid soap on human skin, *J. Soc. Cosmet. Chem. Jpn.* **26**:107–112.

Fulmer, A. W., and Kramer, G. J., 1986, Stratum corneum lipid abnormalities in surfactant-induced dry scaly skin, *J. Invest. Dermatol.* **86**:598–602.

Grubauer, G., Feingold, K. R., Harris, R. M., and Elias, P. M., 1989, Lipid content and lipid type as determinants of the epidermal permeability barrier, *J. Lipid Res.* **30**:89–96.

Hannuksela, A., and Hannuksela, M., 1995, Irritant effects of a detergent in wash and chamber tests, *Contact Dermatitis* **32**:163–166.

Horii, I., Nakayama, Y., Obata, M., and Tagami, H., 1989, Stratum corneum hydration and amino acid content in xerotic skin, *Br. J. Dermatol.* **121**:587–592.

Jemec, G. B. E., and Serup, J., 1992, Scaling, dry skin and gender. A bioengineering study of dry skin, *Acta Derm. Verereol. (Stockh.) Suppl.* **177**:26–28.

Kawai, M., and Imokawa, G., 1984, The induction of skin tightness by surfactants, *J. Soc. Cosmet. Chem.* **35**:147–156.

Keswick, B. H., Ertel, K. D., and Visscher, M. O., 1992, Comparison of exaggerated and normal use techniques for assessing the mildness of personal cleansers, *J. Soc. Cosmet. Chem.* **43**:187–193.

Lahti, A., Pylvänen, V., and Hannuksela, M., 1995, Immediate irritant reactions to benzoic acid are enhanced in washed skin areas, *Contact Dermatitis* **33**:177–182.

Laufer, A., and Dikstein, S., 1996, Objective measurement and self-assessment of skin-care treatments, *Cosmet. Toilet.* **111**:91–98.

Lodén, M., 1990, The simultaneous penetration of water and sodium lauryl sulfate through isolated human skin, *J. Soc. Cosmet. Chem.* **41**:227–233.

Long, C. C., and Marks, R., 1992, Stratum corneum changes in patients with senile pruritis, *J. Am. Acad. Dermatol.* **27**:560–564.

Lukacovic, M. F., Dunlap, F. E., Michaels, S. E., Visscher, M. O., and Watson, D. D., 1988, Forearm wash test to evaluate the clinical mildness of cleansing products, *J. Soc. Cosmet. Chem.* **39**:355–366.

Malten, K.E., 1981, Thoughts on irritant contact dermatitis, *Contact Dermatitis* **7**:238–247.

Mcardle, A., 1997, Overview of the U.S. cosmetics and toiletries market, *Drug Cosmet. Ind.* **160**:28–36.

Middleton, J. D., 1968, The mechanism of water binding in stratum corneum, *Br. J. Dermatol.* **80**:437–450.

Muhkerjee, S., Prowell, S., Hoyberg, K., Gursky, R., Davies, M., Meyers, C. L., Ananth, K. and Aronson, M., 1995, Cleanser induced structural changes in corneum, American Academy of Dermatology 53rd Annual Meeting, New Orleans, LA.

Nassif, A., Chan, S. C., Storrs, F. J., and Hanifin, J. M., 1994, Abnormal skin irritancy in atopic dermatitis and in atopy without dermatitis, *Arch. Dermatol.* **130**:1402–1407.

Oddo, L. P., Willard, M., Bowman, J. P., and Fochtman, D., 1997, Method for measuring skin moisturization with personal care cleansing products, American Academy of Dermatology 55th Annual Meeting, San Francisco, CA.

Patil, S., Singh, P., Sarasour, K., and Maibach, H., 1995, Quantification of sodium lauryl sulfate penetration into the skin and underlying tissues after topical application—pharmacological and toxicological implications, *J. Pharm. Sci.* **84**:1240–1244.

Prall, J. K., Theiler, R. F., Bowser, P. A., and Walsh, M., 1986, The effectiveness of cosmetic products in alleviating a range of skin dryness conditions as determined by clinical and instrumental techniques, *Int. J. Cosmet. Sci.* **8**:159–174.

Pugliese, P., Hines, G., and Wielinga, W., 1990, Skin protective properties of a cationic guar derivative, *Cosmet. Toilet.* **105**:105–111.

Rajka, G., 1986, Atopic dermatitis: correlation of environmental factors with frequency, *Int. J. Dermatol.* **25**:301–304.

Rhein, L. D., Robbins, C. R., Fernee, K., and Cantore, R., 1986, Surfactant structure effects on swelling of isolated human stratum corneum, *J. Soc. Cosmet. Chem.* **37**:125–139.

Rieger, M., 1995, Surfactant interactions with skin, *Cosmet. Toilet.* **110**:31–50.

Shukuwa, T., Kligman, A. M., and Stoudemayer, T. J., 1997, A new model for assessing the damaging effects of soaps and surfactants on human stratum corneum, *Acta Derm. Verereol. (Stockh.)* **77**:29–34.

Simion, F. A., Rhein, L. D., Grove, G. L., Wojtkowski, J. M., Cagan, R. H., and Scala, D. D., 1991, Sequential order of skin responses to surfactants during a soap chamber test, *Contact Dermatitis* **25**:242–249.

Treffel, P., and Gabard, B., 1996, Measurement of sodium lauryl sulfate-induced skin irritation, *Acta Derm. Verereol. (Stockh.)* **76**:341–343.

Tupker, R. A., Pinnagoda, J., Coenraads, P. J., and Nater, J., 1990, Susceptibility to irritants: role of barrier function, skin dryness, and history of atopic dermatitis, *Br. J. Dermatol.* **123**:199–205.

van der Valk, P. G. M., Nater, J. P., and Bleumink, E., 1985, Vulnerability of the skin to surfactants in different groups of eczema patients and controls as measured by water vapour loss, *Clin. Exp. Dermatol.* **10**:98–103.

Wolf, R., 1994, Has mildness replaced cleanliness next to godliness, *Dermatology* **189**:217–221.

Wolf, R., 1996, Entering the 21st century: Future perspectives, *Clin. Dermatol.* **14**:129–132.

Wortzman, M. S., Scott, R. A., Wong, P. S., Lowe, N. J., and Breeding, J., 1986, Soap and detergent bar rinsability, *J. Soc. Cosmet. Chem.* **37**:89–97.

Part IV

Pharmacologic Assays

Chapter 24

Small Plaque Assay

ANTTI I. LAUERMA AND HOWARD I. MAIBACH

1. Introduction

Psoriasis is a disease that is unique to man. Therefore no suitable animal models are yet available, despite the advent of transgenic animals that have raised hopes for such an eventuality. Further, as there is no good *in vitro* model, e.g., a cell culture system, that can replace the *in vivo* situation in studying medications for psoriasis, screening methods are needed in order to find drugs for the treatment of topical psoriasis.

2. Small Plaque Assay

2.1. Why Perform Small Plaque Assay?

As no laboratory models for psoriasis are available, the only way to determine whether or not a drug is efficacious in psoriasis is to study it in humans *in vivo*. The psoriasis small plaque assay (SPA) screens for topical drug activity. Small areas of psoriatic lesions are exposed to different topical preparations under occlusion which enhances drug delivery. Alternatively, open application may be employed if it is considered more useful. SPA permits the study of several different drug preparations at the same time, and there is little likelihood of systemic side effects from absorbed drugs.

SPA has been utilized in comparing the efficacy of corticosteroids (Dumas and Scholtz, 1972), calcipotriol (Bangha and Elsner, 1996), and topical immunosuppressants SDZ 281-240 (Rappersberg *et al.*, 1996) and tacrolimus (Remitz *et al.*, 1999). Additionally a multicenter study proved its efficacy in detecting new potential

Antti I. Lauerma • Department of Dermatology, University of Helsinki, 00250 Helsinki, Finland. *Howard I. Maibach* • Department of Dermatology, School of Medicine, University of California, San Francisco, California 94143-0989.

Cutaneous Biometrics, edited by Doris A. Schwindt and Howard I. Maibach, Kluwer Academic/Plenum Publishers, New York, 2000.

antipsoriatic topical drugs (Weinstein *et al.*, 1981). Its use as a standard may also enable development of laboratory models for psoriasis.

2.2. Selection of Patients, Washout Period, and Pretreatment

Patients with diagnosed plaque-type psoriasis are suited for the assay. On the other hand, there is little experience with the assay in other psoriasis types, including pustular psoriasis, erythrodermic psoriasis, and inverse psoriasis. *Psoriasis guttata* patients have lesions that are too small for the assay. Study patients should have had a stable disease condition for a month or more. The plaque studied should be homogeneous with respect to erythema and infiltration. A wash-out period of 4 weeks is needed for systemic or UV radiation therapy and 1 week or more for topical treatment on study plaques. Plaques other than those of interest may be topically treated during the study.

Selected target lesions should be treated with keratolytics (e.g., 2% salicylic acid in white petrolatum) prior to starting study medications in order to remove psoriasis scales. On the next visit the keratolytic preparation is removed.

2.3. Treatment Period

The microplaque assay method can be performed with 60-uL Finn chambers (Epitest Ltd., Hyrylä, Finland) with an outer diameter of 17 mm and an inner diameter of 11 mm. Finn chambers are normally used in patch-testing for contact allergy, but other patch test or occlusive systems can be used as well.

At the end of the second visit the chambers are filled with study medications and attached to paper tape, which is then applied to the plaque. The distance between chambers should be at least 1.5 cm, as such a distance does not allow horizontal spread of drugs to adjacent chamber sites (Bruze *et al.*, 1995). Without changing the position of the tape and chambers on plaque, a plastic bandage (Tegaderm^{R}) is applied on top of the plaque, tape, and chambers, which allows the patient to wash without wetting the tape or chambers. Assessments (see below) can be done after removal of the bandages, tape, chambers, and assay materials. After assessment a new set of medication chambers, tape, and plastic bandage are applied. The preparations are applied to target lesions with the microplaque assay for at least 5 days.

2.4. Follow-Up

For reasons of safety, as well as to see possible prolonged efficacy of the study medications, it is recommended that a patient be seen 7–14 days after the study so that the studied plaque areas can be assessed with the same methodology that was used during the medication period.

3. Small Plaque Assay Efficacy Parameters

Assessment can be done on the plaques by clinical scoring (Bangha and Elsner, 1996; Rappersberg *et al.*, 1996; Remitz *et al.*, 1999; Weinstein *et al.*, 1981), with laser

Doppler flow measurements of the superficial blood flow of the plaques (Remitz *et al.*, 1996), with chromameter to measure skin color (Bangha and Elsner, 1996), with visiometry to measure skin surface roughness (Bangha and Elsner, 1996), with ultrasound to measure skin thickness (Bangha and Elsner, 1996; Rappersberg *et al.*, 1996; Remitz *et al.*, 1999), and/or by histological examination from punch biopsies (Rappersberg *et al.*, 1996; Remitz *et al.*, 1996) with possible *in situ* immunostainings (Rappersberg *et al.*, 1996) or hybridizations.

3.1. Clinical Scoring

Clinical scoring may be done for erythema, infiltration, and scaling according to a following scale: 0 = none; 1 = mild; 2 = moderate; 3 = severe; 4 = very severe. Clinical scoring is reliable when experienced scientists (e.g., dermatologists) perform it, and the bioengineering methods described below have not proved to be better. However, as clinical scoring involves decision-making, it must always be done in a double-blind setting.

3.2. Chromametry

Chromametry is a method of measuring skin color. It is done with chromameter (e.g., Chroma Meter CR-200, Minolta, Osaka, Japan). The color of the skin is measured using $L^*a^*b^*$ scale. L^* reflects the luminance of the color on a gray scale, a^* reflects redness vs. greenness, and b^* blueness vs. yellowness. In SPA the a^* value is useful because psoriatic skin is erythematous, which is shown as redness. Although chromametry does not provide superior sensitivity to the human eye, it is a highly objective and reproducible method (Bangha and Elsner, 1996).

3.3. Surface Roughness Assessment

Skin surface roughness is increased in psoriatic skin and can be measured with visiometry. In visiometry descaled and cleaned plaque surface is covered with silicon gel, which is removed after hardening. Consequently the silicone replica is photographed and analyzed with special camera-computer equipment (e.g., Visiometer SV40, Courage & Khazaka, Cologne, Germany) (Bangha and Elsner, 1996).

3.4. Transepidermal Water Loss

Laser Doppler measurement is done with standardized equipment after a rest of at least 10 min. The measurement should take place in a controlled temperature setting (20–23°C) to avoid the effects of temperature on skin blood flow, and on sweating (Remitz *et al.*, 1999).

3.5. Skin Ultrasound

Ultrasound measurements are done with specific skin ultrasound devices (e.g., Dermascan A, Cortex Technologies, Hadsund, Germany, or DUB20-S, Taberna Pro

Medicum, Lüneburg, Germany). Ultrasound measurements enable determination of skin thickness; additionally, thickness of epidermis and dermis as well as some morphology can be studied (Bangha and Elsner, 1996; Remitz *et al.*, 1999).

3.6. Histopathology

Histopathology can be studied from biopsies of SPA lesions. In addition to routine histopathology (Remitz *et al.*, 1999), immunostainings (Rappersberg *et al.*, 1996), and/or *in situ* hybridizations can be utilized to study the effect of SPA medicaments on psoriasis in more detail.

4. Alternative Small Plaque Assay Methods

Psoriasis plaques can be used in a similar manner as psoriasis microplaque assay when open application is studied, as was shown by Baadsgaard *et al.* (1995). In the study, different concentrations of tacalcitol (a vitamin D derivative) were applied once a day. The study preparations were rubbed onto psoriasis plaque through a 20-mm-diameter template hole, ensuring that the surrounding plaque area was not treated. The study helped to find an optimal tacalcitol concentration for psoriasis treatment for once-a-day application (Baadsgaard *et al.*, 1995).

5. Conclusions

SPA provides a relatively rapid method for assessing the efficacy of topical drugs to heat for psoriasis. The assay provides information on the occluded condition. Therefore, it cannot be known whether an efficacious drug in occluded SPA would also be efficacious as an open application until such an application is performed.

In the absence of animal and *in vitro* models for psoriasis it seems that SPA provides the safest way to predict the antipsoriatic potential of new compounds or the effect of delivery systems (vehicles). SPA can also serve as a standard for development of alternative methods. It can also be used to study the effect of topical biomodulators on psoriasis to gain insights into the disease processes themselves.

References

Baadsgaard, O., Traulsen, J., Roed-Petersen, J., and Jakobsen, H. B., 1995, Optimal concentration of tacalcitol in once-daily treatment of psoriasis, *J. Dermatol. Treat.* **6**:145–150.

Bangha, E., and Elsner, P., 1996, Evaluation of topical antipsoriatic treatment by chromametry, visiometry and 20-MHz ultrasound in the psoriasis plaque test, *Skin Pharmacol.* **9**:298–306.

Bruze, M., Isaksson, M., and Dooms-Goossens, A., 1995, The influence of patch tests with clobetasol propionate on adjacent patch test reactions, *Contact Dermatitis* **32**:167–170.

Dumas, K. J., and Scholtz, J. R., 1972, The psoriasis bio-assay for topical corticosteroid activity, *Acta Derm. Venereol.* (*Stoch.*) **52**:43–48.

Rappersberg, K., Meingassner, J. G., Fialla, R., Fodinger, D., Sterniczky, B., Rauch, S., Putz, E., Stutz, A., and Wolf,, K., 1996, Clearing of psoriasis by a novel immunosuppressive macrolide, *J. Invest. Dermatol.* **106**:701–710.

Remitz, A., Reitamo, S., Erkko, P., Granlund, H., and Lauerma, A. I., 1999, Tacrolimus ointment improves psoriasis in a microplaque assay, *Br. J. Dermatol.* **141**:103–107.

Weinstein, G. D., McGullough, J. L., Eaglestein, W. H., Golub, A., Cornell, R. C., Stoughton, R. B., Clendenning, W., Zackheim, H., Maibach, H. I., Kulp, K. R., King, L., Baden, H. P., Taylor, J. S., and Deneau, D. D., 1981, A clinical screening program for topical chemotherapeutic drugs in psoriasis, Arch. Dermatol. **117**:388–393.

Chapter 25

Application of the Minolta Chromameter to the Assessment of Corticosteroid-Induced Skin Blanching

RODERICK B. WALKER, JOHN M. HAIGH, AND ERIC W. SMITH

1. Introduction

Technological advances in both analytical methodology and statistical analysis have refined the science of bioequivalence testing. In recent years, the design, performance, and evaluation of bioequivalence studies have received much attention from the industry, academia, and, more recently, from regulatory agencies. The drug delivery field that has possibly received the least attention in this regard is the bioequivalence assessment of topical formulations. Corticosteroids, especially, have been evaluated for several decades using a visual grading system for subjectively recording the intensity of the drug-induced vasoconstriction. As an alternative, a number of researchers (Pershing *et al.*, 1992; Waring *et al.*, 1993; Clarys *et al.*, 1995) have used a Minolta chromameter to assess skin color following the application of corticosteroid-containing formulations. Some years ago the Food and Drug Administration (FDA) proposed that a chromameter be used to measure blanching intensity when assessing topical corticosteroid activity for bioequivalence purposes (FDA, 1995). With this instrument it is theoretically possible to measure the intensity of skin-blanching as an absolute color, based on the tristimulus analysis of a reflected xenon light pulse. The result is expressed as a three-dimensional coordinate consisting of two color coordinates (the green-red or *a*-scale and the yellow-blue or *b*-scale) and a luminance coordinate (black-white or *L*-scale). These three coordinates (*a*, *b*,

Roderick B. Walker and John M. Haigh • Faculty of Pharmacy, Rhodes University, Grahamstown 6140, South Africa. *Eric W. Smith* • Faculty of Pharmacy, Rhodes University, Grahamstown 6140, South Africa • Present address: College of Pharmacy, Ohio Northern University, Ada, Ohio 45810.
Cutaneous Biometrics, edited by Doris A. Schwindt and Howard I. Maibach, Kluwer Academic/Plenum Publishers, New York, 2000.

and L) should define the absolute color of a skin site at each observation time as the vasoconstriction phenomenon varies.

Presently the FDA-recommended assessment of blanching intensity utilizing the chromameter suggests the use of only the *a*-scale parameter for evaluation of the response. A more realistic evaluation of corticosteroid-induced blanching intensity using the chromameter data should be possible using all three coordinates produced by the instrument as color vision is known to be trichromatic, suggesting that a single perceived color results from the effect of three separate stimuli on the visual cortex (Hunter and Harold, 1987). An evaluation (Demana *et al.*, 1997) of the FDA Guidance document for bioequivalence assessment using only the *a*-scale parameter revealed a number of weaknesses in the proposed methodology. In particular, the use of only one application site for each dose duration tested per subject in the pilot trial and the pharmacodynamic modeling of the *a*-scale results were highlighted as areas of concern. In a recent study (Smith *et al.*, 1998), Euclidean distances (ED) calculated using all three chromameter values were used to assess blanching in six subjects that were dosed with two different corticosteroid formulations at multiple application sites. The results indicated that ED may prove more useful for the evaluation of skin-blanching than evaluation with only the *a*-scale parameter. The ED profiles reported in this study were similar in shape and rank order to those generated from the visual assessment of blanching.

In an attempt to further evaluate the application of ED calculation in the assessment of skin blanching, this analytical methodology was applied to data collected during a pilot study conducted as described in the FDA Guidance in which the intensity of skin blanching was evaluated both visually and by using the chromameter.

2. Methodology

2.1. Materials

Two betamethasone 17–valerate-containing creams (0.12%), Lenovate (Pharmacare Lennon, South Africa) and Betnovate (Glaxo-Wellcome, South Africa) were used in this pilot study to evaluate the Guidance proposals. Betnovate was selected as the standard formulation as it has been repeatedly used in our laboratory and an extensive database of results exists for this product (Smith *et al.*, 1989, 1990, 1991; Haigh *et al.*, 1997*a*, *b*). A Minolta CR-200 chromameter (Minolta Corporation, Ramsey, NJ) that had been calibrated immediately prior to the study with a white calibration plate (CD-A223) was used to assess the blanching responses.

2.2. Methods

Twelve healthy Caucasian volunteers with normal forearm skin participated in the study. Ethical approval was obtained from the Rhodes University Ethical Standards Committee in compliance with the 1964* Declaration of Helsinki and its

*Declaration of Helsinki, I, II, III, and IV, Recommendations guiding physicians in biomedical research involving human subjects. Adopted by the 18th World Medical Assembly, Helsinki, Finland, June 1964 (I) and amended by the 29th World Medical Assembly, Tokyo, Japan, October 1975 (II), the 35th World Medical Assembly, Venice, Italy, October 1983 (III) and the 41st World Medical Assembly, Hong Kong, September 1989 (IV).

subsequent amendments. Written informed consent was obtained from each subject. All volunteers were processed on the same day, at intervals of approximately 5 min, in order to minimize any possible effects of environmental variables such as temperature and humidity. Six adhesive labels, from which two 7 mm × 7 mm squares had been punched, were applied to the flexor aspect of both forearms to demarcate a total of 12 application sites per arm of each volunteer. Approximately 3.2 mg (Haigh *et al.*, 1997*a*) of each formulation was applied and spread in a double-blind, randomized manner to each of two designated sites on the right arm (short dose durations: 0.25, 0.5, 0.75, 1, 1.5, and 2 hr) and on the left arm (long dose durations: 4, 5, 6, 7, 8, and 10 hr). This is the dose of formulation normally used in this laboratory for bioequivalence testing. Three sites on each arm were not treated and were used as control sites. Both arms of each volunteer were left unoccluded but were protected by porous perspex frames. The formulations were applied at staggered times with synchronized removal as per the directive in the Guidance. At the designated time, the protective covers and adhesive labels were removed and the application sites were gently washed with soap and distilled water using cotton-tipped buds and patted dry.

(a) Visual Assessment of Blanching Response. The blanching responses were assessed independently by four experienced observers at the time of product removal and thereafter at 1, 2, 3, 4, 5, 6, 7, 8, 9, 10, 11, 19, 22, 24, and 26 hr. Responses were graded using a 0 to 4 scale where 0 = no blanching, 1 = slight blanching, 2 = more intense blanching, 3 = general even and distinct blanching and 4 = marked and very intense blanching. The skin blanching response, expressed as the percentage of the total possible score (% TPS), was calculated (Barry and Woodford, 1978; Smith *et al.*, 1992) and plotted against time in hours after product application to produce effect profiles for each formulation.

(b) Chromameter Assessment of Blanching Response. Blanching responses at all 12 sites on each arm were assessed using all three values (*a*, *b*, and *L*-scale) generated by the chromameter at observation times corresponding to the visual assessment sequence. In addition, the skin color was measured prior to product application at the commencement of the study in order to obtain a baseline value for each site. The use of all three values enables a color to be depicted by a point in three-dimensional space, with each color defined by its unique coordinates. As skin color changes so does the location of the color descriptor point in the defined three-dimensional space. Therefore, an increase in blanching intensity from baseline values as a direct result of corticosteroid absorption manifests as different descriptor points in the *a*-, *b*-, and *L*-scale quadrant. The ED between the normal skin and medicated response points is a direct indication of the blanching intensity and can be calculated as the length of a line joining the points in three-dimensional space using Eq. (1):

$$\mathrm{ED} = [(\Delta L)^2 + (\Delta a)^2 + (\Delta b)^2]^{1/2} \tag{1}$$

The mean ED between the control sites and the medicated sites for each preparation at each observation time was determined for each subject. The resultant ED data for all 12 subjects were then averaged for presentation.

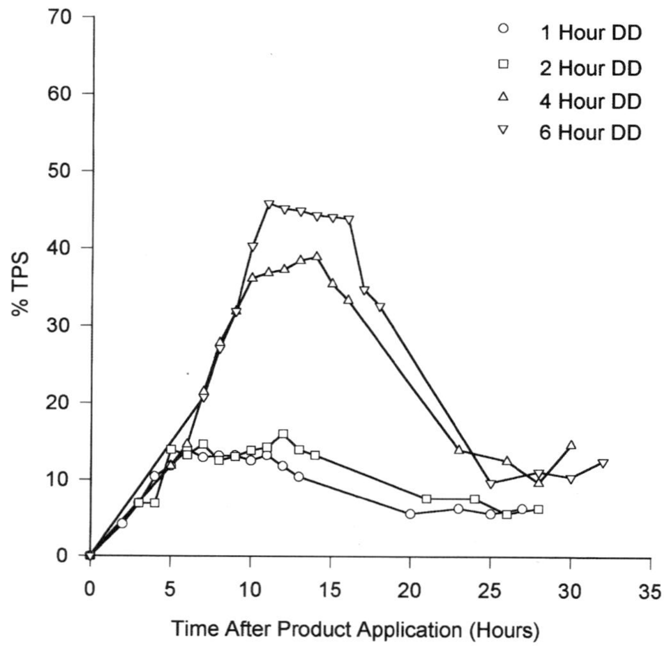

FIGURE 1. Visual blanching profiles for the 1-, 2-, 4-, and 6-hr dose durations (DD) for Betnovate.

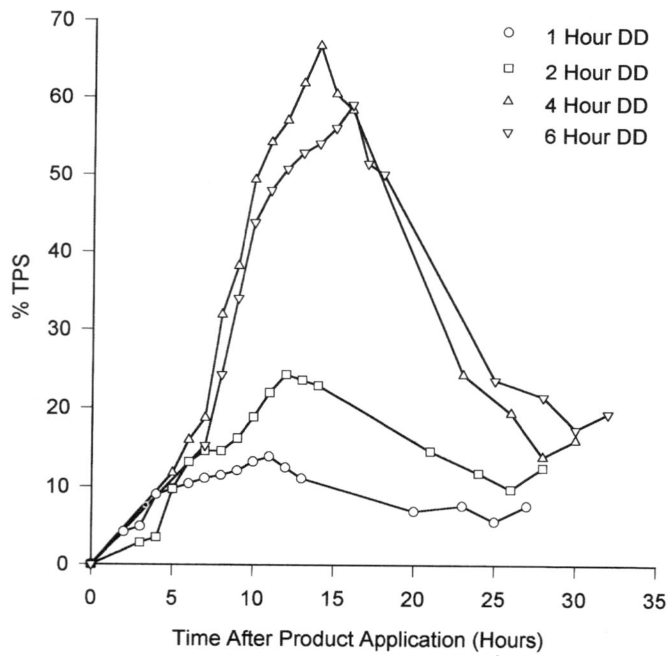

FIGURE 2. Visual blanching profiles for the 1-, 2-, 4-, and 6-hr dose durations (DD) for Lenovate.

3. Results

3.1. Assessment of Visual Blanching Profiles

Figures 1 and 2 represent the blanching response profiles for the 1-, 2-, 4-, and 6-hr dose durations, for Betnovate and Lenovate, respectively, plotted as %TPS vs. time in hours after product application. The profiles show an increase in blanching to a maximum at approximately 15 hr after product application followed by a subsequent decrease in intensity. These profiles are typical of the response normally seen with visual assessment, although they are not as smooth as previously reported (Smith *et al.*, 1991) probably because of the relatively few sites per arm, per dose duration used in this study. As would be expected from previous studies (Magnus *et al.*, 1980), the profiles clearly show that blanching intensity increases as dose duration increases. In addition, the greater response of the Lenovate product compared to the pharmaceutically equivalent Betnovate formulation has been recorded in several previous studies. There is a slight increase in blanching at the later observation times, which may be attributed to a reservoir effect in the stratum corneum (Feldmann and Maibach, 1965).

3.2. Assessment of Chromameter Blanching Profiles

(a) Chromameter a-Scale Data. Figures 3 and 4 represent the mean, uncorrected chromameter *a*-scale blanching response profiles for the 1-, 2-, 4-, and 6-hr dose durations for Betnovate and Lenovate, respectively. The imprecision of the data is demonstrated in Fig. 3 by the standard deviation bars that have been drawn for the last data point of each curve. The magnitude of this variance is typical for all mean data points and, thus, standard deviation bars have not been shown in all figures for the purposes of clarity. This imprecision of data obtained by the chromameter in a pilot study has been reported previously (Demana *et al.*, 1997).

The chromameter data obtained in this study are highly variable and are difficult to interpret, implying that the instrument may be incapable of measuring the blanching response precisely at low dose durations. The results obtained for Betnovate show that there is a rank order difference between the blanching for the 1- and 2-hr dose durations and the 4- and 6-hr dose durations, even though lower *a*-scale values would have been expected for the more intense blanching profiles of the longer dose durations. In contrast, the Lenovate results show that there is no significant difference between the four dose durations tested. In addition, it is not clear that the response for the Lenovate formulation is greater than that of the Betnovate as would be expected from the visual data.

The data collected using the chromameter were subjected to the arithmetic manipulation suggested by the FDA in their Guidance document. Both the medicated and unmedicated site values were initially corrected for the baseline reading at each site and then the baseline-corrected unmedicated values were

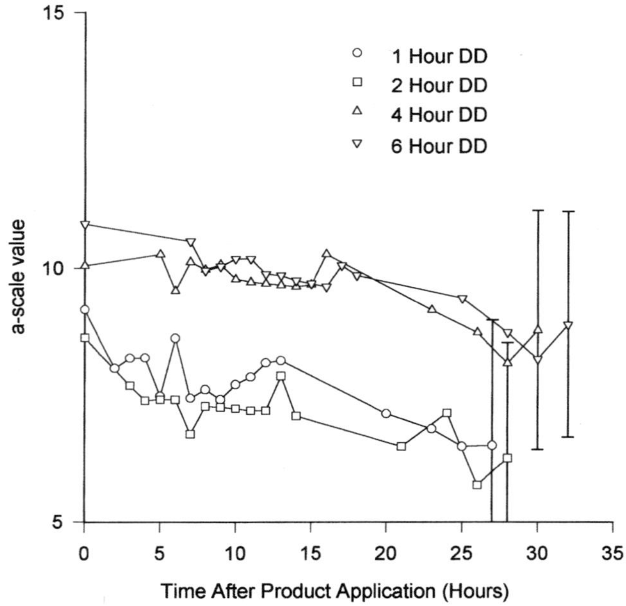

FIGURE 3. Uncorrected *a*-scale chromameter blanching profiles for the 1-, 2-, 4-, and 6-hr dose durations (DD) for Betnovate. Mean $\pm$ 1SD values are shown at the last observation time for each DD. The variance for all other mean points is of similar magnitude.

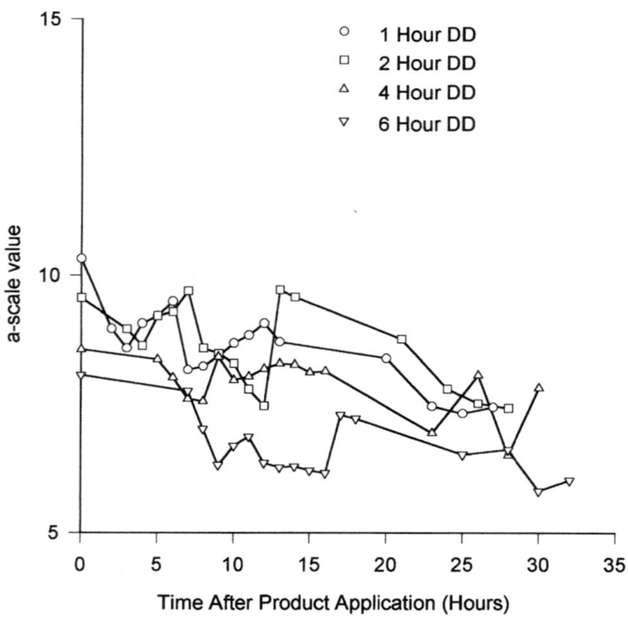

FIGURE 4. Uncorrected *a*-scale chromameter blanching profiles for the 1-, 2-, 4-, and 6-hr dose durations (DD) for Lenovate.

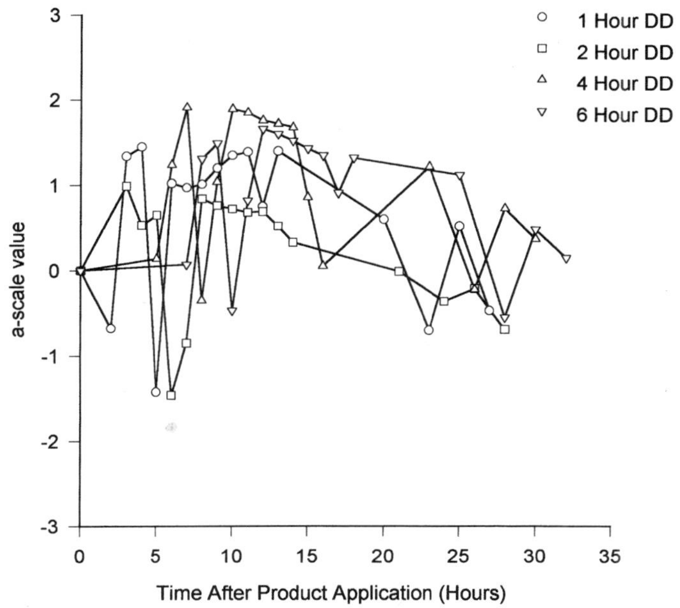

FIGURE 5. Baseline-corrected and unmedicated-corrected *a*-scale chromameter blanching profiles for the 1-, 2-, 4-, and 6-hr dose durations (DD) for Betnovate.

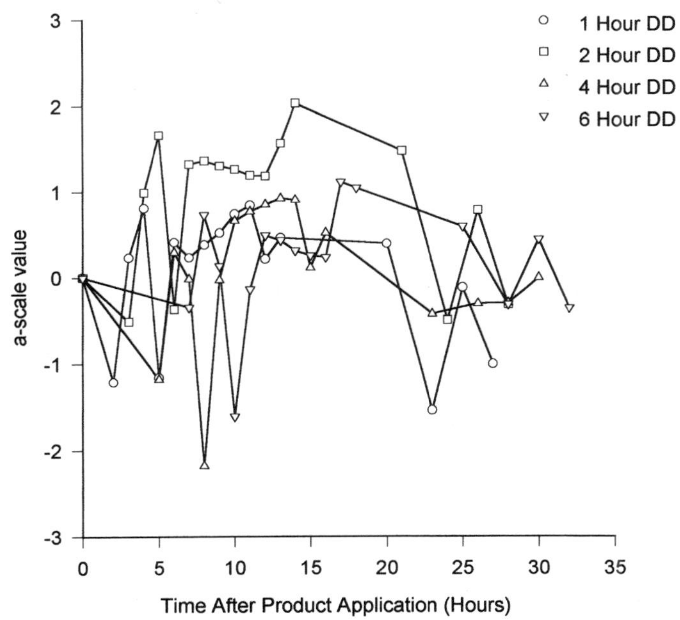

FIGURE 6. Baseline-corrected and unmedicated-corrected *a*-scale chromameter blanching profiles for the 1-, 2-, 4-, and 6-hr dose durations (DD) for Lenovate.

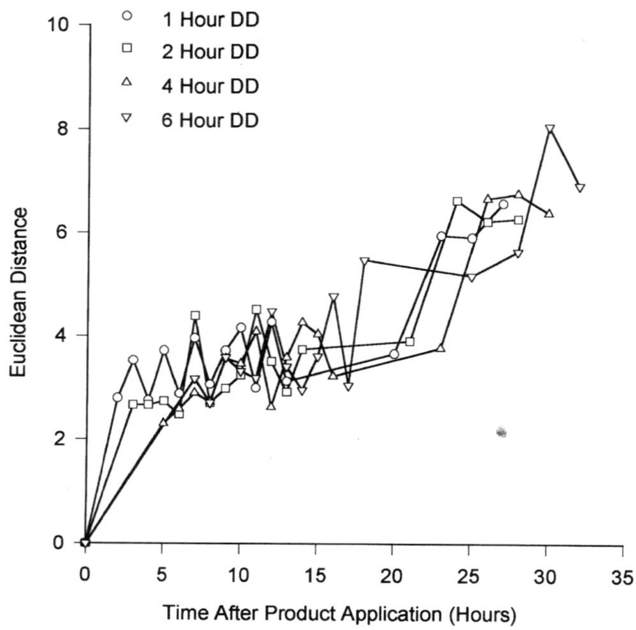

FIGURE 7. Euclidean distance blanching profiles for the 1-, 2-, 4-, and 6-hr dose durations (DD) for Betnovate.

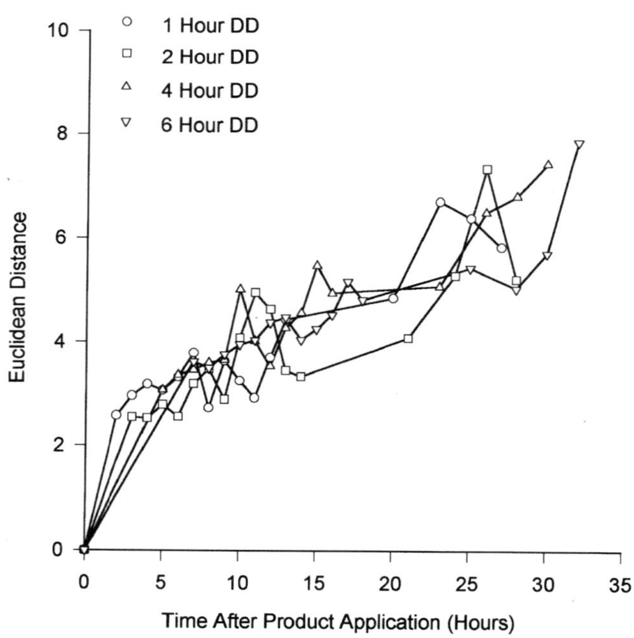

FIGURE 8. Euclidean distance blanching profiles for the 1-, 2-, 4-, and 6-hr dose durations (DD) for Lenovate.

subtracted from the baseline-corrected medicated values. Figures 5 and 6 depict the corrected *a*-scale data for Betnovate and Lenovate, respectively. It is clearly evident that the corrected *a*-scale data are highly variable for both preparations, encompassing both positive and negative corrected *a*-scale values.

(b) Euclidean Distances. Figures 7 and 8 represent the ED response profiles for the 1-, 2-, 4-, and 6-hr dose durations for Betnovate and Lenovate, respectively. The data are highly variable and the profiles do not show an obvious increase in blanching response with increasing dose duration as seen with the visual assessment. The results of a previous study (Smith *et al.*, in press) would suggest that the ED vs. time profiles should mirror the visual response profiles; this is not seen in this study.

4. Conclusions

There are no clear response trends in the chromameter data obtained in this pilot study whereas the visual data produces typical blanching response-time curves, which increase to a maximum and then decline. In addition, the rank order of the observed visual response was typical of that recorded previously.

The expected response trends were not obtained for the uncorrected *a*-scale data (Figs. 3 and 4). As blanching intensity increases, the *a*-scale values should decrease and therefore the longer dose-duration curves should have lower values than the shorter dose-duration curves. This was not recorded for the Betnovate blanching response in this pilot trial. In contrast, the Lenovate results generally show this trend, in rank order, despite the imprecision of the curves.

The *a*-scale data manipulated according to the FDA Guidance document show no clear trends. The validity of this manipulation has already been questioned (Smith *et al.*, in press) and these results show that the data in this particular study are confounded rather thanimproved by the manipulation. Moreover, the resultant positive and negative corrected *a*-scale values hinder interpretation of the response profiles.

The use of ED calculations showed potentially better results in a previous study (Smith *et al.*, in press) in which two different corticosteroids were evaluated using multiple sites per arm; the results from this pilot study did not show a clear ED response profile. These results are almost certainly a function of using only one application site per arm for assessing the dose–response relationship of topically applied corticosteroid-containing formulations. Confidence in the data generated in these studies may be enhanced by generation of a larger data set for analysis. The use of ED calculations in these bioequivalence assessments may have potential but requires further investigation.

In contrast, the visual assessed blanching profiles clearly show a trend with more potent blanching occurring at the 4- and 6-hr dose durations for both Betnovate and Lenovate than at the 1- and 2-hr dose durations. These results correlate well with previous studies using these two preparations, despite the fact that only one site per arm was used for each dose duration. In addition, the rank order of Lenovate and Betnovate blanching intensity was as expected from previous studies conducted in this laboratory. The visual assessment of skin blanching continues to appear more robust than current chromameter techniques when evaluating different formulations containing the same corticosteroid, especially when relatively few application sites per arm are employed. The recommendation from these data would be that pilot studies on which pivotal bioequivalence evaluations are based should use multiple sites per arm for each dose duration so that a scientifically acceptable data set is obtained for analysis.

References

Barry, B. W., and Woodford, R., 1978, Activity and bioavailability of topical steroids: *In vivo/in vitro* correlations for the vaso-constrictor test, *J. Clin. Pharm.* **3**:43–65.

Clarys, P., Wets, L., Barel, A., and Gabard, B., 1995, The skin blanching assay with halcinolide, influence of halcinolide concentration and application time, *J. Eur. Acad. Dermatol. Venereol.* **5**:250–257.

Demana, P. H., Smith, E. W., Walker, R. B., Haigh, J. M., and Kanfer, I., 1997, Evaluation of the proposed FDA pilot dose–response methodology for topical corticosteroid bioequivalence testing, *Pharm. Res.* **14**:303–308.

FDA, 1995, Guidance: Topical dermatologic corticosteroids: *in vivo* bioequivalence, Division of Bioequivalence, Office of Generic Drugs, Food and Drug Administration, Rockville, MD., June 1995.

Feldmann, R. J., and Maibach, H. I., 1965, Penetration of ^{14}C-hydrocortisone through normal skin, *Arch. Dermatol.*, **91**:661–666.

Haigh, J. M., and Kanfer, I., 1984, Assessment of topical corticosteroid preparations: the human skin blanching assay, *Int. J. Pharm.*, **19**:245–262.

Haigh, J. M., Meyer, E., Smith, E. W., and Kanfer, I., 1997*a*, The human skin blanching assay for *in vivo* topical corticosteroid assessment: I. Reproducibility of the assay, *Int. J. Pharm.* **152**:179–183.

Haigh, J. M., Meyer, E., Smith, E. W., and Kanfer, I., 1997*b*, The human skin blanching assay for *in vivo* topical corticosteroid assessment: II. Subject- and observer-dependent variation in blanching responses, *Int. J. Pharm.* **152**:185–192.

Hunter, R. S., and Harold, R. W., 1987, *The Measurement of Appearance*, 2nd Ed., Wiley-Interscience, New York.

Magnus, A. D., Haigh, J. M., and Kanfer, I., 1980, Assessment of some variables affecting the blanching activity of betamethasone 17-valerate cream, *Dermatologica* **160**:321–327.

Pershing, L. K., Lambert, L. D., Shah, V. P., and Lam, S. Y., 1992, Variability and correlation of chromameter and tape-stripping methods with the visual blanching assay in the quantitative assessment of topical 0.05% betamethasone dipropionate bioavailability in humans, *Int. J. Pharm.* **86**:201–210.

Smith, E. W., Meyer, E., Haigh, J. M., and Maibach, H. I., 1989, The human skin blanching assay as an indicator of topical corticosteroid bioavailability and potency: An update, in: *Percutaneous Absorption*, Vol. 2 (R. L. Bronaugh and H. I. Maibach, eds.) Marcel Dekker, New York, pp. 443–460.

Smith, E. W., Meyer, E., and Haigh, J. M., 1990, Blanching activities of betamethasone formulations: The effect of dosage form on topical drug availability, *Drug Res.* **40**:618–621.

Smith, E. W., Meyer, E., Haigh, J. M., and Maibach, H. I., 1991, The human skin blanching assay for comparing topical corticosteroid availability, *J. Dermatol. Treat.* **2**:69–72

Smith, E. W., Meyer, E., and Haigh, J. M., 1992, Accuracy and reproducibility of the multiple-reading skin blanching assay, in: *Topical Corticosteroids* (H. I. Maibach and C. Surber, eds.), Karger, Basel, pp. 65–73.

Smith, E. W., Haigh, J. M., and Walker, R. B., 1998, Analysis of chromameter results obtained from corticosteroid-induced skin blanching: I. Manipulation of data. *Pharm. Res.* **15**:280–285.

Waring, M. J., Monger, L., Hollingsbee, D. A., Martin, G. P., and Marriott, C., 1993, Assessment of corticosteroid-induced skin blanching: evaluation of the Minolta Chromameter CR200, *Int. J. Pharm.* **94**:211–222.

Chapter 26

Assaying for Bioavailability in Acne

OTTO H. MILLS, JR, CAROLINE W. CARDIN, AND MITCHELL WORTZMAN

1. Introduction

Although the skin is completely available for clinical evaluation, dermatopharmacology assessments are limited when one looks for studies demonstrating the presence and level of drugs or metabolites at the targeted disease site. Blood levels of drugs have not been meaningful when discussing therapy for skin diseases.

Ideally topical applications for dermatologic conditions would result in a therapeutic impact. Orally administered agents are commonly used to treat skin conditions. In both routes of administration, one would like to know not only if the drug and/or metabolites of the drug are found at the target site but some measure of the amounts present would be helpful for comparative purposes.

The pathology of acne vulgaris is centered in the sebaceous follicle, and three abnormalities are usually cited for the disease: (1) excessive sebum, (2) hyperkeratinous impaction in the sebaceous follicle (microcomedone), and (3) rupture of the follicular epithillium. *Propionibacterium acnes* (*P. acnes*) is said to play a key role in the third abnormality. Of these three, the hyperkeratinization of the sebaceous follicle and the formation of a microcomedone is considered by many to be the initial step in the pathology. Some years ago Mills and Kligman (1980, 1983), described a sampling technique using a modification of the method of Marks (Marks and Dawber, 1971) that Holmes (Holmes *et al.*, 1972) had used to extract this hyperkeratinous impaction (microcomedone) from the sebaceous follicle without invasive methods. This approach opened up the way to study this microcomedone in a variety of ways (Webster and Kligman, 1979; Mills *et al.*, 1984; Weinstein *et al.*, 1988; Groh *et al.*, 1992; Mills, 1995). It also allowed for the study of the penetration of

Otto H. Mills, Jr. • University of Medicine and Dentistry of New Jersey, Robin Wood Johnson Medical School, New Brunswick, New Jersey 08901. *Caroline W. Cardin* • The Proctor & Gamble Company, Sharon Woods Technical Center, Cincinnati, Ohio 45241. *Mitchell Wortzmann* • Medicis Pharmaceutical Corporation, Scottsdale, Arizona 85258.

Cutaneous Biometrics, edited by Doris A. Schwindt and Howard I. Maibach, Kluwer Academic/Plenum Publishers, New York, 2000.

drugs into the microcomedone. This line of investigation using high-pressure liquid chromatography (HPLC) began in the early 1980s.

2. Pilot Studies of Benzoyl Peroxide and Topically Applied Erythromycin

Some 15 years ago, one of us (Wortzman *et al.*, 1984) gave an oral presentation at the annual Society for Investigative Dermatology meeting. The studies described at that time involved taking follicular biopsy samples from the face following treatment with benzoyl peroxide (BPO) or erythromycin (ERY) and removing 10 to 20 moderately sized microcomedones using no. 15 disposal scalpel. These microcomedones were placed in a 5-ml test tube and diluted to a 1.0 final volume with tetrahydrofuran (Burdick-Johnson Labs, Muskegon, Michigan, HPLC Grade). The tubes were sonicated for 1 hr in the absence of light in order to completely extract the content of the lesions for analysis.

For BPO, the extract was analyzed using a HPLC reverse phase ODS column (Waters Associates, Milford, Mass.) With the mobile phase consisting of acetonitrile: water, 65:35 (Ph 4.0), the BPO was detected by absorbance at 280 nm. Benzoic acid, the expected degradation product of BPO, was assayed using the same system with detection at 254 nm.

In the ERY assay the comedone extract was analyzed using a reverse phase ODS column, with a mobile phase of methanol, water, 2-propanol, acetic acid, and diethanolamine (80:15:5:.05:.05). The ERY was detected by its UV absorbance at 215 nm.

In order to determine penetration of topically applied antiacne drugs into the microcomedone these studies focused on ERY and BPO with measurement of the densities of *P. acnes* using the scrub cup sampling method of Williamson and Kligman (1965) (modified by wiping the skin with a solution of the detergent triton X-100 in sterile water, then drying it). This was done in order to ensure that the surface of the skin was clean, and a stabilization period of 1 hr was observed. The prospective test site was now covered with perforated plastic that was slightly elevated so that the skin was protected but not occluded. At the time of sampling, the covering was removed and the glass cup was held in place in the center of the protected area (3.8 cm^2). The organisms were drop-plated on brain–heart medium and identified by colony morphology and bacteriophage susceptibility.

The clinical applications and sampling schedule were as follows for two studies with BPO: In the first study, four volunteer subjects reported to the lab twice daily for applications of a 2.5% BPO gel for 2 weeks. Samples were taken prior to starting medication and at the end of 2 weeks. In a second study, eight volunteer subjects reported to the lab twice daily for 1 week to participate in a half-face study. On the right side of each subject's face a mask containing 5% BPO was applied on one of these visits for 20 min and then washed off. On the left side of the face a 5% BPO gel was applied twice daily. Sampling was done prior to and at 1 week of applications.

Topically applied ery was studied in the following way. Seven volunteer subjects reported to the laboratory for twice-daily applications of a 2% ERY (hydroalcohol vehicle) solution for 6 weeks. Sampling was done at 0 and 6 weeks.

TABLE 1. Benzoyl Peroxide Therapy[a]

Subject	BPO/comedone (μg)	BZA/comedone (μg)	*P. acnes* reduction (log cm^2)
A	7.19 ± 0.21	0.70 ± 0.12	−1.72
B	6.49 ± 0.33	0.66 ± 0.10	−1.49
C	1.40 ± 0.18	4.98 ± 0.72	−0.24
D	7.63 ± 0.35	0.90 ± 0.18	−1.84

[a]2.5% benzoyl peroxide, applied twice daily for two weeks; BPO = benzoyl peroxide, BZA = benzoic acid.

The results of these pilot studies were as follows: Table 1 shows the results from a 2.5% BPO product applied twice daily for 2 weeks. BPO concentration averaged 5.5 μg comedone. The amount of benzoic acid, the primary breakdown product of BPO, per comedone varies greatly—one subject with low levels of active drug and high levels of benzoic acid had only minor reduction of *P. acnes* levels. Table 2 shows the result of a comparison study using two currently marketed 5% BPO products, one an alcoholic gel and other a Kaolin mask. Despite the differences in vehicle and number of applications, the retention of BPO within the comedone was rather close in all trials and the *P. acnes* reduction in 1 week averaged between 1 and 2 logs. This result was similar to the 2.5% BPO product after 2 weeks.

Table 3 shows the results of a 6-week trial of 2.0% ERY in a hydroalcoholic vehicle twice a day. ERY levels averged between 1 and 2 μg/comedone. This resulted in an approximate 2 to 3 log reduction of *P. acnes* concentration. In one case, the ery concentration of about 0.5 μg per comedone resulted in a reduction of *P. acnes* of less than 1 log.

TABLE 2. Benzoyl Peroxide Product Comparison[a]

Subject	BPO/comedone (μg)		BZA/comedone (μg)		*P. acnes* reduction (log cm^2)	
	Left	Right	Left	Right	Left	Right
A	7.9 ± 0.14	3.2 ± 0.11	0.84	1.47	−1.85	−1.10
B	9.7 ± 0.14	2.9 ± 0.27	0.97	1.35	−1.29	−0.32
C	13.4 ± 0.17	5.7 ± 0.15	1.14	0.59	−2.41	−0.82
D	19.2 ± 0.29	4.9 ± 0.33	0.89	0.94	−3.14	−1.23
E	6.8 ± 0.14	4.4 ± 0.15	0.75	0.82	−1.64	−1.21
F	5.7 ± 0.09	4.1 ± 0.11	0.80	0.77	−1.13	−0.65
G	3.9 ± 0.30	5.1 ± 0.28	1.32	0.93	−0.82	−1.40
H	8.2 ± 0.33	8.2 ± 0.31	0.79	0.73	−1.26	−1.16

[a]Left = 5% benzoyl peroxide gel used twice daily for 1 week; right = 5% benzoyl peroxide mask used once daily for 1 week.

TABLE 3. Topical Erythromycin Therapy[a]

Subject	Comedone μg ERY	*P. acnes* reduction (log cm^2)
A	0.92 ± 0.02	-1.61
B	0.85 ± 0.03	-1.40
C	0.99 ± 0.06	-2.02
D	0.48 ± 0.02	-0.85
E	1.86 ± 0.06	-3.25
F	1.50 ± 0.04	-2.51
G	2.09 ± 0.04	-3.16

[a]2%ERY in a hydroalcoholic vehicle, applied twice daily for 6 weeks.

3. *Comparative Study of Salicylic Acid Levels*

3.1. *Clinical Applications and Sampling Schedule*

Eleven years after the initial report (Wortzman, 1984) there was a presentation at another meeting of the Society for Investigative Dermatology in the form of a poster (Mills *et al.*, 1995). This study focused on salicylic acid and changed the sampling site from the face to the back, allowing for direct comparison of formulations. The study was 12 days in duration, single-blind, using 16 male and female acne vulgaris sufferers (18 to 50 years of age). Prior to enrollment, subjects who had given informed consent were screened for their eligibility to participate on the basis of having at least 15 microcomedones/cm^2 on the back and being in good general health apart from the facial acne. Subjects were excluded if they had a significant number of papulopustules of scarring of the back or recent sun exposure that would interfere with evaluations. Also excluded were subjects who had used any topical medication on the back, or had been on systemic contraceptive therapy, or had participated in any investigational drug study within the previous 3 months.

Use of topical or oral medications containing salicylates, use of an oral medication that might affect the course of acne, or a history of very sensitive skin or a known allergy to any of the active ingredients found in the antiacne products were also reasons for exclusion. Subjects with a positive Woods lamp response at the treatment sites, indicating the presence of salicylates, were excluded as well.

Each subject had eight (1 inch $\times$ 1 inch) treatment sites across the right and left scapulae (four sites on each scapula) to which six treatments and two untreated controls were randomly assigned using a replicated Latin square design. Each of the six active treatment sites was treated twice daily with 10 μl of product, for the specified number of days, by laboratory personnel. The treatments were dispensed using an automatic displacement volume microliter pipette onto the designated area of the black of each subject. After application, the sites were allowed to dry. None of the sites was occluded. At the designated times—on days 2, 7, and 12 for the

TABLE 4. Follicular Biopsy Schedule[a]

Treatment	Days of treatment
2% SA in hydroalcoholic vehicle	2
2% SA in hydroalcoholic vehicle	7
2% SA in hydroalcoholic vehicle	12
No treatment (baseline)	0
2% SA in hydroalcoholic gel	2
2% SA in hydroalcoholic gel	12
No treatment (12 days)	0

[a]SA = salicylic acid.

active-treatment sites and days 0 and 12 days for the no-treatment sites—the sites were biopsied using the cyanoacrylate glue slide technique (Table 4). A total of 15 to 16 microcomedones was harvested from the designated sites. The harvested microcomedones were placed in 1-dram glass vials for salicylic acid analyses.

3.2. Analytical Method for Salicylic Acid Analysis (Isocratic HPLC Using UV Detection

(a) Preparation of the Calibration Standard

1. Approximately 10 mg of salicylic acid was accurately weighted into a 100.0-ml flask.
2. Methanol (50 ml) was added and mixed to dissolve the salicylic acid.
3. Methanol was added to the mark of the flask. This solution was mixed well.
4. A 2.0-ml aliquot was transferred to a 100.0-ml flask and diluted to volume with eluent.
5. A 5.0-ml aliquot was transferred to a 25.0-ml flask and diluted to volume with eluent. This calibration standard has a concentration of 0.4 μg of salicylic acid per 1.0 ml, if 10.0 mg of salicylic acid was weighted in step 1 (10 mg/100 ml $\times$ 2 ml/100 ml $\times$ 5 ml/25 ml $\times$ 1000 g/mg = 0.4 μg/ml or 0.4 ng/μl).
6. A portion of the salicylic acid standard was transferred into an autoinjector vial and injected into the HPLC under the conditions described above. Note: Linearity was confirmed over the range of interest; single-point calibration was used for calculation purposes; the calibration standard was reinjected after every eight samples to verify system suitability and precision.

(b) Sampling Preparation. (Samples were excised microcomedones in glass vials):

1. A 2.0-ml aliquot of the eluent was transferred into the vial and capped.
2. The vial was shaken vigorously for 30 sec to wet the sample.
3. The microcomedone mixture was sonicated for 1 min.

4. The mixture was swirled occasionally and allowed to stand for at least 10 min.
5. One ml of the contents was poured into a 1-ml tuberculin syringe attached to a Millex filter unit (4 mm) and the filtrate was collected in an autoinjector vial.
6. The contents were injected into the HPLC under the conditions described above.

(c) Chromatographic Conditions

1. Column: Shandon Hypersil ODS, 5u, 15 cm × 4.6 mm I.D. or equivalent.
2. Guard Column: 2 cm × 2 min I.D. packed with *BONDAPAK* C_{18} (37-50U).
3. Column temperature: 25°C.
4. Eluent: methanol/distilled water/phosphoric acid (45:55:0.1, v/v).
5. Flow rate: 1.0 ml/min.
6. Detector: UV at 308 nm, Att: 64 (Lcl-100 at 1.0 V full scale output).
7. Chart speed: 2 mm/min.
8. Injector volume: 200 u.
9. Retention time: 7.5 min (may vary due to column or setup).
10. Analysis time: 10 min.

(d) Calculation

$$\frac{\text{Peak height of sample}}{\text{Peak of calibration}} \times C \times 2 = \mu\text{g of salicylic acid found per vial}$$

where $C = \mu$g of salicylic acid in 1.0 ml of calibration standard (0.4 μg).

The salicylic acid concentration was subjected to an analysis of variance for treatment effect with factors for subject and site. The equality of the salicylic acid concentration for the three treatments was compared using least squares means analyses (Fig. 1). A P-value $\leqslant 0.05$ was used as the criterion for significance.

Significantly more salicylic acid (about three times) was delivered to the follicles treated with a 2% salicylic acid hydroalcoholic solution (clinically effective formulation) after 24 treatments (day 12), than a 2% hydroalcoholic gel product (clinically ineffective formulation). The salicylic acid delivery to the follicle appeared to reach a plateau after 14 treatments on day 7. Sites treated with 2% salicylic acid formulations had significantly more salicylic acid in the follicles after 4 treatments (day 2) in comparison to the no-treatment sites. The no-treatment sites did not show any meaningful level of salicylic acid at day 0 or day 12.

Analysis of human back microcomedones for salicylic acid active content following treatment provides a means to assess the follicular delivery of salicylic acid from salicylic acid formulations. The amount of salicylic acid delivered to the follicle has been demonstrated in this study to be associated with the clinical activity of these formulations. This assay can be used as a screening tool for assessing the relative antiacne efficacy potential of salicylic acid formulations.

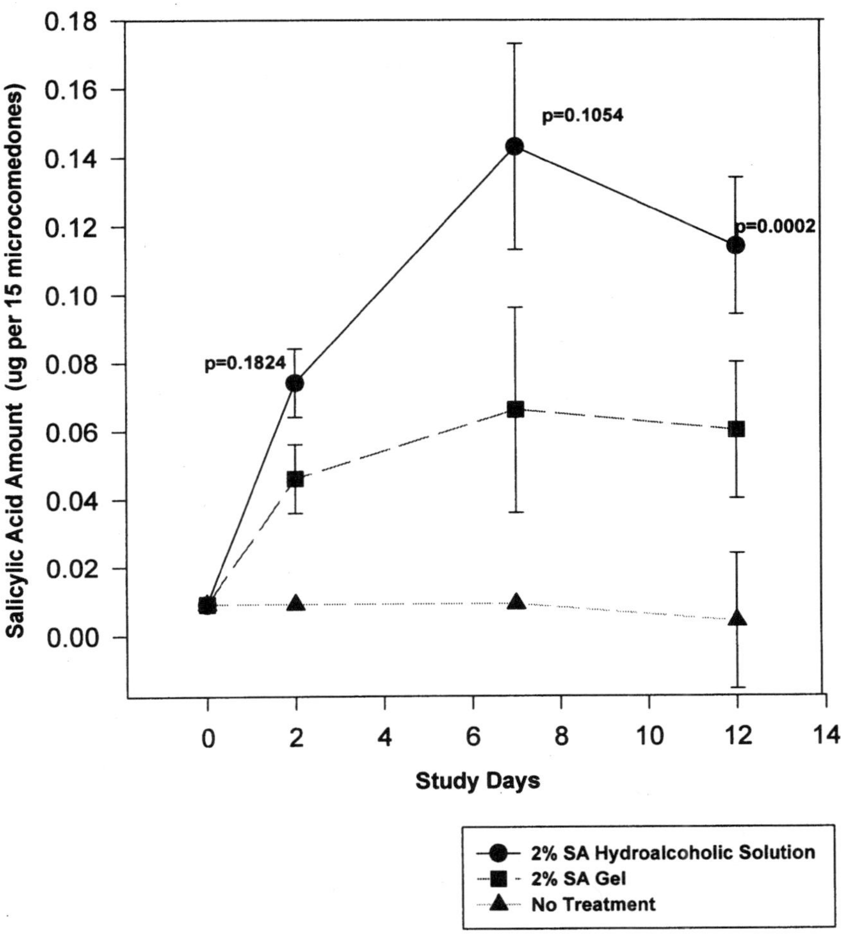

FIGURE 1. Effect of vehicle on follicular delivery of salicylic acid (SA).

4. Conclusion

This chapter has described a line of investigation centering on analysis of drug levels in the microcomedone. From early pilot studies one was able to see correlations between active forms of BPO and ERY and *P. acnes* reductions. In addition, these early trials showed that the method may also be able to identify potential treatment failures. As shown in the BPO study, certain subjects had more BZA than BPO in their follicles. Speculating, one can think of biochemical characteristics or certain resistant bacteria capable of altering the metabolism of the drug.

In the more recent work with this method, looking at salicylic acid active content in the microcomedones, the amount of salicylic acid delivered to the follicle was associated with the clinical activity of the formulations tested.

The ability to quantify active and inactive forms of drugs directly in the microcomedone *in vivo* offers an advance over *in vitro* percutaneous penetration assays for antiacne drugs. Through other pilot studies (unpublished data) it appears this HPLC–microcomedone approach can be used to measure the amounts and metabolites of orally administered antiacne drugs. Also, with this method, it is now possible to determine the active forms of drugs with positive therapeutic histories. Information gathered with this technique should be very helpful to both those studying the pathology of acne as well as those designing new therapies for the disease.

ACNOWLEDGMENTS: Over the years, this line of investigation has been supported by Hill Top Research, Inc., Neutrogena Corporation, and Proctor & Gamble Company.

References

Groh, D. G., Mills, O. H., and Kligman, A. M., 1992, The quantitative assessment cyanoacrylate follicular biopsies by image analysis, *J. Soc. Cosmet. Chem.* **43**:101–112.

Holmes, R., Williams, M., and Cunliffe, W. F., 1972, Pilosebaceous duct obstruction and acne, *Br. J. Dermatol.* **87**:327–332.

Marks, R., and Dawber, R. P. R., 1971, Skin surface biopsy: An improved technique for the examination of the horny layer, *Br. J. Dermatol.* **84**:117–123.

Mills, O. H., 1995, The follicular biopsy, in: *Handbook of Non-Invasive Methods and the Skin* (J. Serum and B. E. Jemec, eds.) CRC Press, Boca Raton, pp. 511–515.

Mills, O. H., and Kligman, A. M., 1980, The follicular biopsy: Methodology and application, *Clin. Res.* **23**:251A.

Mills, O. H., and Kligman, A. M., 1983, The follicular biopsy, *Dermatologica* **167**:57–63.

Mills, O. H., McGinley, K. J., and Kligman, A. M., 1981, The follicular biopsy in the study of acne, Program, 40th Annual Meeting, American Academy of Dermatology, 117, San Francisco.

Mills, O. H., Berger, R. S., Cardin, C., Turner, C., Smiles, K., 1995, Analysis of salicylic acid in the sebaceous follicle. *J. Invest. Dermatol.* **104**:661.

Webster, G. B., and Kligman, A. M., 1979, A new method for the assay of inflammatory mediators in follicular casts, *J. Invest. Dermatol.* **73**:266–268.

Weinstein, M., Mills, O. H., Berger, R. S., Dammers, K., and Baker, M., 1988, Follicular localizations of *Propionibacterium Acnes* by ultraviolet examination, *Clin. Res.* **36**:101.

Williamson, P. E., and Kligman, A. M., 1965, A new method for the quantitative investigation of cutaneous bacteria, *J. Invest. Dermatol.* **45**:498–503.

Wortzman, M., Scott, R., Wong, P., and Mills, O. H., 1984, A quantitative method for analysis of drug levels in the microcomedone, *J. Invest. Dermatol.* **82**:413.

Chapter 27

In Vitro/In Vivo Correlations

ERIC BEYSSAC, J-M. CARDOT, AND C. BONNABRY

1. Introduction

While the main function of the skin is to protect an organism against dehydration and external stress, it is more or less permeable to some chemical substances and may allow the permeation of certain drugs for therapeutic purposes. The observation of professional intoxications challenged by this permeability has triggered research in this field. Depending on the type and level of therapeutic activity sought in the drug and excipients used in the formulation, one finds it possible to class the different dermatological preparations into three groups:

1. Epidermal preparations: drugs are used to treat the epidermis, and only a superficial penetration limited to the epidermis is designed in order to achieve a local action. This type of application is used, for instance, in cosmetology (e.g., sunscreen preparations) and in therapy in the case of antifungal powders.
2. Endodermal preparations: penetration of drugs must be deeper and pass through the stratum corneum and attain the dermis without reaching the general circulation, which, e.g., is the case with anti-inflammatory preparations containing corticoids.
3. Transdermal preparations: in this case, the skin is used as a route of administration for drugs. A local action on the muscle underlying the skin may be achieved. However, as a general rule, a systemic action is achieved after the permeation of drugs in the general circulation: transdermal systems currently represent the most characteristic example. For this case, we will only discuss general bioavailability with respect to the other formulations for which it will only be question of topical availability.

Eric Beyssac, J.-M. Cardot, and C. Bonnabry • Biopharmaceutic Department, Faculty of Pharmacy, Auvergne University, 63000 Clermont-Ferrand, France.

Cutaneous Biometrics, edited by Doris A. Schwindt and Howard I. Maibach, Kluwer Academic/Plenum Publishers, New York, 2000.

It is therefore obvious that the *in vitro* or *in vivo* study of each one of these preparations will be specific for both the methods used and the interpretation of results. Therefore, correlations, or more exactly, *in vitro/in vivo* relationships, will also be specific.

2. Topical Availability or Bioavailability

Assessing the bioavailability of topical formulations is a complex process since many factors are involved. First, the drug must be released from the preparation, and this process may involve various steps. Then, it must reach the target tissue either in one of the skin levels or far from the site of drug administration on the surface of the skin. The process of permeation of molecules though the different layers of the skin involves a passive diffusion mechanism, with the following limiting steps and factors:

1. Release and diffusion of the drug from the vehicle of the dosage form (where it is dissolved or made soluble) up to the surface of the skin. This step depends on the excipients present in the formulation, on the solubility of the drug, and finally on the partition coefficient of the drug, between the excipient and the stratum corneum (Zatz, 1983).
2. Diffusion and penetration of the drug through the different layers of the skin (stratum corneum, epidermis, dermis and endoderm).
3. Probable permeation (absorption) of the drug from cutaneous structures into the blood and lymph circulation. Depending on the physicochemical characteristics of the drug, once it is in the viable epidermis, it diffuses to the dermis before reaching the systemic circulation.

Factors influencing or limiting this last step are linked to the drug and excipients on one hand and to the skin on the other. The drug intervenes by its solubility in the vehicle of the preparation or in the components of the skin (sebum, intercellular or internal medium of cells). The stratum corneum, owing to its structure and composition, constitutes a limiting factor on the diffusion, the penetration level, and the intensity of action of the drug. Live epidermis and dermis act as an aqueous gel and do not have a significant barrier function (Baker, 1979; ECETOC, 1993; Kligman, 1983; Zatz, 1983).

The skin is also the center of numerous metabolism phenomena (20% of the amount of nitrous derivative reaching the dermis are metabolized). It is an essential immunological organ (sensitization with clonidine in 30% of cases with reactions of inflammation), it plays an endocrine role and participates in the metabolism of certain hormones and vitamin D (Ademola and Meibach, 1995; Guy *et al.*, 1987).

Finally, skin plays the role of a reservoir for numerous substances that are stored before being slowly released, with delay, as e.g., corticoids, some antibiotics, and even scopolamine (Kligman, 1983; Katz, 1983).

3. In Vitro and In Vivo Methods for Studying Drug Release: Diffusion and Permeation of Drugs from Dermatological Preparations

3.1. In Vitro Study Methods

The aim of these studies was twofold: highlighting the release of the drug from the preparation (used as a quality control test in order to ensure the compliance and the reproducibility of the dosage form) and visualization of the availability of the drug or formulation to pass through a membrane (natural or synthetic), thus simulating skin. In all cases, this assessment depends on quantification of the drug dissolved in an appropriate solvent.

(a) Release of the Drug. Methodologies used in this case depend mainly on the composition of the vehicle. If it is not miscible with the solvent, the preparation is directly in contact with it and the drug is directly assessed in this solvent. If the drug is miscible, it will be necessary to separate the two parts by a membrane as permeable as possible, in order to avoid an additional error factor. The Poulsen model with isopropyl myristate as the solvent (for the study of corticosteroids) or the Yotsuyanagi model using a cellophane membrane are good examples of these approaches (Poulsen, 1970; Poulsen *et al.*, 1986; Paire, 1977). Obviously, these models are only useful for epidermal and endodermal preparations, since, in order to actually be therapeutic, the drug that is administered has to be maintained in these cutaneous layers on the strength of its own physicochemical properties.

The *in vitro* release study of the drug from cutaneous dosage forms could be performed using the methods described in the Pharmacopoeas (EP, 1997, USP, 1995) in the section "Dissolution test for transdermal patches." These techniques usually apply to transdermal systems, but the cell method can be used for overall cutaneous dosage forms.

The different types of apparatus used are the following:

1. The disk assembly method or paddle over disk (USP, 1995, and EP, 1997): A stainless steel disk assembly in the shape of a net designed for holding the transdermal system is used with the paddle and the vessel assembly from the paddle apparatus (Fig. 1). A distance of 25 ± 2 mm is maintained between the paddle blade and the surface of the disk assembly during the test. The disk assembly holds the system flat, with the release surface parallel to the bottom of the paddle blade. This system is suitable only for transdermal patches that can be cut (matrix systems) because the patch must not overlap the borders of the disk assembly and, in this case, an appropriate measured piece of patch must be cut. Owing to the aperture of the net (125 mm), this method cannot be applied to the various cutaneous formulations.

2. The cell method (EP, 1997): An extraction cell, made of chemically inert materials, consisting of a support, a cover, and if necessary a membrane, is used with the paddle and vessel assembly from the paddle apparatus (Fig. 2). The central part of the support forms a cavity with a depth of 2.6 mm and a defined diameter. The cover has a central opening with a diameter selected according to the support used or the size of the patch to be examined. A membrane can be inserted between the

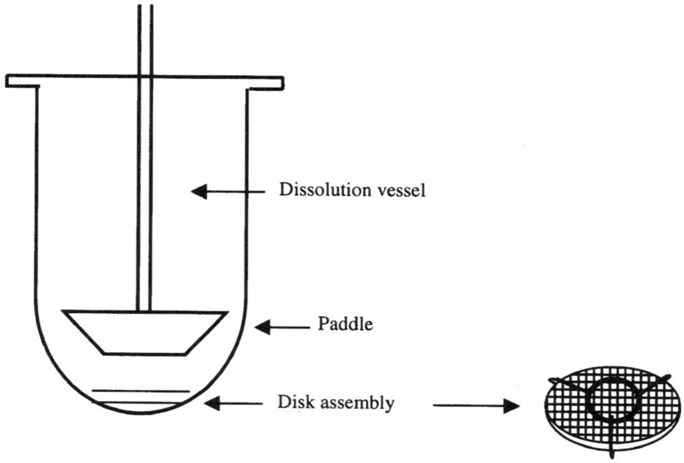

FIGURE 1. Paddle over disk.

support, and the cover is held in place by screws. The extraction cell is introduced flat into the bottom of the vessel and a distance of 25 ± 2 mm is maintained between the paddle blade and the surface of the disk assembly during the test. This cell method can be applied to all cutaneous formulations, whatever the device.

In the case of a cream, ointment, or gel, a membrane is used to isolate the dosage form from the medium.

3. The rotating cylinder method (USP, 1995; EP, 1997): a stainless steel cylinder stirring element replaces the paddle and is used with the vessel from the paddle method (Fig. 3). The patch is placed on the cylinder at the beginning of the test with a suitable adhesive, so that the release surface is in contact with the dissolution medium. The distance between the bottom of the vessel and the cylinder is

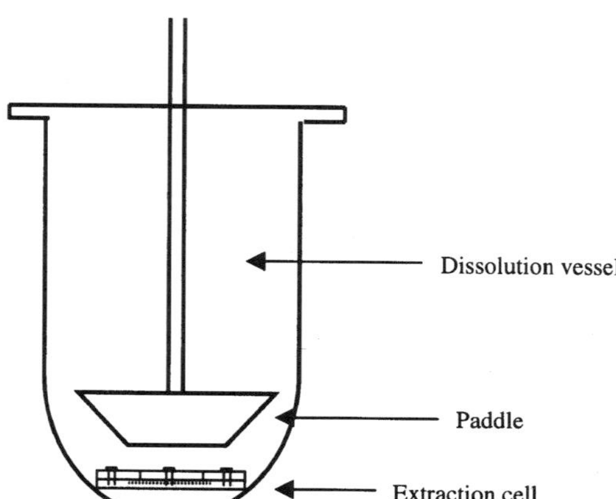

FIGURE 2. The cell method.

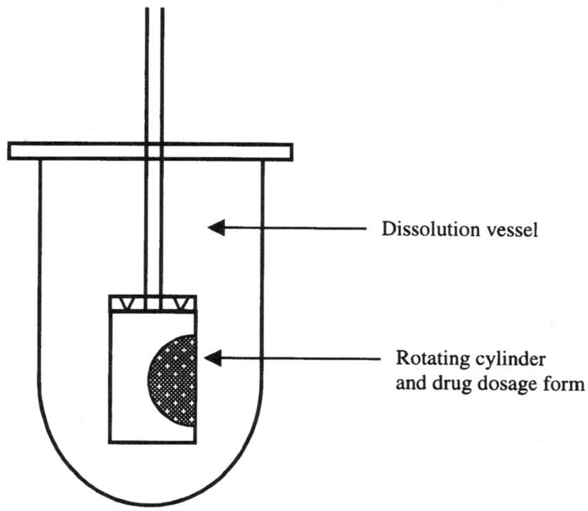

FIGURE 3. The rotating cylinder method.

maintained at 25 ± 2 mm during the test. This method is suitable for all types and shapes of patches (matrix system, membrane system) and the surface in its entirety is in contact with the dissolution medium, but with some matrix patches there can be erosion of the matrix (Aiache *et al.*, 1989; Cappel and Kreuter, 1991).

Whichever method is used, samples are withdrawn at predetermined intervals and drug release is assayed by a suitable analytical technique. These studies allow the determination of the percentage or amount released per surface unit as a function of time, i.e., the flow of the drug, which is the essential parameter of transdermal systems.

(b) Release and Diffusion of the Drug. Studies on drug release and diffusion are essentially research, development, and formulation trials; they are rarely quality control tests, since they are particularly difficult to perform and the experimental protocol is not yet completely standardized. There are many different diffusion cells in use for such studies, but as a general rule, all apparatuses, vertical or horizontal, are composed of two compartments: the donor compartment and the receptor compartment, separated from each other by a membrane. This membrane is either representative of the skin or a model of skin, in order to mimic what may happen *in vivo*, after the application of a preparation (Friend, 1992).

A good *in vitro* diffusion cell can be defined by its ability to discriminate the permeation of drug from very similar donor formulations with *in vivo* differences. These diffusion studies not only allow the assessment and quantification of the permeation of the drug through a "skin," but also enable study of the penetration level through the different layers of this skin and the rate of such penetration. Generally, studies performed on the dosage form are always preceded by a study of the pure drug, in solution or in suspension, in order to determine its own capability for permeation and/or absorption through the membrane.

The cells currently available can be classified into two different types, namely static and flowthrough, and are generally based on the original Franz static cell model. Only two of them, which reflect the basic principles, will be described.

The Franz cell (1975), designed in 1975–1978, is vertical and has a small donor compartment and a receptor chamber. The lower portion communicates with the upper portion of the chamber by a narrow cylindrical tube that widens near the area in contact with the membrane. The temperature of the solution in the receptor chamber is maintained by circulating water at a appropriate temperature through a jacket that surrounds the chamber. Portions of the receptor chamber (sampling port) and the donor compartment (top) are open to room temperature. As most of the vertical systems, the receptor medium is stirred by a Teflon-covered magnet. Several important modifications have been introduced to the system: rings, a second exit in the donor compartment in order to allow circulation of fluids and to transform this static cell into a dynamic cell, and a donor compartment that can be sealed according to a broad range of diameters.

The modified Franz diffusion cell or Keschary and Chien cell (1984) is another type of vertical cell, which was developed in 1990, in response to observations concerning the original Franz cell (Fig. 4). The lower compartment, straight and cylindrical, is entirely surrounded by a waterbath. The two compartments are separated by a membrane. The dermal compartment (receptor) is continuously renewed, completely or partially. Numerous parameters influence the *in vivo* diffusion of the drug, mainly:

1. The donor compartment and in particular the temperature and stirring rate.
2. The donor medium more or less complex depending on the membrane used; it should allow maintainence of the potency of the membranes and vitality

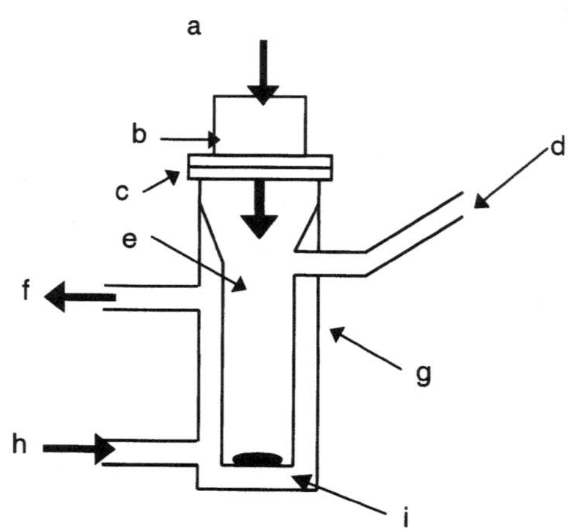

FIGURE 4. Modified Franz diffusion cell: (a) open to air, (b) donor cell cap, (c) membrane, (d) media injection port, (e) receptor chamber, (f) water bath out, (g) water jacket, (h) 37°C water bath in, (i) stirring bar.

in the case of a living tissue, but should also allow for solubilization of the drug if it is absorbed.

3. Type and nature of the membrane used.

At present, different types of membranes are being used (Bonina and Montenegro, 1992; Bronaugh and Stewart, 1986; Bronaugh and Franz 1986), such as excised animal skins, full thickness excised or dermatomed human skin, artificial membranes, and more recently reconstituted human epidermis.

According to the FDA and AAPS guidelines (Skelly *et al.* 1987), human skin should be used for *in vitro* percutaneous absorption studies, and it can be in the form of fine dermatomed sections (< 0.5 mm thickness) or epidermal sections isolated by heating or other methods.

Faced with the problems presented by the recuperation and the variability of the human skin, many studies (Bonina and Montenegro, 1992; Grissom *et al.*, 1987) have been done using animal skin models (nude mouse, rat, guinea pig, rabbit) and bovine udder. Apart from work with these animal models, studies were done using snake skin, owing to its similarity to human stratum corneum in terms of structure, composition, lipids, and water permeability (Itoh *et al.*, 1990).

A new approach to the study of cutaneous absorption involves using cell cultures as the diffusion cell membrane (Slivka *et al.*, 1993). Cutaneous substitutes correspond only to one or two layers of skin; epidermis, epidermis and dermis, dermis. Spheres of application of reconstituted epidermis are *in vivo* as replacement (or covering) materials for healing and *in vitro*, as a sophisticated, very differentiated dermal model or dermoepidermal model. This reconstituted epidermis is essentially composed of keratinocytes of human origin and fibroblasts in some cases, cultivated at the air–liquid interface on a support. After culture, the tissue culture architecture is near to the one found in the stratum corneum of the epidermis *in vivo*: germinative, prickel, granular layers and the stratum corneum, composed of 10 layers of compact cells (Rosdy *et al.*, 1993). Its production is reproducible and standardized and limits the problems of variability linked to the use of live tissue.

In order to find alternative experimental models and to eliminate disadvantages linked to the use of live tissues, different synthetic membranes, simple homogeneous one-layer systems, were used to study the cutaneous penetration of products with topical application (Behl *et al.*,1990; Nastruzzi *et al.*, 1993; Neurbert *et al.*, 1991). In order to separate percutaneous absorption from subsequent distribution, biotransformation, and excretion processes, the use of artificial membranes comparable to human and animal skin can be used. The main artificial membranes are grouped together in Table 1.

Since absorption is a passive diffusion process, penetration flows depend on the thickness of the barrier and of the diffusion constant of the drug. However the type of formulation and the release of the product play an important role in the diffusion. Preparation of the *in vitro* barrier membrane must take into account the thickness of the *in vivo* barrier. The barrier layer of human skin can be estimated at about 200 μm. This is about the same distance that a substance must traverse before reaching the dermal microcirculation. Human skin coming from an anatomical site close to the site of clinical application and prepared with a dermatome adjusted at

TABLE 1. Artificial Membranes Used Most Widely in Permeation Studies

Silastic® (polydimethylsiloxane)
Cellulose acetate membrane
Polyurethane membrane
Supor® (modified polysulfone)
Zeolite (aluminosilicates)
Multimembrane system (cellulose acetate : Silastic® cellulose acetate)
Diaflo-ultrafiltration membrane
Miscellaneous: collagen, egg shell
Liquid membranes (organic fluids, isopropyl myristate soaked on filter membranes)

200 μm seems to be the ideal membrane for percutaneous absorption studies (Hadgraft *et al.*, 1990).

Although human skin remains the most reliable way to study *in vitro* skin diffusion, a fairly good *in vitro/in vivo* correlation has been found for a number of animal models described in the literature (ECETOC, 1993; Grissom *et al.*, 1987). An animal model can be used for screening a series of compounds, a series of penetration enhancer principals, or a series of formulations, if one is certain that the route of penetration and the skin barrier for both the compound and excipients used are similar in animal and human skin.

The results obtained from diffusion studies allow the quantification of the *in vitro* permeation of the drug and the transfer rate, the presence or absence of a lag time, and the transfer flow.

(c) Other Types of In Vitro Techniques to Measure Drug Penetration. The use of radiolabeled molecules and the measurement of radioactivity in the receptor compartment as well as in the different layers of the skin used in the cell allows the assessment of the degree of permeation of the drug and the reservoir effect on the horny layer. This type of study can be completed with the use of a fluorescent tracer, whose behavior is similar to that of the drug and allows the visualization of the *in vitro* absorption rates of the drug by laser scanning confocal microscopy.

3.2. In Vivo Methods

In vivo studies are performed in human or intact animals or animal grafted with human skin (grafted nude rats). Measurements can be classified either by the nature of the preparation or by the method of measurement.

(a) Direct or Indirect Methods. In direct methods the effect of the preparation is assessed directly and is quantifiable. This category includes all methods in which drug concentrations are evaluated by cutaneous assay in tissues, in an organ, or in the blood. (1) In the stripping method: after drug application and skin surface washing, a strip of adhesive tape is placed on each application area to determine the amount of drug in the stratum corneum and drug absorption can thus be evaluated (Dupuis *et al.*, 1984; Tojo and Lee, 1989). (2) Determination of the plasma levels after drug administration allows quantification of the permeation of the molecule. (3) In

animals, the distribution in the cutaneous tissues may be determined by histologic methods such as biopsies, microdialysis, use of radiolabeled elements and measurement of the radioactivity in histologic sections.

The indirect methods include the following: (1) In some cases, the residual amount of drug deposited on the skin after a determined time of application is assessed in order to highlight nonpenetration. (2) The evaluation of a preparation can be performed by measuring the pharmacological or pharmacodynamical effect on the dermis. This type of measurement is widely used in the case of noninflammatory corticoids, where the vasoconstriction provoked by these drugs and the resulting blanching is correlated to the quantity of the drug present in the skin.

(b) Type of Preparation. Epidermal preparations can be dealt with by both indirect and direct measurement methods. Therefore, for preparations containing sunscreen substances intended to protect the skin from sun exposure, the *in vivo* method will be an assessment after exposure to UV radiation. The method involves the application on a healthy human or animal skin of this type of preparation, followed by an exposure delivering a previously fixed dose of radiation. The protective activity is determined indirectly by visual comparison with an area near the non treated skin or treated by a placebo or a reference product. The value and duration of the protection obviously depend on the drug used but also on its ability to remain on the skin surface and its release rate from the excipient.

The skin penetration profile of UV filters has been studied *in vitro* and *in vivo*. The *in vitro* penetration measurements were performed with static diffusion Franz cells. After experimentation on human skin, biopsy punches were performed and UV filter was extracted from the epidermis and dermis compartment. *In vivo*, horny layer concentrations were measured directly (after stripping) and the sun protection factor (SPF) was evaluated. Significant *in vivo* and *in vitro* differences between the vehicles were observed. The *in vivo* findings (stratum corneum and SPF values) are in agreement with the *in vitro* findings (Treffel and Gabard, 1996)

Highlighting *in vivo* activity for endodermal preparations is slightly more delicate, as their site is located in the deep epidermis or endoderm. Measurement will often be more indirect. The most common example is the one of dosage forms containing corticoids. MacKenzie's blanching test is the universally accepted method, but involves a controversy in regard to the assessment method used. It can be visual by judges or by automatic colorimeters or by various other systems described in various publications (FDA, 1995). The experimental protocol is well established. Drugs or reference are applied double-blinded on the flexor aspect of the forearm in different arrangements, and each application is chosen randomly. Defined samples of drugs are applied on 12-mm-diameter sites on each forearm. Occlusion is generally attained by application of chambers (inox) and adhesive labels. The residual cream, adhesive label, and chamber are all removed after application (6 hr). Assessment of the blanching responses is performed independently by three observers using a 0–4 scale or by means of a colorimeter (Haigh and Kanfer, 1984).

In vivo assessment of transdermal preparations depends on direct methods such as those used for the other routes of administration, i.e., determination of drug levels in the blood or in certain tissues after application of the therapeutic system. Blood level curves present a profile similar to that obtained after use of conventional routes of administration for sustained release drug dosage forms. Values of plasma and tissue levels are generally lower since they are a function of the quantity that passes through the different layers of the skin, the application surface, and the degree of skin hydration.

4. In Vivo/In Vitro Correlations or Relationships

Mathematically, the term correlation means the interdependence of two variables. In the case of *in vivo/in vitro* correlation, it means establishing a relationship between a parameter determined from any *in vitro* assessment method and one obtained after *in vivo* experimentation. Depending on the data, the type of correlation obtained may be a simple relationship between two parameters or a true correlation defined from a mathematical point of view by a function (Cardot and Beyssac, 1993).

For dosage forms designed to have a systemic action and in particular in the case of oral forms, the term *in vitro/in vivo* correlation applies to study the relationship between the *in vitro* release (or a parameter derived from these data) and the parameters observed in biological fluids. *In vitro/in vivo* correlations are defined in the USP as well as the methods for their determination: the correlations are classified as follows:

1. Level A: Relationship between *in vitro* dissolution and the input rate into the organism.
2. Level B: Relationship based on statistic moments.
3. Level C: Point-to-point relationship between a dissolution parameter and a pharmaceutical parameter.

The *in vitro* data are generally provided by the release kinetics performed according to numerous methods described. The *in vivo* data are derived from a series of plasma samplings carried out at various times. These *in vitro/in vivo* correlations are reasonable when drug release is the limiting factor for drug input into the general circulation, i.e., in the case of controlled release formulations (Aiache *et al.*, 1997; Cardot and Beyssac, 1993).

For cutaneous preparations, it is not possible to limit the *in vitro/in vivo* correlations of this type of testing, requiring dosage in biological media. For dosage forms intended to be superficial treatments or designed for the internal skin layers, the drug is neither absorbed nor found in the general circulation. However, other types of studies and relations might be sought.

For dermal and transdermal dosage forms, *in vitro/in vivo* relationships may be classified into two grades, depending on the degree of drug penetration into the skin, i.e., its locale or general action: (1) Grade 1 is the study of the relationships between *in vitro* release and diffusion and penetration in the different skin structures. (2) Grade 2 is the relationship between *in vitro* release and systemic concentrations.

4.1. Relationships between In Vitro Release and Diffusion and In Vivo Penetration

In research and development, it becomes more and more important to have access to experimental models in the laboratory, which allow for the prediction of the dermal and transdermal transport in humans after exposure to xenobiotics. These relationships are used for dosage forms and drugs that will have a local action and for the study of penetration of substances intended to have a systemic action.

Comparison of the *in vitro/in vivo* diffusion results allows the verification of the validity of the proposed *in vitro* model and its use in predicting the behavior of the drug or of a pharmaceutical dosage form. This type of approach is essential in order to handle a specific galenic problem, such as the use of absorption promoters, testing of the qualitative and quantitative influence of different excipients, or research on the release of drugs.

For endodermal preparations, the diffusion measured *in vitro* is correlated to indirect measurement such as the pharmacological effect observed *in vivo*. This type of study allows the classification of formulations according to the vehicles and excipients used and the comparison of the *in vitro* and *in vivo* rank obtained. Thus for preparations containing corticoids, a relationship has been developed between *in vivo* skin blanching and *in vitro* release rate. Both the blanching assay and *in vitro* experiments measure the rate and extent of corticosteroid absorption by the skin. It is theoretically possible to correlate findings from both experimental studies, i.e., the rank order *in vitro* release rate with the blanching score of the *in vivo* study. A number of comparative experiments have been conducted using bethametasone-17 valerate, bethametasone-17 benzoate and hydrocortisone (Shah *et al.*, 1992). In most cases, the *in vitro* test was carried out using a Franz cell with artificial membranes or animal (pig, mouse) or human skin.

The rank order correlation of pharmacodynamic data with the *in vitro* release rate provides additional validity of the *in vitro* quality control test. Direct quantitative comparison of results is not possible as the blanching assay generally makes use of a subjective assessment method. However, if the *in vitro* results show that a specific membrane (animal or artificial) adequately simulates the *in vivo* permeation of a drug for a particular class of topical dosage form, then the use of this membrane can be used to validate comparative findings for testing other products—in the same formulation class and containing the same corticosteroid.

Correlation obtained between the *in vitro* release and the *in vivo* performance provides additional validity for the *in vitro* quality control test. In this case, the term relationship must be used rather than correlation since the diffusion phenomenon measured *in vitro* is related to that observed *in vivo*. This validation of the *in vitro* model allows its further use in order to initiate the design of a pharmaceutical form or to compare similar types of products.

The use of *in vitro* diffusion cells has highlighted the numerous *in vitro/in vivo* relationships from the diffusion rate measured *in vitro* on the one hand and cutaneous permeation *in vivo* rate on the other. These relationships, assessed by the relation between *in vivo* absorption data in human beings or animals and the *in vitro* data, show, the existence of a relationship between the data for numerous drugs. Various reviews of the literature survey the assessment of tests performed and the *in vitro/in vivo* relationships established. Most of the time, these relationships have been realized with the Franz cell and different *in vivo* animal models. This ability of the *in vitro* method to reflect *in vivo* absorption was extensively studied for a range of chemicals with different physicochemical properties (Bronaugh and Franz, 1986; ECETOC, 1993; Fujimara *et al.*, 1996).

If the cells used are appropriate for studying the diffusion of preparations as solutions, aqueous or oily suspensions, gels, or creams, their application to transdermal systems is limited. In the transdermal field, system-control input has been achieved by both membrane and matrix devices. Diffusion cells, in view of their size, may not be used with transdermal system reservoirs. In addition, drug release from transdermal systems is widely influenced by the absolute value of cutaneous permeability, which is not always found *in vitro*. This type of approach does not allow the determination of the drug after penetration, if no further biological fluid sampling is performed.

4.2. Relationship between In Vitro Release and Systemic Occurrence of the Product

Unlike the previous approach, the relationship between *in vitro* release and systemic occurrence of the product concerns only transdermal forms with a systemic action. The aim is not only to implement efficient control of production, but primarily to optimize development of the transdermal form (Hadgraft et al., 1993).

Here, the *in vivo* studies are normally a series of biological fluid samplings carried out at various times. From these data, the common pharmacokinetics parameters C_{max}, T_{max}, equilibrium concentration, AUC are determined as well as the input kinetics of the drug in the general circulation (percent of the fraction of the dose absorbed as function of time) by different mathematical methods, such as deconvolution (Aiache *et al.*, 1997; Cardot and Beyssac, 1993).

Numerous authors also study, as an additional parameter, the amount of drug remaining in the transdermal system after application. The transdermal system releases only a fraction of the drug and the apparent dose is calculated as the difference between the transdermal device initial potency and final potency. This allows the calculation of the true amount of drug delivered per unit time from the absorption kinetics expressed in percentage (Banerjee and Robinson, 1991; Catterall, 1997; Friend, 1992).

There are three possible correlations (Van Buskirk *et al.*, 1997):

1. *In vitro* release/*in vivo* skin permeation: the results obtained allow the quantification of the rate of transfer to assess the reservoir effect of the horny layer and to define release parameters of the system according to the cutaneous permeation rate.
2. *In vitro* permeation/*in vivo* bioavailability: comparison of the *in vitro* dissolution results with the *in vivo* diffusion results and plasma levels allows the determination of the relative contributions of the system and the skin in the total absorption of the drug; however, this type of study is limited in view of the restrictions previously mentioned concerning the *in vitro* diffusion measurements with transdermal systems.

3. *In vitro* release/*in vivo* bioavailability: correlation between *in vitro* dissolution kinetics and drug input kinetics in the general circulation. The aim of the study is the search for different correlation levels (A, B, or C); it is obvious that one learns the most from level A relationships.

The mechanism of action of transdermal devices is based on diffusion. Skin acts as a homogeneous membrane; flow of the active drug should be continuous; skin surface concentration should therefore be constant, depending on the steady state resulting from drug release from the system and cutaneous permeation (Guy, 1996).

If release from the system is greater than absorption, the superficial concentration attains a saturation level. Permeation is constant but is a function of the cutaneous permeability. This is the case for certain types of nitroglycerine transdermal matrix systems: the skin regulates the permeation of the drug at 87% and the system at only 13%. The rate of release of the drug from the transdermal system must be far below the saturation level in order to ensure the permeation of a dose independent of the conditions of permeability. Most of the substance contained in this type of system serves as an internal reservoir, in order to maintain a concentration gradient sufficient to allow transcutaneous permeation, thus supplying the reservoir located in the horny layer.

Whatever the galenic form of the device (matrix or reservoir), the amount, and the *in vitro* release rate, blood level curves will show the same profile since the skin controls the penetration (Boyd *et al.*, 1996; Hadgraft *et al.*, 1993). Under these conditions, correlation between *in vitro* and *in vivo* data is quite difficult since release is not the main factor limiting the drug input into the general circulation.

Different studies have been published on *in vitro/in vivo* relationships for the drugs commercially available at present for the transdermal route. Results are variable according to the type of system, the particular drug, and *in vitro* study method used. These relationships cannot always be considered as level A.

5. Conclusion

The biopharmaceutical assessment of cutaneous dosage forms by the means of *in vitro* experiments is a crucial step during development. This type of study enables an understanding and an evaluation of the way that different dosage forms diffuse through the skin. The facility and rapidity with which *in vitro* cutaneous absorption experiments have been conducted recently are indicative of the current interest of this type of study.

The comparison between *in vitro* and *in vivo* diffusion results allows verification of the validity of the *in vitro* model and its utility in predicting the behavior of the drug or a pharmaceutical dosage form. This type of approach is essential in order to handle a specific galenic problem, such as the use of absorption promoters, testing the qualitative and quantitative influence of different excipients, or technologies on drug permeation.

References

Ademola, J., and Meibach, H., 1995, Cutaneous metabolism and penetration of methoxypsoralen, betamethasone, retinoic acid, nitroglycerin and theophyllin, *Current Prob. Dermatol.* **22**:201–213.

Aiache, J.-M., Cardot, J.-M., and Aiache, S., 1989, Comparative study of release kinetics of trinitrine from TTS, *Int. J. Pharm.* **55**:147–155.

Aiache, J.-M., Beyssac, E., and Cardot, J.-M., 1997, Les corrélations *in vitro/in vivo*, in: *Traité de biopharmacie et de pharmacocinétique*, PUM and Lavoisier, Montreal and Paris.

Baker, H., 1979, The skin as a barrier, in: *Textbook of Dermatology* (A. Rook, D. S. Wilkinson, and F. S. G. Eblin), Blackwell, Oxford.

Banerjee, P., and Robinson, J., 1991, Novel drug delivery systems, *Clin. Pharmacokinet.* **20**:1–14.

Behl, C., Kumar, S., Malick, A. W., Patel, S. B., Char, H., and Piemontese, D., 1990, Choice of membranes for *in vitro* skin uptake studies and general experimental techniques, in: *Methods for Skin Absorption.* (B. W. Kemppainen, and W. G. Reifenraft, eds.), CRC Press, Boca Raton.

Bonina, F. P., and Montenegro, L. 1992, Penetration enhancer effects on *in vitro* percutaneous absorption of heparin sodium salt, *Int. J. Pharm.* **82**:171–177.

Boyd, R., Yang, B., Abel, R., Eldon, M., Sedman, A.,and Forgues, S., 1996, Pharmacokinetics of a 7 day 17β estradiol transdermal delivery system: Effect of application site and repeated applications on serum concentrations of oestradiol and estrone, *J. Clin. Pharmacol.* **36**:998–1005.

Bronaugh, R. L., and Stewart, R. F., 1986, Methods for *in vitro* percutaneous absorption studies: VI. Preparation of the barrier layer, *J. Pharm. Sci.* **75**:487–491.

Bronaugh, R. L., and Franz, T. J., 1986, Vehicle effects on percutaneous absorption: *in vivo* and *in vitro* comparisons with human skin, *Br. J. Dermatol.* **115**:1–2.

Cappel, M. J., and Kreuter, J., 1991, Effect of nonionic surfactants on transdermal drug delivery: II. Poloxamer and poloxamine surfactants, *Int. J. Pharm.* **69**:155–167.

Cardot, J. M., and Beyssac, E., 1993, *In vitro/in vivo* correlations: Scientific implications and standardisation, *Eur. J. Drug Meth. Pharmacokinet.* **18**:113–120.

Caterall, R. A., 1997, Problems of sweating and transdermal fentanyl, *Palliative Med.* **11**:169–170.

Dupuis, D., Rougier, A., Roguet, R., Lotte, C., and Kalopissis, G., 1984, *In vivo* relationship between horny layer reservoir effect and percutaneous absorption in human and rat, *J. Invest. Dermatol.* **82**:353–356.

ECETOC (European Centre for Ecotoxicology and Toxicology of Chemicals), 1993, Percutaneous absorption, *Monograph No. 20.*

EP: European Pharmacopoeia, 1997, Dissolution test for transdermal patches, 3rd Ed., Council of Europe, Strasbourg, pp. 131–133.

FDA Guidance, 1995, Topical dermatologic corticosteroids: *In vivo* bioequivalence, Rockville, MD.

Franz, T. J., 1975, *In vitro* vs. *in vivo* percutaneous absorption: The relevance of *in vitro* data, *J. Invest. Dermatol.* **64**:190–195.

Friend, D. R., 1992, *In vitro* skin permeation techniques, *J. Control Rel.* **18**:235–248.

Fujimara, A., Sasaki, M., Harada, K., Kamugai, Y., Ohashi, K., and Ebihara, A., 1996, Influence of bathing and hot weather on the pharmacokinetics of new transdermal clonidine, M-5041-T, *J. Pharm. Clin.* **36**:892–896.

Grissom, R. E., Brownie, C., and Guthrie, F. E., 1987, *In vivo* and *in vitro* dermal penetration of lipophilic and hydrophilic pesticides in mice, *Bull. Environ. Contam. Toxicol.* **38**:917.

Guy, R. H., 1996, Current status and future prospects of transdermal drug delivery, *Pharm. Res.* **13**:1765–1769.

Guy, R. R., Hadgraft, J., and Buck, D., 1987, Transdermal drug delivery and cutaneous metabolism, *Xenobiotica* **17**:325–343.

Hadgraft, J., Wolff, M, Bonn, R., and Cordes, G., 1990, Development of biphasic transdermal nitroglycerin delivery systems *Int. J. Pharm.* **64**:187–194.

Hadgraft, J., Beutner, D., and Wolff, H., 1993, *In vivo-in vitro* comparisons in the transdermal delivery of nitroglycerin, *Int. J. Pharm.* **89**:R1–R4.

Haigh, J. M., and Kanfer, I., 1984, Assessment of topical corticosteroid preparations: The human skin blanching assay, *Int. J. Pharm.* **19**:245–262.

Itoh, T., Magavi, R., Casady, R. L., Nishihata, T., and Rytting, J. H., 1990, A method to predict the percutaneous permeability of various compounds: shed snake skin as a model membrane, *Pharm. Res.* **7**(12):1302–1306.

Keshary, P. R., and Chien, Y. W., 1984, Mechanisms of controlled nitroglycerin administration (I): Development of a finite-dosing skin permeation system and transport of model compound, *Drug Dev. Ind. Pharm.* **10**:883–913.

Kligman, A. M., 1983, A biological brief on percutaneous absorption. *Drug Dev. Ind. Pharm.* **9**:521.

Nastruzzi, C., Esposito, E., Pastesini, C., Gambari, R., and Menegatti, E., 1993, Comparative study on the release kinetics of methylnicotinate from topic formulation. *Int. J. Pharm.* **90**:43–50.

Neurbert, R., Bendas, C., Wohlrab, W., Gienau, B., and Furst, W., 1991, A multilayer membrane system for modelling drug penetration into skin, *Int. J. Pharm.* **75**:89–94.

Paire, M., 1977, Contribution à l'évaluation biopharmaceutique des préparations à usage dermatologique, thèse de Doctorat en Pharmacie, Clermont-Ferrand, France.

Poulsen, B. J., 1970, The use of models in estimating vehicle effects on the activity of topical corticosteroid formulations. *Br. J. Dermatol.* **82** (Suppl. 6):49–52.

Poulsen, B. J., Young, E., Coquilla, V., and Katz, M., 1986, Effect of topical vehicle composition on the *in vitro* release of fluocinolone acetonide and its acetate ester, *J. Pharm. Sci.* **57**:928–933.

Rosdy, M., Pisani, A., and Ortonne, J.-P., 1993, Production of basement membrane components by a reconstructed epidermis cultured in the absence of serum and dermal factors, *Br. J. Dermatol.* **129**:227–234

Shah, V. P., Elkins, J., and Skelly, J. P., 1992, Relationship between *in vivo* skin blanching and *in vitro* release rate for betamethasone valerate creams, *J. Pharm. Sci.* **81**(1):104–106

Skelly, J. P., Shah, V. P., Maibach, H. I., Guy, R. H., Wester, R. C., Flynn, G., and Yacobi, A., 1987, FDA and AAPS report of the workshop on principles and practises of *in vitro* percutaneous penetration studies: Relevance to bioavailability and bioequivalence, *Pharm. Res.* **4**(3):265–267.

Slivka, S. R., Landeen, L. K., Zeigler, F., Zimber, M., and Bartel, R. L., 1993, Characterisation, barrier function, and drug metabolism of an *in vitro* model, *J. Invest. Dermatol.* **100**:40–46.

Tojo, K., and Lee, A. C., 1989, A method for predicting steady-state rate of skin penetration *in vivo*; *J. Invest. Dermatol.* **92**:105.

Treffel, P., and Gabard, B., 1996, Skin penetration and sun protection factor of ultra violet filters from two vehicles, *Pharm. Res.* **13**:770–774.

USP: *United States Pharmacopia*, 1995, USP 23, 1995, Drug release ⟨724⟩, 1796–1799.

Van Buskirk, G. A., Gonzales, M. A., Sha, V., Barnhardt, S., Barrett, C., Berge, S., Cleary, G., Chan, K., Flynn, G., Foster, T., Gale, R., Garrison, R., Gochnour, S., Gotto, A., Govil, S., Gray, V. A., Hammar, J., Harder, S., Hoiberg, C., Hussain, A., Karp, C., Llanos, H., Mantelle, J., Noonan, P., Swanson, D., and Zerbe, H., 1997, Scale-up of adhesive transdermal drug delivery systems, *Pharm. Res.* **14**(7):848–852.

Zatz, J. L., 1983, Factors in percutaneous absorption, in: *Pharmacokinetics and Topically Applied Cosmetics* (J. McNemey and A. Curry, eds.), CFTA Scientific Monograph Series No. 2, Cosmetic, Toiletry and Fragance Association, Inc., Washington DC.

Part V

Pharmacoeconomics

Chapter 28

Pharmacoeconomics and Outcomes Research

RUKMINI RAJAGOPALAN

1. Introduction

Over the past decade, as pressures to control health care spending have escalated, the term "cost-effective" has become increasingly commonplace in environments as diverse as the Congress, the business community, managed care organizations, and the pharmaceutical industry. Various factions use this term to serve their own purpose:

1. Purchasers of health care use the term to convey a careful assessment of the relative value of different health care services.
2. Producers of health care technologies and programs use the idea to support marketing claims.
3. Advocates of special interest groups for particular illnesses and constituencies use the term to garner resource investments.

2. Audience for the Economic Evaluation and Economic Messages

It is important that the economic evaluation results are conveyed to various audiences internally within the organization as well as externally. The internal audience includes personnel in corporate marketing, operational marketing, strategic planning, and corporate R & D. The external audience includes general practitioners, patients, health care researchers such as epidemiologists, economists, insurers, managed care executives, hospital executives, formulary managers, and the general public. Table 1 shows various analysis models appropriate for various audiences and different stages of drug development.

Rukmini Rajagopalan • Glaxo Wellcome, Inc., Research Triangle Park, North Carolina 27709, and University of North Carolina, Chapel Hill, North Carolina 27599.

Cutaneous Biometrics, edited by Doris A. Schwindt and Howard I. Maibach, Kluwer Academic/Plenum Publishers, New York, 2000.

TABLE 1. Studies and Audiences

Stage	Study type	Audiences Internal	External
Preclinical	Cost of illness	Stragegic R&D planning	Government
Phase I	Preliminary model	Strategic R&D planning	
Phase II	Detailed model	R&D marketing	Clinicians, healthcare managers
Phase III	Piggyback study, validated model, management trial	Marketing R&D	Clinicians, healthcare managers
Phase IV	Comparative management trials, comparative models	Marketing	Clinicians, healthcare managers, formulary managers

Economic messages will be related to net social benefit, cost-effectiveness, cost-minimization, prices, and tax costs.

3. Evaluation Methodology

Various economic evaluation methods are illustrated in Fig. 1.

3.1. Cost-Benefit Analysis

This is a general method applicable for evaluation of almost any drug, device, or service with the broadest perspective. The evaluation is in terms of monetary value and the results are "absolute" and not "relative." The outcomes may be expressed in terms of a ratio (benefit to cost) or in terms of net cost or benefit (benefit minus cost, i.e., ($\Delta B - \Delta C$). By converting outcomes to dollar values, cost–benefit analysis helps the decision-maker to compare the program alternatives directly and choose the program that gives the maximum benefit.

One disadvantage of cost-benefit analysis is the ambiguity that lies in converting nonmonetary units, such as, e.g., the quality of life improved or deteriorated, or adverse events. Cost-effectiveness analysis is preferred to avoid such difficulties.

3.2. Cost-Utility Analysis

This method is useful for evaluating drugs, devices, or services from the social perspective. Utility measurements can use any one of the following methods to integrate patient preferences and satisfaction. Cost-utility analysis not only takes into account the associated monetary costs for a drug, procedure, or device but also looks at the added costs of patient discomfort or change in capacity to function or level of satisfaction.

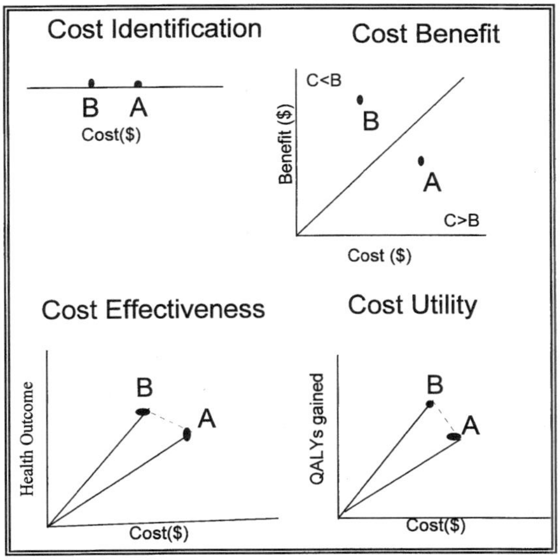

FIGURE 1. Graphic illustration of various economic evaluations.

Utility translates as preference for or desirability of a particular state of health. Life-years added to one's life because of a treatment and quality-adjusted life-years added are used as a measures of utility. However, while evaluating chronic diseases that are not really debilitating but are the cause of really annoying symptoms, such as skin diseases, it is very difficult to measure utility of a treatment using life-years, or quality-adjusted life-years added, because in such circumstances, the improvement due to the treatment has to be measured by more subtle evaluation tools such as quality-of-life instruments that are sufficiently sensitive to detect the improvement gained by the intervention.

- *Standard gamble* requires the evaluator to collect data on the values of substances and services the consumer would be willing to give up for set of available alternatives, even when they do not know the chances of obtaining the preferred alternative.
- *Time Trade-off* is almost like the old-fashioned barter as to what you can give up for something you want.
- *Rating scale* is assigning some value to the alternatives that are available and rating them in an order.
- *Multiattribute index:* when the outcome is multifaceted, an index is created combining all the factors appropriately weighted so that the index is used for the evaluation of the outcome.

Sources of valuations are various costs, direct and indirect, and the benefits of the consequences.

3.3. Cost-Effectiveness Analysis

Cost-effectiveness analysis (CEA) is performed for specific disease/patient groups. The results are always relative and are more easily understood by the general public. The main purpose of cost-effectiveness analysis is to compare the relative value of different interventions in improving health and/or prolonging life. The results of these evaluations can be summarized by the "cost-effectiveness ratio," in which the denominator reflects the gain in terms of years of life gained, or improvement in the health status, etc., and the numerator stands for the costs involved in obtaining those gains (Gold *et al.*, 1996). The information from the CEA help decision-makers comprehend the intricacies of each of the alternatives they have at hand and enable them to decide on an appropriate alternative that would suit their programs and meet their financial constraints. These decision-makers may very well be federal, state, or local in the public sector or they may be in the private sector. CEA provides a framework within which decision-makers may ask questions and look for answers methodologically.

Steps in studying cost-effectiveness consist of calculating the expected cost of treatment, cost ratio, rates of mortality, major and minor disabilities, quality-adjusted life years (QALY), duration of disease-free period, and finally the relative cost-effectiveness (Arikian *et al.*, 1994).

In the problem identification stage, the perspective should be identified, and appropriate cost analysis including the cost of the treatment regimen, expected cost, and factors of cost-effectiveness should be specified.

Next is the clinical management analysis, which includes data on clinical success rates, relapse rates, adverse event rates, mortality rates, and major and minor morbidity rates. With all this information, a decision analysis model including therapeutic pathways, outcomes, probabilities, and costs associated with each outcome can be constructed.

By using the above pieces of information together in a decision-analysis model, an economic analysis will be done to obtain cost-effectiveness ratio using the cost of treatment regimen, the expected cost of therapy, and the rates of various factors of outcome.

Finally, the uncertainty of the results of economic evaluation is estimated by identification of the range of variation for the input parameters, calculation of parameter elasticities, and calculation of 95% confidence intervals for treatment success rates, relapse rates, and adverse event rates. Other methods of evaluating uncertainty are sensitivity analysis and threshold analysis.

3.4. Cost-Minimization Analysis

Cost-minimization analysis is a special case of CEA. Unless equal effectiveness of the comparators is proven, this is not a legitimate economic approach. Most decision-makers use this methodology, explicitly or implicitly. When considerable clinical evidence suggests equality among alternatives, cost-minimization analysis lends itself to compare just the costs of two or more of the interventions (Bootman *et al.*, 1996). For instance, if we can prove that the outcomes of the use of two antibiotics happen to be found equal in terms of time to cure, success rate for cure, etc., the costs of interventions can be compared so that the least expensive intervention can be chosen. In cases where equality among treatment outcomes is questionable, cost-effectiveness analysis is the preferred method.

4. Benefits of Outcomes Research Studies

Medical interventions have various consequences, such as direct health care resource use, direct nonhealth care resource use, indirect societal costs, intermediate clinical success, and impact on patient quality of life and survival.

When outcomes research studies are done, the results provide the information to build better methods of managing the respective diseases and provide optimal care. A combination of science describing the disease mechanisms and the action of the drug in treating of the disease and health economics providing data on efficacy/effectiveness and costs of disease and treatment will provide a knowledge-based advantage in disease management.

5. Examples of Cost-Effectiveness Studies

5.1. An Economic Analysis of Oral Therapies for Onychomycosis

This study was performed using the statistical data input from various countries and the confidence interval for success rates for all the comparator treatments were ranked for both fingernail and toenail diseases. Meta analysis was used to pool the data and decision analysis model was used for the economic evaluation of various treatments to identify and rank their cost-effectiveness (Arikian, 1994).

5.2. An Economic Evaluation of Patch-Testing in Diagnosis and Management of Allergic Contact Dermatitis

This study (Rajagopalan *et al.*, 1997) was performed recently, and using a decision analysis model the ranking of the risk groups for whom the diagnostic process was cost-efficient was identified (Figs. 2–5). The decision analysis model

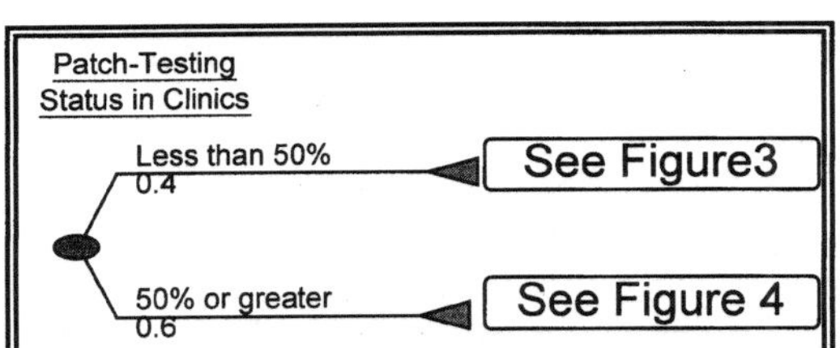

FIGURE 2. Decision analysis model (Part I)—total costs.

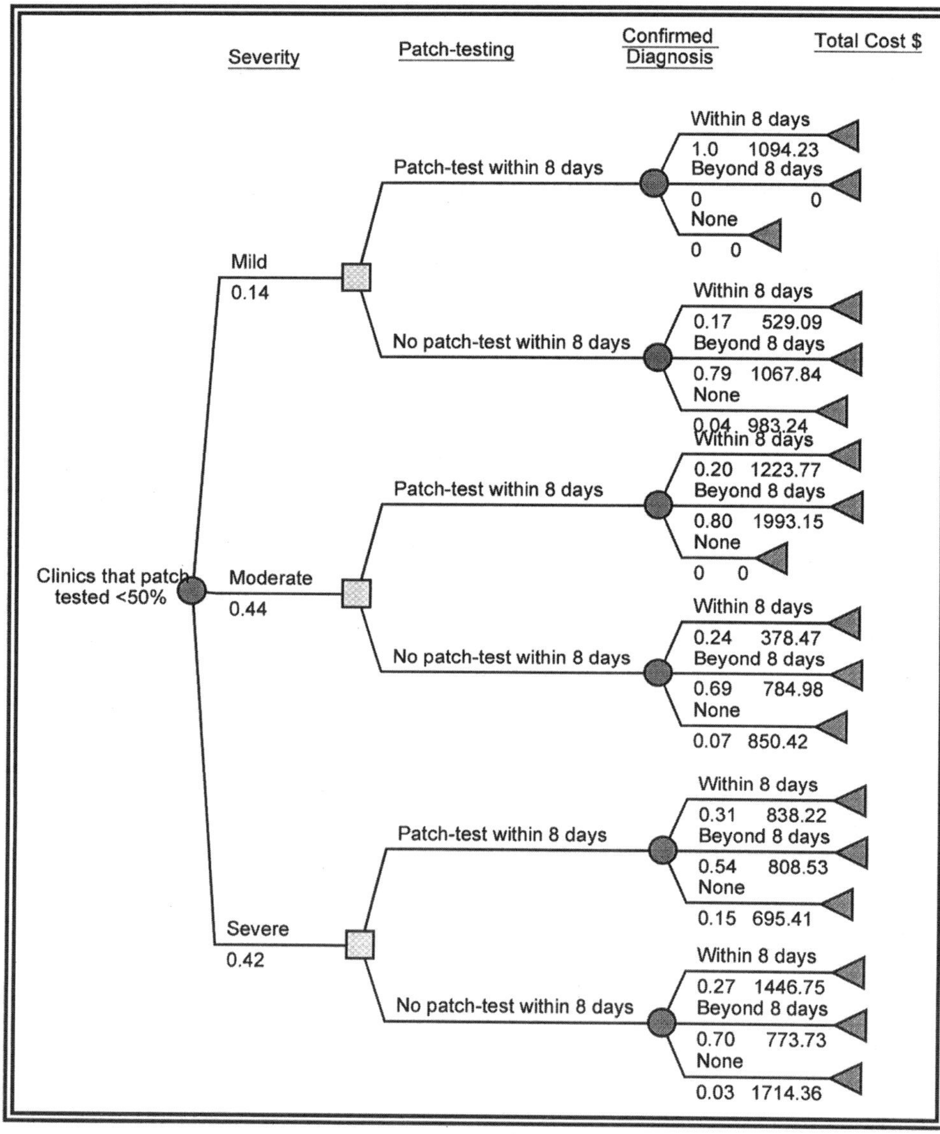

FIGURE 3. Decision analysis model (Part II)—total costs.

systematically evaluates the decision alternatives and their consequences and provides the audience information on the cost and effectiveness of each decision alternative by comparing their expected values. The expected value calculation is based on the probabilities and utilities of the decision outcomes derived from the data collected in a multicenter study. In this particular study, it was clearly evident

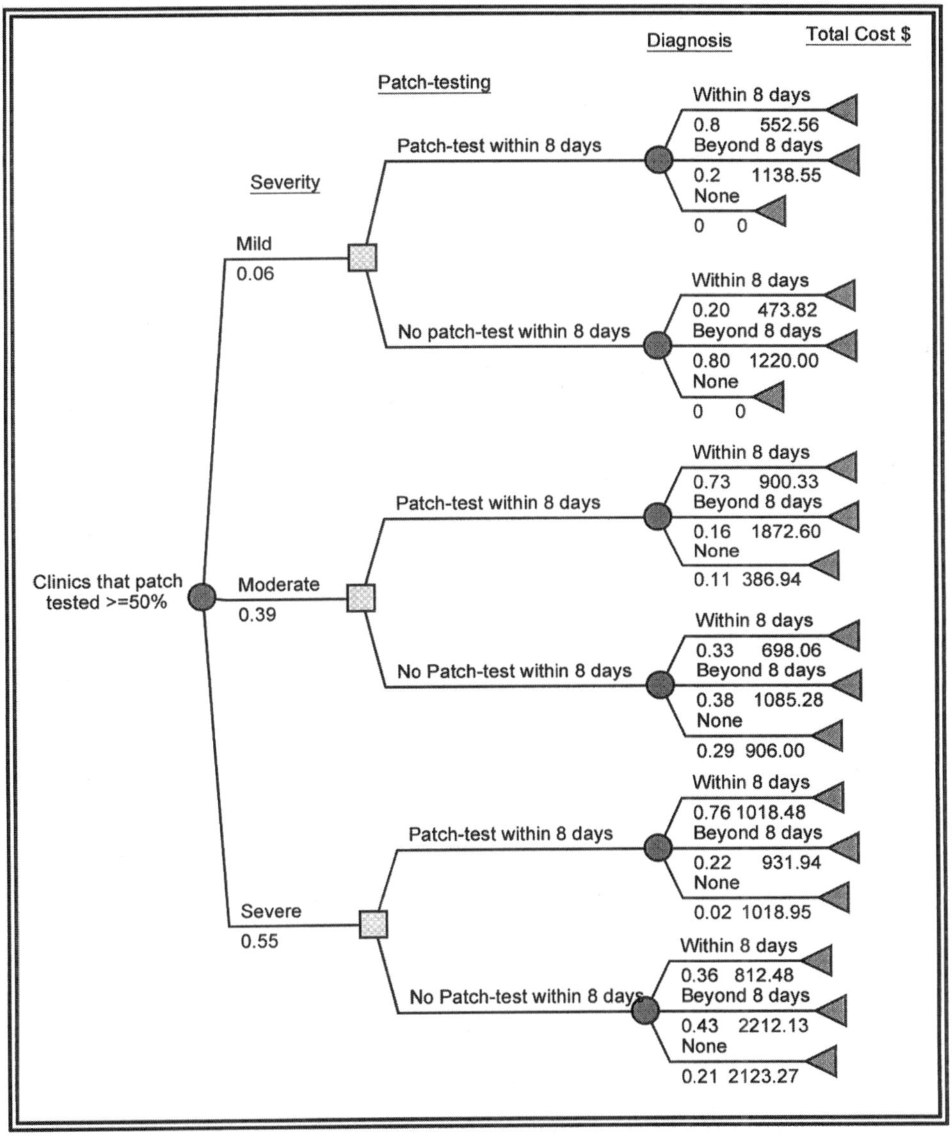

FIGURE 4. Decision analysis model (Part III)—total costs.

that for the total costs involved in the diagnostic process with patch-testing, the effectiveness in terms of diagnosis and treatment success was most pronounced in the group of patients with severe allergic control dermatitis. The model also ranked other groups of patients for whom the procedure was cost-effective, in a descending order.

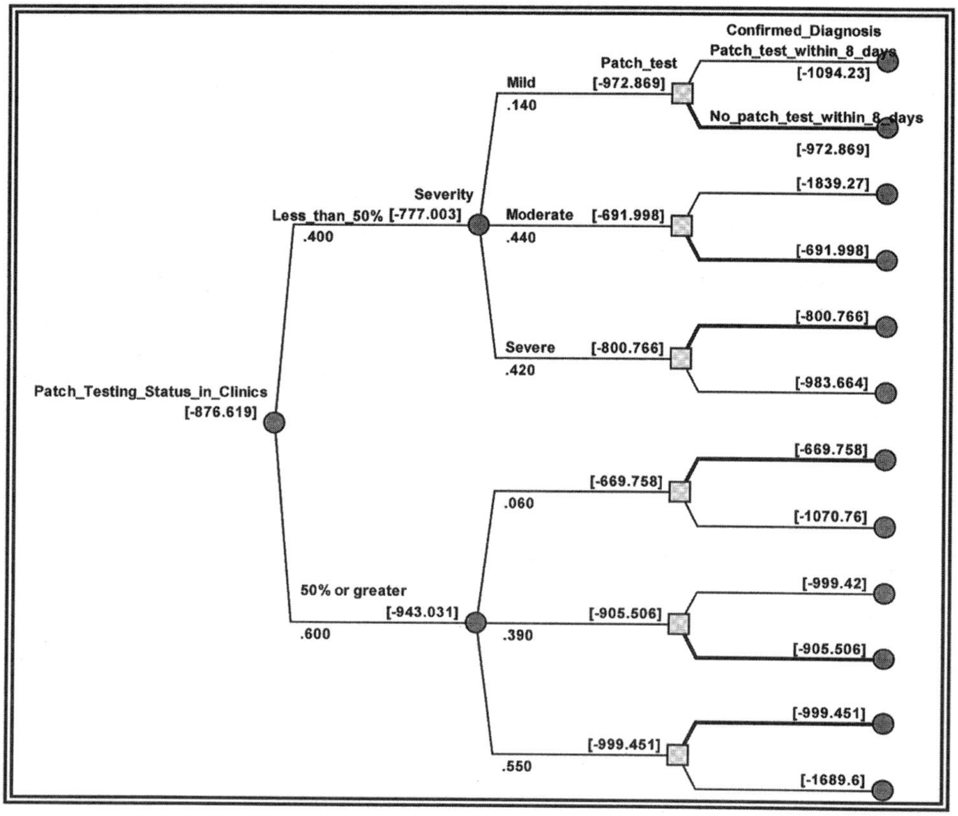

FIGURE 5. Decision preferences based on study data—total costs.

5.3. Systematic Analysis of Components of Compounds and Outcomes Data

The role of the complement system in shock and tissue injury induced by tumor necrosis factor and endotoxin examined and demonstrated by Hsueh *et al.* (1990) describes how systematically studying various components of a compound in a physiological process helps understand their activities individually and collectively, and by modeling their activities using step-by-step elimination and inclusion processes, scientists can obtain a thorough knowledge of the characteristics of the compound and make for valuable contributions to drug development. It is worth noting that a consolidated effort on the part of the professionals from the whole spectrum of health care fields from computational chemistry to pharmacoeconomics and from gene mapping to disease intervention simulations (Fig. 6) can add great value to health care outcomes (Evans, 1996).

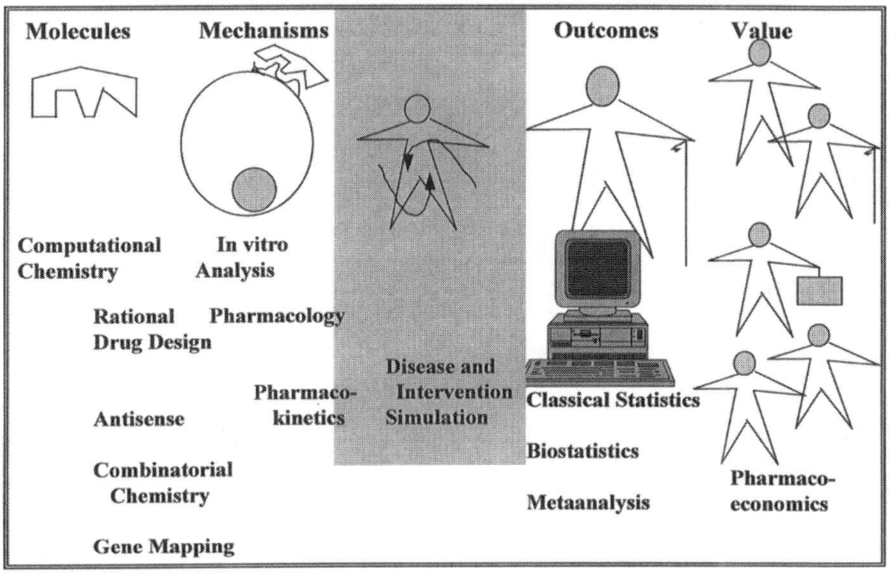

FIGURE 6. Dynamic simulation fills a gap in tools available to learn at the systems level.

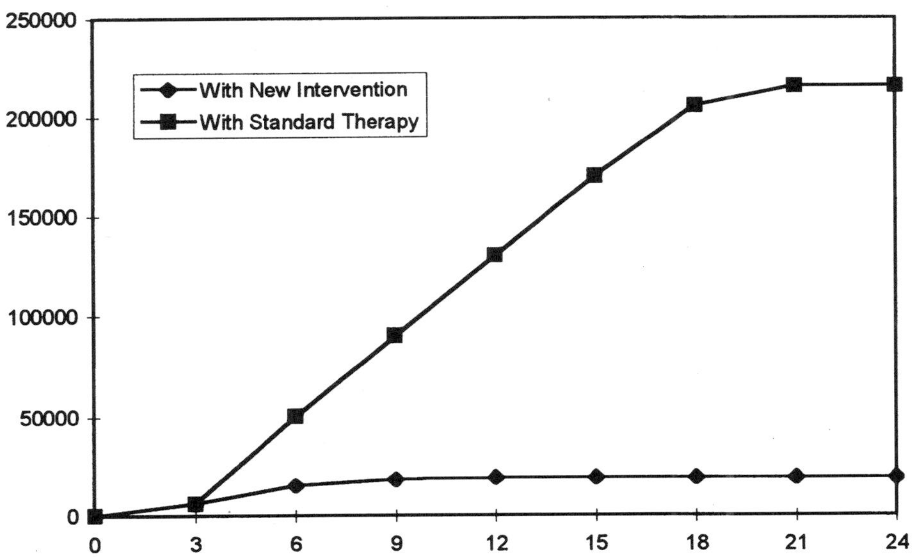

FIGURE 7. Pharmacoeconomic consequences of specific interventions.

5.4. A Comparison of Outcomes between the Standard Treatment and the Treatment Based on Iterative Feedback and Simulation Modeling

By using the components of the outcomes data such as underlying disease conditions, onset of clinical symptoms, the progress of disease, the categories of patient populations for whom the treatment is more effective, etc., in iterative feedback (Hsueh *et al.*, 1990) and simulation modeling, the costs of effective treatment can be considerably reduced as shown in Fig. 7.

6. Conclusion

As the health care industry progresses, increasingly conscious of cost-effectiveness, studies evaluating outcomes are going to be extremely important and providers will be required to base their determination of which drug to prescribe on the results of such outcome studies. Hopefully, material such as what is presented in this chapter will help readers to acquire some expertise to comprehend and interpret such study results and use them efficiently.

References

Ariikian S. R., Einarson, T. R., Nguyen, G. K. T., Schubert, F., and the Onychomycosis Study Group, 1994, A multinational pharmacoeconomic analysis of oral therapies for onychomycosis, *Br. J. Dermatol.* **130** (Suppl.43):35–44.

Bootman, J. L., Townsend, R. J., and McGhan, W. F.(eds.), 1996, *Principles of Pharmacoeconomics*. 2nd Ed., Harvey Whitney Books, Cincinnati.

Evans, P., 1966, Using health economics to gain a knowledge-based advantage in disease management, presented to DIA, Barcelona, Feb. 1996.

Gold, M. R., Siegel, J. E., Russell, L. B., and Weinstein, M. C. (eds.), 1996, *Cost Effectiveness in Health and Medicine*, Oxford University Press, New York, pp. xvii–xxiii.

Hsueh, W., Sun, X., Rioja, L. N., Gonzalez-Crussi, F., 1990, The role of the complement system in shock and tissue injury induced by tumor necrosis factor and endotoxin, *J. Immunol.* **70**:309–314.

Rajagopalan, R., Anderson, R. T., Sarma, S., Kallal, J., Ritchin, C., Jones, J., Fowler, J. F. Jr., and Sherertz, E. F., 1997, Allergic contact dermatitis: cost effectiveness of early confirmed diagnosis and patch testing, in: *Care Management of Skin Diseases: Economic and Life Quality Impact* (R. Rajagopalan, R. T. Anderson, and E. F. Sherertz, eds.), Marcel Dekker, New York, pp. 299–316

Chapter 29

The Cost of Atopic Eczema to the Society: An International Perspective

RUKMINI RAJAGOPALAN

1. Introduction

Skin reacts to chemical, mechanical, and immunological injuries in a variety of ways depending on the intensity, time course, and spatial distribution of the stimulus. Eczema is one such pattern of response. The intraepidermal edema and inflammatory reaction are together responsible for the clinical picture of eczema and an enormous amount of discomfort, disability, and economic loss (Marks, 1992).

2. Clinical Definition

Atopic eczema or dermatitis is a most complex condition for which even a common clinical definition is very difficult to formulate. If we are not certain that all populations concerned have a similar definition of the disease condition and we are dealing with different population groups, the economic evaluation of one cohort may not purport to hold valuable information for use by decision-makers for another. The choice of a suitable definition is more often governed by practical constraints rather than ideals. It is, nevertheless, important to consider what makes a good epidemiological definition for atopic dermatitis (AD). Williams (1997), in his pursuit of defining diseases has made great strides to build a hierarchical algorithm to define any complex condition. According to this algorithm, a good epidemiological disease definition should be valid (sensitive and specific), repeatable (between and within observers), acceptable to the population, rapid and easy to use by field workers, coherent with prevailing clinical concepts, a reflection of some degree of morbidity, comprehensive in its applications, and compatible with other studies.

Rukmini Rajagopalan • Glaxo Wellcome Inc., Research Triangle Park, North Carolina 27709, and University of North Carolina, Chapel Hill, North Carolina 27599.

Cutaneous Biometrics, edited by Doris A. Schwindt and Howard I. Maibach, Kluwer Academic/Plenum Publishers, New York, 2000.

It is important to note that different studies have different requirements of disease definitions, especially in terms of the relative importance of sensitivity and specificity. Therefore, for comparative prevalence studies, Williams *et al.* (1996) recommends the following format of definition recorded as a 12-month period prevalence measure, to obtain optimum results for prevalence estimates of a magnitude typical in developed countries (Table 1).

What we recognize as a clinical syndrome of atopic dermatitis today may in time be shown to be composed of three or four different diseases, based on genetic, immunological, or etiological discoveries (Imayama *et al.*, 1992; Roth, 1987). Almost half-a century ago, D. Williams (1951) compared the occupational blindness of dermatologists groping around the elephantine colossus of atopic dermatitis (Fig. 1) to the blind men in John Godfrey Saxe's famous "Hindoo fable," which relates how

> "It was six men of Indostan
> To learning much inclined
> Who went to see the Elephant
> (Though all of them were blind),
> That each by observation
> Might satisfy his mind."

Each in his blindness "saw" the elephant differently — one saw it as a wall, another as a spear, a third as a snake, and so on — as a tree, a fan, and a rope. Similarly, various authorities from various vantage points have viewed this dermatological "elephant" and gained various impressions of it (Fig. 2). The visible nature of skin diseases is a clear advantage over "internal" diseases for developing valid and reliable diagnostic measures. Unless the definitions of the diagnosis are comparable among world populations, the outcomes of a study done in Boston cannot provide effective treatment measures for populations in Bombay or Borneo.

3. Prevalence

Eczema is the most common of all inflammatory dermatoses, particularly in children. Recent studies of children suggest that the prevalence ranges from 2 to 20%

TABLE 1. UK Refinement of the Hanifin and Rajka Diagnostic Criteria for Atopic Dermatitis[a]

To qualify as a case, the child must have:

1. An itchy skin

Plus three or more of the following:

2. Onset under age 2 years
3. History of rash in skin creases
4. Personal history of asthma or hay fever
5. A history of generally dry skin
6. Visible flexural dermatitis

[a]Hanifin *et al.*, 1980.

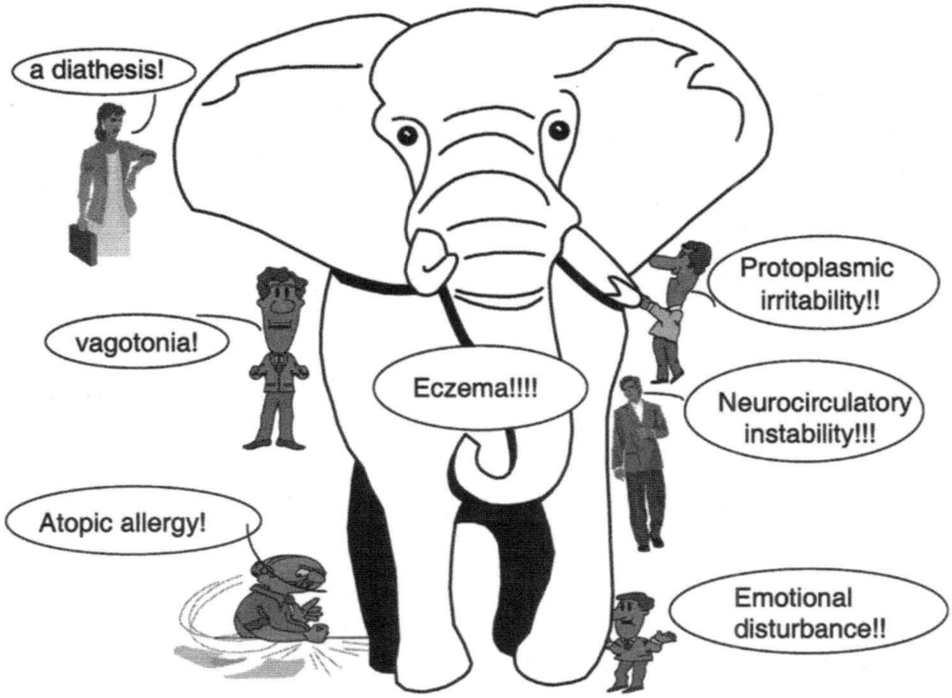

FIGURE 1. Elephantine colossus of atopic dermatitis: diathesis — constitutional predisposition to certain disease condition, vagotonia — hyperirritability of parasympathetic nervous system.

by the age of 7, particularly in developed countries (Dottered *et al.*, 1995; Peat *et al.*, 1994; Buser *et al.*, 1993; Williams *et al.* 1995; Schafer *et al.*, 1993; Saval *et al.*, 1993; Bobak *et al.*, 1995; Okuma, 1994., Leung and Ho, 1994; Henderson, 1995) (Table 2).

Prevalence estimates among adults suggest an overall frequency of AD of between 1 to 2%. Although AD is considered a "childhood" disorder, because adults make up around 80% of the population in developed countries, adult AD cases may constitute around a third of the "AD problem" in absolute terms for a given population.

For example, in a population of 100,000 where the adult population constitutes 80% and the prevalence of AD in children under the age of 7 is 5%, there are 20,000 children and the number of children with AD is 1000. If the overall prevalence is, say, 1.5% the total number of AD subjects is 1500. Therefore a third of the AD subjects are adults. If we consider only the severe cases of AD, this proportion will be even greater. In the recent years, the disease burden due to AD has been studied in depth in the United States (Lapidus *et al.*, 1993), United Kingdom (Herd *et al.*, 1995), and Australia (Su *et al.*, 1997). Dermatologists in all the three countries express similar concerns regarding the inadequate processes followed to diagnose and treat AD adults and children, suboptimal therapies available to treat them, and incomplete records on the occurrence and prevalence of AD. The details of these three studies will be discussed here.

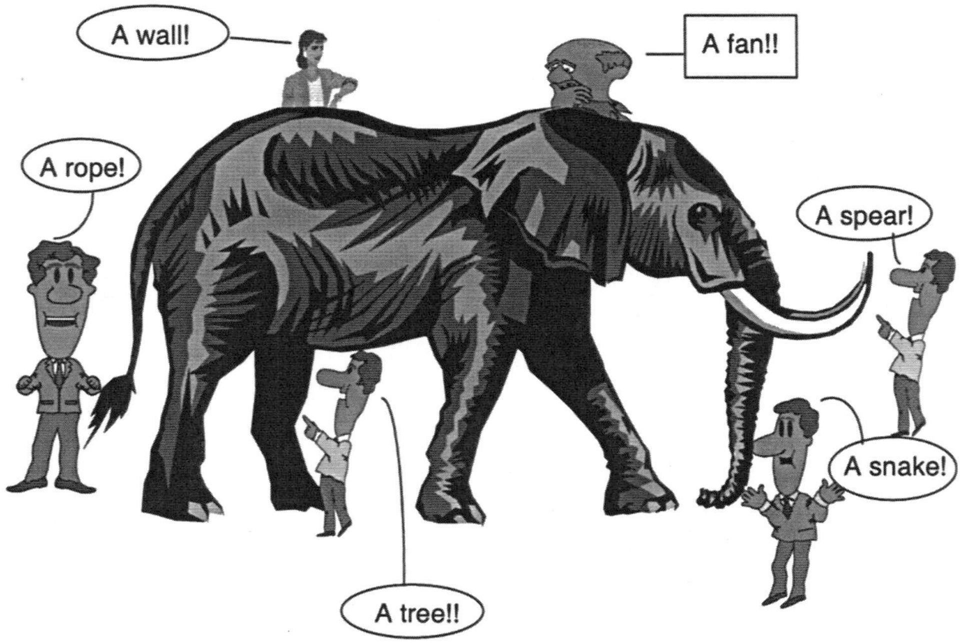

FIGURE 2. The elephant and the six blind men.

4. International Scenes

4.1. United States

Lapidus *et al.* (1993) have expressed great concern over the fact that despite the escalating prevalence and associated morbidity of AD particularly in urban children, no systematic approach has yet been documented to deliver care and manage the disease.

In United States, 60% of AD cases occur in the first year of life and 90% in the first 5 years (Rajka, 1975). The prevalence of AD over the past three decades has more than tripled from 3 to 10% and seems to be even higher in some heavily populated urban areas (Larsen *et al.*, 1986; Hanifin, 1987). In the absence of documentation for care delivery, local institution-specific data provided information on service use and expense of care for children with AD and national data sets were used to estimate the direct cost for care of children with AD in the United States.

The hospital data revealed that 63% of the outpatient visits occurred in the emergency department setting and 37% in other practice setting through the hospital. Further, it was interesting to note that 60% of emergency department visits for AD occurred during regular clinic hours. Just for one urban children's hospital, a sum of approximately $410,000 was incurred in annual medical reimbursement cost for AD.

TABLE 2. Prevalence Surveys of Atopic Dermatitis from Selected Developed Countries around the World

Country/reference	Prevalence	Definitions	Age group	Year	Comments
Norway (Dottered *et al.*, 1995)	26% ($N = 424$)	12-month prevalence (history and examination)	7–12 years	1993	Detailed study of an isolated rural community in Northern Norway
Australia (Peat *et al.*, 1994)	24.4% ($N = 873$)	Lifetime prevalence (questionnaire)	8–10 years	1992	Serial cross-sectional studies in two towns
Germany (Buser *et al.*, 1993)	11.8% ($N = 4651$)	Current symptoms (questionnaire)	Primary school	1993	Part of validation study in Hannover
Germany (Schafer *et al.*, 1993)	22.2% ($N = 1091$)	Life time prevalence (questionnaire and some examination)	5–6 years	1993	Survey of seven areas in Bavaria
United Kingdom (Williams *et al.*, 1995)	11.7% ($N = 693$)	Point prevalence	9–11 years	1994	Survey of London schoolchildren
Denmark (Saval *et al.*, 1993)	7% ($N = 4952$)	Symptoms last 12 months (questionnaire)	5–16 years	1990	Large survey of Viborg
Czech Republic (Bobak *et al.*, 1995)	3.9% boys 6.3% girls ($N = 2060$)	12 month prevalence (questionnaire)	3–5 years	1995	Survey of atopy and pollution in the "Black Triangle"
Japan (Okuma, 1994)	9.5% ($N = 10137$)	Current AD (questionnaire)	Primary school	1993	Large questionnaire survey of atopic diseases in Okinawa
Hong Kong Malaysia, China (Leung and Ho, 1994)	20.1% 7.1% 7.2% ($N = 2060$)	History of flexural rash ever in the past	Secondary school	1993	Large survey of atopy in three southeast populations
Tanzania (Henderson, 1995)	0.7% ($N = 548$)	Point prevalence based on doctor's examination	7–18 years	1994	Survey of skin diseases in two rural villages

The total estimated cost for treatment of AD in the United States is $364 million annually. This includes inpatient and outpatient care (office visits and emergency department visits). AD definitely carries a high cost for the nation. As its prevalence continues to increase and funding for research and health care continues to decrease, the cycle of unending issues can be broken only by organizing the management of diagnosis and care in innovative and efficient ways, such as directing these children to dermatology clinics, and family guidance including home care.

4.2. United Kingdom

According to Herd *et al* (1995), atopic eczema affects 2–3% of the UK population. The objective of the Lothian atopic dermatitis study was to estimate the costs associated with treatment atopic eczema (AE) in patients of all ages in the United Kingdom. They studied a sample of 155 patients randomly selected from the computerized records of eight general practitioners in Livingson, West Lothian. Diagnostic coding for various forms of eczema was used for easy identification. Patients were selected in stages: (1) all subjects that satisfied the diagnostic criteria; (2) all families in the practice who were known to have children under the age of 2 years were questioned and examined; (3) a random sample of families was chosen and contacted on the phone, and each adult in the family was interviewed to get information on any child in the family known to have a complaint or history of itchy skin; (4) all members in a family of six which denied the presence of an itchy member at stage 3 were individually examined to avoid oversight of an AE-afflicted subject.

To calculate the cost of disease and treatment prospectively, clear instructions were given to the patients to provide data on direct costs, such as purchase of drugs and medical office visits and procedures, costs of nonmedical purchases due to the disease, indirect costs such as loss of salary, absence from school or work, and any other expenses directly or indirectly resulting from the skin condition. The costs were monitored for a 2-month period. The findings were:

Personal Cost

1. The mean personal expenditure by patients over the 2 month period was (25.90 and the maximum spent was £546, 81% of which was due to salary loss.
2. 45% of patients under the age of 16 and 26% over the age of 16 had no expenditure because they were in remission.
3. Patients aged 2–15 years incurred significantly fewer expenses than those over the age of 16, and those under the age of 2 had significantly fewer expenses than the former group. Note: Patients below the age of 16 do not pay for prescriptions.

Health Service Cost

4. The mean health service cost was £16.20 and the maximum attributable to one patient was £177.07. The median cost for the group 2–15 years of age was significantly higher than that for the group over 16 years of age. The group under the age of 2 had even significantly higher expenses than the former group.

5. Most of the health service costs were for treatment: 38% on emollients or bath additives, 32% on topical steroids, 10% on bandages, and the remaining 20% on antihistamines, shampoos, antibiotics, and evening primrose oil.

HOSPITAL COHORT

6. The 2-month cost to the health service of the more severely affected cohort of patients attending hospital was upto £1500 per patient (mean £415), 63% of which was incurred by the hospital and 34% by the GPs. The maximum personal cost was £1225 (mean £325) over 2 months, 75% of which was due to loss of salary.

Extrapolation from the above data collection and analysis projected a total annual expenditure in the UK of £297 million by patients, an estimated annual cost to the health service of £125 million, and the cost to the society in lost working days of £43 million for a grand total of £465 million. The authors have cautioned readers regarding some limitations of the data. The final estimate of the cost is likely to be an underestimate for the following reasons:

1. The prevalance of AE in Scotland is less than in the rest of the United Kingdom.
2. The GPs in the study practice are low prescribers.
3. None of the patients in the study population had required hospital treatment during the study.
4. Patients lived within walking distance of the health center, and that reduced travel costs.
5. The cost of time spent by caretakers of dependents is not included in the calculation.

The data do give us an indication of the magnitude of the problem in United Kingdom with a national cost of £465 million. Patients' readiness to spend large sums of money for relief from their skin disease is demonstrated by 10 patients in the sample paying over £100 in 2 months. In an attempt to compare the burden of the disease on the society of few diseases, the annual per capita cost of treating AE (£7.38) was examined against treating venous ulcers (£6.73), stroke in Scotland (£18.78), and benign prostatic hypertrophy (£1.04–£1.53). It is totally astounding that the economic burden of these conditions so closely resembles that of AE, even though admittedly the comparison is not very precise.

The financial impact of severe AE undoubtedly has considerable repercussions on the sufferers. The burden is heavy and the ceiling of expenditure high.

4.3. Australia

Su *et al.* (1997) found that some 25–50% of AD children develop asthma and 30% develop allergic rhinitis. Atopic eczema is the single most common skin condition in children younger than 11 years of age, and it is also an important cause of morbidity when it affects adults. (Hanifin and Rajka, 1980). They decided to evaluate the impact of childhood atopic eczema of varying severity on families and assess the personal financial cost of its management by cross-sectional survey.

Infants and children presenting with AD to the dermatology clinic at the Royal Children's Hospital between March and August 1995 and their families were studied. Forty-eight children were selected sequentially and their czema was categorized as mild, moderate, or severe by assessing the percent body surface area involved, the course of the illness, and intensity of itching.

The impact-on-family questionnnaire of Stein and Riessman (1980) was administered by a single researcher. This questionnaire, which has been used to assess the impact of several other chronic conditions, has four subscales assessing: (1) financial burden; (2) familial/social impact; (3) personal strain; and (4) mastery, as perceived by the parent.

A control population of 46 children with insulin-dependent diabetes also enrolled sequentially from the diabetes clinic at the same hospital was also studied using the same questionnaire. This group was compared with eczema group by a two-sample t-test. Analysis of covariance was performed to adjust for age as a covariate.

Impact-on-family scores were compared between the diabetes group and the eczema group (mild, moderate, and severe eczema children separately).

IMPACT-ON-FAMILY SCORES: Children with moderate or severe eczema had higher scores than those with diabetes after adjusting for age differences. Higher scores means higher impact of the disease.

COMMUNITY COSTS: The annual cost to the community of medical consultations for each child in the mild, moderate, and severe groups (calculated from the medical benefits schedule book) was Aus$209, 389, and 642, respectively. The annual cost to the community of hospitalization for each child in these three groups was Aus$453, 1523, and 2912, respectively.

DIRECT FINANCIAL COSTS: The mean annual cost of eczema was related to the severity of eczema. The total mean costs of medication, dressings, diet, other eczema management strategies, and medical consultations for the mild, moderate, and severe groups were Aus$330, 818, and 1255, respectively. The major costs were for medication and dressings. Other expenses included change of carpets, purchase of dust mite free covers, and purchase of extra nonirritant clothing. Comparable costs for 22 diabetic children averaged Aus$444.

INDIRECT COSTS: Only time off from work to care for the sick child was included. The potential income loss due to employment reduced, ceased, or not started for parents was not estimated. The cost and time to travel to and from the provider's office or clinic were not determined. Therefore the indirect costs were quite underestimated.

The social, family, and emotional effects and personal financial costs of eczema have never been quantified. The impact on family scores was higher in the moderate and severe eczema groups, compared with scores for the type I diabetes group. It is very clear that the profound impact of this common disorder on families, a fact that is underappreciated in view of the lack of support for families of children with eczema compared with that for families of children with diabetes or asthma. Even in the mild eczema group, the family score was equivalent to that for children with insulin-dependent diabetes.

Management of atopic eczema in children is quite complex and costly, often requiring a well-planned multifaceted approach for optimal care. There is a great impact on the quality of life in children and families. There is a real need for doctors and health care managers to recognize that eczema is not merely a minor skin disorder, but a major handicap with a considerably personal, social, and financial burden on the family.

5. Discussion

Literature from over half-a-century back up to the present demonstrates repeatedly that hardly the surface has been scratched with respect to finding the etiology of atopic eczema. Various drugs such as corticosteroids and antihistamines have only a palliative effect to minimize the signs and symptoms and none of them cure the disease totally. Researchers believe that normal looking skin in a subject who suffered atopic eczema is never normal and the skin even long after the disappearance of the rash still contains inflammatory cell infiltrate (Hamid *et al*, 1994 and Leung *et al*, 1983). Research has a long way to go before the immunological factors are identified to totally prevent relapse of the disease episodes. Until such time, the costs of treating diseases such as atopic eczema will keep increasing.

These three studies in combination give us a comprehensive picture of the burden of atopic eczema—financial, family, and emotional. We have a clear sense of the expenditure for the society, the stress and dilemma on the family, and the plight of the patients, adults and children suffering equally. The disruption caused in the quality of life of the sufferers and the family dynamics is too heavy a price in addition to the financial burden. The authors of all the three publications suggest unequivocally that atopic eczema is by no means any less of a disability than asthma or diabetes and that more attention should be devoted to finding an optimal remedy for the sufferings resulting from this disease and more support provided for its victims.

6. Conclusion

As the population around the world is becoming knowledgeable of the health implications and disease entities, it is very well understood that skin disease should not be taken lightly and it is equally important to keep the skin healthy as we pay attention to the other organs of the body. Further, it is not sufficient to provide short-term relief for the signs and symptoms of the disease, which is going to flare up as soon as the treatment is discontinued. The burden of atopic eczema in terms of financial costs as well as deterioration of quality of life continues to rise in the developed countries very much at the same rate as in the developing countries. Future research should be focused in identifying the immunological factors that could lead into finding a cure for the condition.

References

Bobak, M., Koupilovà, I., Williams, H. C., Leon, D. A., Danova, J., and Kriz, B., 1995, Prevalence of asthma, atopic eczema and hay fever in five Czech towns with different levels of air pollution, *Epidemiology* **6**:S35.

Buser, K., Bohlen F. V., Werner, P., Gernbhuber, E., and Robra, B. P., 1993, Neurodermitis-Prävalenz bei Schulkindren im Landkreis Hannover, *Dtsch. Med. Wachr.* **118**:1141–1145.

Dotterud, L. K., Kvammen, B., Lund, E., and Falk, E. S., 1995, Prevalence and some clinical aspects of atopic dermatitis in the community of Sør-Varanger, *Acta. Derm. Venereol.* **70**:50–53.

Hamid, Q., Boguniewicz, M., and Leung, D. Y. M., 1994, Differential cytokine gene expression in acute vs. chronic atopic dermatitis. *J. Clin. Invest.* **94**:870–876.

Hanifin, J. M., Epidemiology of atopic dermatitis, *Monogr. Allergy*. **21**:116–131.

Hanifin, J. M., and Rajka, G., 1980, Diagnostic features of atopic dermatitis, *Acta. Derm. Venereol. (Stockh.)* **Suppl. 92**:44–47.

Henderson, C. A., The prevalence of atopic eczema in two different villages in rural Tanzania, *Br. J. Dermatol.* **133** (Suppl. 45):50.

Herd, R. M., Tidman, M. J., Prescott, R. J., and Hunter, J. A. A., 1995, The cost of atopic eczema, *Br. J. Dermatol.* **135**:20–23.

Imayama, S., Hashizume, T., Miyahara, H., Tanabashi, T., Takeishi, M., Kubota, Y. *et al.*, 1992, Combination of patch test and IgE for dust mite antigens differentiates 130 patients with atopic eczema into four groups, *J. Am. Acad. Dermatol.* **27**:531–538.

Lapidus, C. S., Schwartz, F. D., and Honig, P. J., 1993, Atopic dermatitis in children: Who cares? Who pays? *J. Am. Acad. Dermatol.* **28** (5):699–703.

Larsen, F. S., Holm, N. V., Henningsen, K., 1986, Atopic dermatitis, *J. Am. Acad. Dermatol.* **15**:487–494.

Leung, D. Y. M, Bhan, A. K., Schneeberger, E. E., Geha, R. S., 1983, Characterization of the mononuclear cell infiltrate in atopic dermatitis using monoclonal antibodies, *J. Allergy Clin. Immunol.* **71**(1, pt. 1):47–56.

Leung, R., and Ho, P., 1994, Asthma, allergy, and atopy in three southeast Asian population. *Thorax* **49**:1205–1210.

Marks, R., 1992, Introduction, in: *Eczema* (R. Marks, ed.), Martin Dunitz, London.

Okuma, M., 1994, Prevalence rate of allergic diseases among school children in Okinawa, *Jap. J. Allergol. (Arerugi)* **43**:492–500.

Peat, J. K., van der Berg, R. H., Green, W. F., Mellis, C. M., Leeder, S. R., and Woolcock, A. J., 1994, Changing prevalence of asthma in Australian children, *Br. Med. J.* **308**:1591–1596.

Rajka, G., 1975, Atopic dermatitis, in: *Major Problems in Dermatology* (A. Rook, ed.), W. B. Saunders, London, p. 3.

Roth, H. L., 1987, Atopic dermatitis revisited, *Int. J. Dermatol.* **26**:139–149.

Saval, P., Fuglsang, G., Madsen, C., and Osterballe, O., 1993, Prevalence of atopic disease among Danish school children, *Pediatr. Allergy Immunol.* **4**:117–122.

Schafer, T., Przybilla, B., Ring, J., Kunz, B., Greif, A., and Überla, K, 1993, Manifestation of atopy is not related to patient's month of birth, *Allergy* **48**:291–294.

Stein, R. E. K., and Riessman, C. K., 1980, The development of an impact-on-family scale: Preliminary findings, *Med. Care.* **18**:462–472.

Su, J. C., Kemp, A. S., Varigos, G. A., and Terence, M. N., 1997, Atopic eczema: Its impact on the family and financial cost, **76**:159–162.

Williams, D., 1951, *AMA Arch. Dermatol. Syphil.*, **63**(5)545–560.

Williams, H. C., 1997, Defining cases, in: The Challenge of Dermato-Epidemiology (H. C. Williams and D. P. Stratchen (eds.), CRC Press, New York, pp. 13–23.

Williams, H. C., Pembroke, A. C., Forsdyke, H., Boodoo, G., and Hay, R. J., 1995, London-born black Caribbean children are at increased risk of atopic dermatitis, *J. Am. Acad. Dermatol.* **32**:212–217.

Williams, H. C., Pembroke, A. C., Burney, P. G. J., Forsdyke, H., Boodo, G., and Hay, R. J., 1996, Community validation of the UK working party's diagnostic criteria for atopic dermatitis in a population setting, *Br. J. Dermatol.* **135**:12–17.

Index

A priori order, 13, 14
Absorption, 321
Acid mantle of skin, 83, 86–87
Acne, assaying bioavailability in, 307–314
Acneiform eruptions of the face
relationship between judged lesion numerosity and actual lesion number, 137–144
scales of measurement, 129–131, 139–147
Acrylates, allergic reactions to, 240
Allergens, eczema sites and, 233–241
Allergic contact dermatitis: *see* Contact dermatitis, allergic
Allergic reactions, 238–240
Amerchol L 101, allergic reactions to, 240
Antibiotics: *see* Erythromycin
Assessment criteria, 57–59
Asymptotic principle and statistical test, 5
Atopic dermatitis (AD), 93, 263, 351; *see also* Dermatitis; Eczema
across various countries, 346–351
clinical definition, 343–344
costs, 348–350
diagnostic criteria for, 47
UK refinement of Hanifin and Rajka, 344
EFA supplementation in, 109–116
epidemiological studies, 103
extent of, 94–95, 98
inclusion criteria for clinical trials, 102–103
localized assessment, 102
methodology of clinical trials in, 114–115
prevalence, 344–345
quality control and routine scoring of, 101–102
subjective items and quality of life, 98–99
Atopic Dermatitis Area and Severity Index (ADASI), 94, 124
Atopic dermatitis severity
biophysical and biochemical evaluation of, 103
and time course, 101
Atopic dermatitis severity/intensity items, 96–97
choosing/defining, 95–96
Atopic dermatitis severity/intensity items (*cont.*)
ETFAD recommendations and definitions for grading, 103–104
recommendations regarding, 126–127
Atopic dermatitis severity scoring/grading, 93, 102, 119–120, 126–127
historical perspective on, 94
interobserver variability and, 96–98
training and, 103
Atopic dermatitis severity scoring systems, 120, 122–123
for clinical practice, 120–121
for clinical research, 121, 124–126
Atopy, 46–47

Barrier function of stratum corneum, 81–82, 86–87
Basic Clinical Scoring System (BCSS), 125
Benzoyl peroxide (BPO) product comparison, 309
Benzoyl peroxide (BPO) therapy, 308–309, 313
Berkson's bias, 32
Bias, 32
Binary data, 8–10
Bioengineering methods/techniques, 81–85
choosing, 85
important noninvasive, 85–87
Bioequivalence testing, 14–15
Biophysical and biochemical parameters, 63–64
Blanching response, visual assessment of, 297
Bleeding time (BT), 206–209
Blood flow, skin, 84, 86–87
Bonferroni and Bonferroni-Holm rules, 12–13
Borage oil, 111–112
Bronopol, allergic reactions to, 240

Calculated perceived values of acneiform lesions (CPVAL), 142, 144, 147
Cancer, skin, 47
Capacity, 82–83, 86
Capillary erythrocyte flow, 84, 86–87
Case-control comparison, objective of, 27

Case-control studies, analysis of, 29–30
Case-control study design, 23, 34
  planning and carrying out
    data analysis and interpretation of results, 31–34
    data collection, 30–31
    developing a study protocol, 25–30
  reasons for selecting, 24–25
  schematic diagram of, 24, 25
Case report forms, 40–41, 59–62
Case:control ratio, 27
Cases
  determining the source of, 27
  identifying and collecting, 30–31
Causality, criteria used to judge, 34
Cell method, 317–318
Chi-square test, 8
Chromametry, 291
Cleansing products: *see* Personal cleansing products
Color of skin, 83–84, 86–87, 262
Colorimetry, 154, 155
Comparison test *vs.* defined resources, 72–76
Computed tomography (CT), 193
Conductivity, 82–83, 86
Confidence interval (CI), 9–12, 15, 33–34
Confidentiality, 30
Confounding, 33
Consistency of skin products, 73
  test procedure for evaluation of, 73, 74
Contact dermatitis, *see also* Dermatitis
  allergic, *see also* Eczema sites
    economic analysis of patch-testing in diagnosis and management of, 337–340
    quaternium-18 bentonite in the prevention of, 269–271
  evaluating surveillance systems in, 247, 251–252
    proposal for establishing guidelines for, 252–253
  frequency of, and the time factor, 244–245
  irritant *vs.* allergic, 243
Control group
  choosing, 27
  types of, 27–28
Controls, matching, 27, 33
Corneocyte cohesion, 81–82, 86–87
Corticosteroid-induced skin blanching, Minolta chromameter used to assess, 295–303
Corticosteroids, 237–238
Cosmetic field, areas of application for sensory assessment in, 78
Cosmetic products, *see also* Skin-care products
  sensory profiles of, 76, 78–80
Cosmetics, 65
  quantification of test results
    seven-point comparison test for, 74–77
  quantitative testing, 71–78
    mathematical aspects, 77
Cosmetics (*cont.*)
  and the relationship between effect and perception, 65–66
  and the significance of language, 67–71
Cosmetics effects, reasons for sensory vagueness in assessing, 66–67
Cost–benefit analysis, 334
Cost-effectiveness, 333
Cost-effectiveness analysis (CEA), 336
Cost-effectiveness studies, 337–341
Cost-minimization analysis, 336–337
Cost–utility analysis, 334–335
Cremor Basalis DAC, 74, 75
Cross-sectional studies, 44
Crusts/oozing, 104
"Cushion"
  interconnection of semantic fields dealing with the word, 71, 72
  score values for the test parameter, 77

Data generation, active *vs.* passive, 245, 248–251
Data handling and documentation, 40–41
Data types, classes of, 6
Decision analysis models, 337–340
Dental fillings, allergy to amalgam in, 240
Dermatitis: *see* Atopic dermatitis; Contact dermatitis
Dermatitis model, water soluble irritant, 257–264
Dermatological preparations, types of, 315
Dermatophyte infections, 221–222
  clinical assessment, 227–230
  clinical trials of antimycotic agents, 222–231
Dermatophytes, 222
Dermatoses, 44–46
  hand, 45–46, 264
Desquamation (D), 150, 154–155
Detection bias, 32
Diagnostic criteria, *see also under specific disorders*
  establishing, 26
Diagnostic words, 69
Diaper rash grading, case report form for, 60–61
Diffusing capacity (DLCO), 193
Diffusion of drugs from dermatological preparations, 319–322
Diseases of unknown etiology, 25
Disk assembly method, 317
Dose–response relationship, 34
Drug release
  diffusion from dermatological preparations, 319–322
  permeation from dermatological preparations, 322, 326
  in vitro methods for studying, 317–322, 324
  in vitro release and systemic occurrence of, 326–327
  in vivo methods for studying, 322–324
Drug *vs.* nondrug clinical studies, 38–39

Drying potential, skin, 281
Dryness, 104, 275, 276; *see also* Moisture

Economic evaluation and economic messages, audience for, 333–334
Economic evaluation methodology, 334–337
Eczema, 46, 343; *see also* Atopic dermatitis
  hand, 45–46, 264
Eczema Area and Severity Index (EASI) system, 46
Eczema sites, allergens and, 233–241
Edema, 96
  reliable, clinically validated measures of, 209–210
Edema/papulation, 104
Effect size measures, 9–10
Eicosapentaenoic acid (EPA), 110, 112
Elasticity, skin, 86–87
Elevation and grooves of skin, deviation in, 83
Eligibility criteria for cases and controls, 26, 33
Endodermal preparations, 315, 325
Eosinophil cationic protein (ECP) levels, 119
Epidemiologic methods, 244
Epidemiological concepts, 43–44
Epidemiological studies, 44
Epidemiology, 244
  clinical, 244
Epidermal preparations, 315
Epithelialization, 211
  reliable, clinically validated measures of, 212–213
Epoxy resin allergy, 239
Errors
  in measurement process, 31–32
  type I and II, 30
Erythema (E), 153
  reliable, clinically validated measures of, 209
  severity of, 58–59, 61, 104, 150, 153–154
Erythromycin (ery), topically applied, 308–310
Essential fatty acid (EFA) supplementation in atopic dermatitis, 109, 115–116
  combined n3 and n6 series, 112–114
  evidence for efficacy, 110–111
  n3 series, 112
  n6 series, 111–112
  rationale for, 110
Essential fatty acids (EFA), 110
European Task Force on Atopic Dermatitis (ETFAD), 94–97, 99, 102–104
Excited skin syndrome, 263
Excoriations, 104
Exposure assessment, 29, 31

Facial photography, high-resolution, 179–180
Final reports, 41
Fish oil supplementation, 111–114
Fisher's exact probability, 8
"Flash foam," test procedure for evaluation of, 73, 75
Fluorescent photography, 180
Forearm controlled application technique (FCAT), 280–281
Fragrance mix, allergic reactions to, 239
Franz diffusion cells, 320, 323, 325

Gamma linolenic acid (GLA), 110–112
Grading scales: *see* Scale grades
Granulation, 211
Granulation phase healing
  reliable, clinically validated measures of, 212–213
Group sequential design, 14

Hand dermatitis/eczema, 45–46, 264
Healing: *see* Wound healing
Health Assessment Questionnaire (HAQ), 184, 191, 197
Hedonic words, 69
Hemostasis, 206, 209
  reliable, clinically validated measures of, 206–208
High-pressure liquid chromatography (HPLC), 308, 311, 312, 314
Histologic assessments, 180–181
Hodges–Lehmann estimator, 9, 10
Hospital controls, 27
Hydration, skin, 262–263
Hyperirritable skin, 263

ICP-GCP guidelines, 38–39
ICP (International Conference of Technical Requirements for Registration of Pharmaceuticals for Human Use), 37, 38
In vitro/in vivo correlations/relationships, 316, 324–327
In vitro methods for studying drug release, 317–322, 324
In vitro release
  diffusion, *in vivo* penetration, and, 325–326
  and systemic occurrence of drug, 326–327
In vivo methods for studying drug release, 322, 324
  direct *vs.* indirect methods, 322–323
  type of preparation and, 323–324
In vivo penetration
  diffusion, *in vitro* release, and, 325–326
Incidence, 43, 244–245
Incidence studies, 43
Infiltration (I), 150, 152–153
Inflammation, 209–210
Inflammatory response, cellular, 210
Information bias, 246–247
Informed consent, 30, 39
Interim analysis, 14
Interrater reliability, 62–63, 96–98
Irritancy of personal cleansing products, 274–281

Irritant contact dermatitis, 243
Irritants, *see also* Dermatitis model, water soluble irritant
  characteristics proposed for irritants used experimentally, 257–258
Irritation, 81
  from sodium lauryl sulfate, 261–264
Iterative feedback, treatment based on, 341, 342

KOH (potassium hydroxide), 223, 224, 226, 227, 229, 230
KOH (potassium hydroxide) examination, 224–225

Language
  correlation of experience and, 67–68
  generation of verbal terms, 68–71
  processing of verbal terms, 69–70
  support of word finding by preformed verbal terms, 70
Laser Doppler velicometry (LDV), 84, 153
Latency periods, diseases with long, 24
Lesions: *see* Acneiform eruptions of the face
Lichenification/prurigo, 96–97, 104
Light reflection, 83–84, 86–87
Linguistic methods, 70–71
Logrank-test, 7
Longitudinal studies, 44

Mann–Whitney (MW) superiority measure, 9–11, 15–17
Mantel–Haenszel test, 7, 30
Matched-pairs signed ranks test, 7, 8
Matched sampling, 27–28
Matching subjects/controls, 27, 33
Maturation and complications (wound healing), 211, 214–215
Measurement data, *see also specific topics*
  errors in, 30–32
  types of, 6–8
    with unknown distribution, 7, 8
Measurement scales, *see also specific scales*
  meaning of, 10
Measures of relevance with relations, 11
Melanoma
  clinical diagnosis, 160
  diagnostic criteria, shifting pathological, 160–161
  early detection and thin, 162
  "epidemic," 159–165
    in Kauai, Hawaii, 163–165
  incidence, 159–160, 164–165
    birth cohort effect and, 162
    decreasing mortality and increasing, 162–163
    increases only in certain geographic areas, 162
    methods of case finding and documentation of, 160–161
  prognostic methods, current, 168
Melanoma (*cont.*)
  prognostic models in, 167–168
    future directions, 171–172
    mixed-effects survival model, 168–171
  tumor registry and underreporting, 160
  types of, 160–162
Mercury, allergic reactions to, 240
Metals, allergic reactions to, 238–239
Michaelis relationship, 141–142
Mini-immersion method, 277–280
Minolta chromameter, used to assess corticosteroid-induced skin blanching, 295–303
Misclassification, 44
MOHL/MOAHL/MOAHLFA indexes, 246
Moisture, skin, 82–83, 86–87; *see also* Dryness
Moisturization from personal cleansing products, 281–282
  examples, 284
  protocol considerations for demonstrating, 282–284
Moisturizing cream, all-purpose
  sensory profile, 78, 79
Multiple points in time, testing at, 14
Multiple subgroups, 14
Multiple testing, and control of α, 11–15
Multiple tests and pretests, 14
Multiple treatment groups, 14
Mycological cure, tinea pedis results using, 230

Nondrug clinical studies
  vs. drug studies, 38–39
  principles for the conduct of, 37–42
Nonresponse bias, 27, 32
Normally distributed measurement data, 6, 8
Null hypothesis, 4, 11

Occlusion, time of, 269–270
Octyl gallat, allergic reactions to, 240
Odds ratio (OR), 11, 23, 29, 34
One-group studies, tests for, 9
Onychomycosis, economic analysis of oral therapies for, 337
Oozing/crusts, 104
Ordinal data with a few categories, 7–8
Outcomes, of standard treatment *vs.* treatment based on iterative feedback and simulation modeling, 341, 342
Outcomes data, systematic analysis of components of compounds and, 340
Outcomes research, pharmacoeconomics and, 341
Outcomes research studies, benefits of, 337

P-value, meaning and interpretation of, 4–5, 9, 33
Paddle over disk, 317, 318
Pain
  reliable, clinically validated measures of, 209

Parallel groups, 6–8
PASI, 93, 94
Patch test grading scale
  clinician's, 270, 271
  expanded, 271
Patch tests, 337–340
  when to grade when reading, 270–271
  when to read, 270
Permeation of drugs from dermatological preparations, 322, 326
Permutational principle and statistical test, 5, 6
Personal cleanser irritation potential, 277–281
Personal cleansing products, 273; *see also* Sodium lauryl sulfate
  irritancy, 274–281
  moisturization from, 281–284
pH value of skin, 83, 86–87
Phenylene diamine, allergic reactions to, 240
Photoaging, 173–174
  biometrics
    applications, 174–181
    rationale for, 174
  diagnosis, 174
Photodamage, 176
Photodocumentation, 62–63
Physicians, differences among, 96; *see also* Interrater reliability
Pilot studies, 30
Polarized light photography, 180
Population base, defining, 26
Population-based studies/designs, 27
Predictive values, 44, 45
Preservatives, allergic reactions to, 239
Prevalence, 244–245
  defined, 43
Prevalence–incidence bias, 32
Prevalence studies, 43
Primin, allergic reactions to, 239
Primrose oil, 110, 112
Profilometry
  mechanical, 178
  optical, 178–179
Propylene glycol, allergic reactions to, 240
Psoriasis, 149, 155
Psoriasis area and severity index (PASI), 149–152, 156
Psoriasis severity scoring using bioengineering techniques, 151–156
Pulmonary computed tomography (CT), 193
Pulmonary function tests (PFTs), 193

Quality of life measures, 99
Quantification of subjective tests, 71
Quaternium-18 bentonite, 269–271
Questionnaire methods used in skin disease epidemiology, 43–48
Questionnaires, 29
  answer categories, 50
  characteristics of good, 49
  designing, 48–51
    minimal requirements before, 48, 49
    order of steps in, 51
Questions
  art of asking, 49–50
  requirements for good, 50

Random sampling, 27
Rare diseases, evaluation of, 25
Rash grades, 61
Rating scales, 7–8
Raynaud's phenomenon in scleroderma, 194–195
Redness, 276
Reference products, strategy for generation of, 75, 76
Reliability, 44
  interrater, 62–63, 96–98
Research, *see also specific topics*
  confirmatory, 11
Research questions, 26
Rotating cylinder method, 318, 319
Roughness of skin, assessment of, 291
  visiometry, 84–87
Rubber, contact allergy to, 238–239

Safety documentation for test products, 40
Salicylic acid levels and analysis, 310–312
Sample size, 5, 30
Samples, correlated, 8–10
Sampling methods, 27
Scale development, 57–59
Scale grades, 57–58
Scleroderma, 183
  diagnosis, 184–187
  Raynaud's phenomenon in, 194–195
Scleroderma clinical trial guidelines, statistics and, 184
Scleroderma clinical trials, 183–184, 188–189, 195–198
  outcome measures, 196
  skin outcome measurements, 187, 190–194
Scoring Atopic Dermatitis (SCORAD) index, 94, 96, 99–103, 105, 124–125
Sebaceous glands, activity of, 86–87
Seborrheic dermatitis, 264
Selection bias, 31, 32, 245–246
Self-assessment, 58–59
Sensitive skin, 263
Sensitivity, statistical, 44, 45
Sensory assessment in cosmetics, 71–72
  areas of application for, 78
Sensory profile sheet, 76
Sensory profiles, structuring of, 78
Sensory properties, procedures for evaluation of, 73
Simulation modeling, treatment based on, 341, 342

Six Area, Six Sign Atopic Dermatitis (SASSAD) score, 114, 116, 126
Skin: *see specific topics*
Skin-care products, 39; *see also* Cosmetic products
 sensory parameters concerning, 70
Skin thickness, 85
Small plaque assay efficacy parameters, 290–292
Small plaque assay methods, alternative, 292
Small plaque assay (SPA), 289–290, 292
 follow-up, 290
 patient selection, washout period, and pretreatment, 290
 reasons for performing, 290
 treatment period, 290
Soap: *see* Personal cleansing products
Sodium lauryl sulfate (SLS), 257–264
 application methods, 257–259
 biologic endpoints, 259–261
 host-related factors and irritation from, 261–264
 purity and carbon length, 258
Specificity, 34, 44, 45, 47
Standard difference (SD), 11
Statistical power, 30
Statistical significance, 4–5, 33
Statistical testing, examples of, 15–19
Statistical tests
 for difference and equivalence, 14–15
 selecting appropriate, 5–9
Statistics in clinical research, 3–19
 in nondrug studies, 41
Stratified sampling, 27
Stratum corneum, barrier function of, 81–82, 86–87
Stratum corneum corneometry, 155
Study base, identifying, 26
Study protocols, 25–30, 39–40
Subgroup analysis, 14
Sun protection factor (SPF), 323
Surfactant based products, sensory parameters concerning, 73
Surfactant interactions with skin, 274–277
Surveillance, defined, 244
Surveillance systems, evaluating, 247, 251–253
Syndet: *see* Personal cleansing products
Systematic sampling, 27

T-test, 7, 10
 one- *vs.* two-sided, 14–15
 paired, 8
Terminology, variation in, 44–45
Test panels, use of, 68–69
TESTIMATE, 16, 18
Thiosalicylic acid, allergic reactions to, 240
Time-to-event data, 7
Tinea pedis, 230
Topical availability *vs.* bioavailability, 316
Transdermal preparations, 315, 326, 327
Transepidermal water loss (TEWL), 81–83, 155, 259–264, 291
Trial subjects, protection of, 39, 42
Tuohilampi questionnaire, 51–52
Two-by-two table, 8, 10
Two parallel groups, tests for, 9–10
Two-stage procedure, 14

Ultrasound imaging, 85, 180, 291–292
 laser Doppler, 85
Ultraviolet radiation (UVR), 173, 180

Validation and calibration of investigators, 62–63
Validity, 44
Viscoelasticity ratio, 86–87
Visiometry, and assessing roughness of skin, 84–87

Wei–Lachin procedure, 13–14
Wilcoxon/Wilcoxon–Mann–Whitney tests, 7–10, 14
World Medical Association Declaration of Helinski, 38
Wound healing, 205
 complications, 211
 quantifying the stages of, 206–211, 215
 reasons for measuring, 205
Wound healing measures, 206